# ORGANISMAL AND MOLECULAR MALACOLOGY

Edited by **Sajal Ray**

**Organismal and Molecular Malacology**
http://dx.doi.org/10.5772/65885
Edited by Sajal Ray

**Contributors**

Genaro Diarte-Plata, Ruth Escamilla-Montes, Antonio Luna-González, Jesús Arturo Fierro-Coronado, Héctor Manuel Esparza-Leal, César Arturo Ruíz-Verdugo, Salvador Granados-Alcantar, Natalia Fokina, Tatiana Ruokolainen, Nina Nemova, Eric Hallerman, Pawel Michalak, Kang Lin, Serena Ciparis, William Henley, Jess Jones, Andrew Phipps, Thanasis Manousis, Sophia Galinou-Mitsoudi, Sofia Galinou-Mitsoudi, Melissa Palos Ladeiro, Iris Barjhoux, Aurelie Bigot, Marc Bonnard, Elise David, Odile Dedourge Geffard, Elodie Geba, Alain Geffard, Emilie Lance, Maxime Lepretre, Gabrielle Magniez, Dominique Aubert, Isabelle Villena, Gaëlle Daniele, Arnaud Salvador, Jean Armengaud, Emmanuelle Vulliet, Damien Rioult, Paulo Henriques, Ricardo Sousa, João Delgado, Minoru Saito, Raul Bettencourt, Inês Barros, Eva Martins, Inês Martins, Teresa Cerqueira, Ana Colaço, Valentina Costa, Domitília Rosa, Sergio Stefanni, Hugo Froufe, Conceição Egas, Paul Dando, Ricardo Serrão Santos, Mudasir Dar, Radhakrishna S Pandit, Kiran D Pawar, Shelley Bhattacharya

**Published by InTech**
Janeza Trdine 9, 51000 Rijeka, Croatia



**Publishing Process Manager** Mirena Calmic
**Technical Editor** SPi Global
**Cover** InTech Design team

First published June, 2017
Printed in Croatia
Legal deposit, Croatia: National and University Library in Zagreb

Additional hard copies can be obtained from orders@intechopen.com

Organismal and Molecular Malacology, Edited by Sajal Ray
p. cm.
Print ISBN 978-953-51-3305-6
Online ISBN 978-953-51-3306-3

For EU product safety concerns:
IN TECH d.o.o., Prolaz Marije Krucifikse Kozulić 3, 51000 Rijeka, Croatia,
info@intechopen.com or visit our website at intechopen.com.

# Meet the editor

Sajal Ray received his MS and MPhil degrees from Calcutta University in Zoology and Environmental Science, respectively, and his PhD degree from Jadavpur University. His thesis reports immunotoxicity of pesticide in an economically important snail of India. As an awardee of Fogarty Visiting Fellowship, Ray carried out his postdoctoral research in cardiac pathology at the National Institutes of Health, USA. His research interest is studying the immunological responses of mollusc, sponge, crab and earthworm exposed to pollutants. His team is engaged in understanding the evolutionary mechanism of immunity in phylogeny. He presented his research in the World Congress of Malacology at Washington DC. Sajal Ray, a professor of Zoology at Calcutta University, has been teaching Zoology for nearly 30 years at the postgraduate level.

# Contents

# Preface

Malacology, the study of molluscs, is an age-old discipline of natural science. Its genesis perhaps dates back to the time of Aristotle. Mollusc is the second largest group of invertebrates after arthropods. The group is represented by snails, oysters, squids, octopuses, slugs, clams and their relatives. Scientists had recorded nearly 200,000 extant species of molluscs on earth occupying diverse ecological niches. They also exhibit a significant paleontological evidence since Cambrian. Molluscs present a wide range of variation with reference to their anatomy, physiology, behaviour, adaptation, food preference and habitat selection. Body plan of mollusc is extremely varied, which ranges from tiny worm–like creatures to giant squids, a relatively less researched group of cephalopod. Molluscs bear several taxonomical classes of organisms with distinct morphofunctional attributes. They are distributed widely in all zoogeographical realms. An advanced level of adaptational efficiency and physiological resilience enabled molluscs to survive and propagate in diverse ecological conditions over the span of millions of years of evolution. Ecologically, they are distributed in almost all types of environments, namely, aquatic, terrestrial and deep sea regions. Their existence from hydrothermal vent was indicative to their adaptational uniqueness, which enabled them to survive in an extreme environmental condition with a minimal level of interspecific competition.

Molluscs, in general, evolved an advanced mode of innate immune system, which, with other factors, might have played a key role in their evolutionary success. Report of efficient phagocytes, cytotoxic molecules, agglutinins and possible Toll-like receptors is the evidence of effective immunoresponsiveness against pathogen and toxin characteristic to this group. Similarity of immunological system between mollusc and mammal is also claimed. Aquatic molluscs, in recent years, have been identified as effective monitoring organisms of the water quality of lotic and lentic ecosystems. Filter feeding mode of digestion enabled them to interact with diverse groups of environmental contaminants and pathogens intimately. Malacologists had identified several cellular, molecular and biochemical markers of ecotoxicity both in the gastropods and bivalve species. Cellular apoptosis, respiratory burst activity, lysosomal membrane fragility and lysozyme activity in the immunocytes of molluscs had already been suggested as effective biomarkers of toxicity. Gene expression analysis under pathogen and toxin exposure is comparatively a new area of malacological research. Being a putative source of ethnomedicines and bioactive molecules, several species of molluscs bear a high degree of biotechnological and pharmaceutical significance. However, molluscan biodiversity and its rational utilisation is a less researched area with limited scientific information.

Molluscs, in general, bear an immense degree of societal and other influences on mankind. It is an established source of food and ethnomedicine for human, culturable fish and poultry. It is a lucrative source of ornamental shell and pearl and thus bears a special commercial importance. In recent years, malacology has been gaining a special scientific significance due to novel discoveries and findings in the areas of physiology, biochemistry, molecular biology and ethology. Development of advanced instrumentation technology enabled the malacologists to gather a new set of information on basic and applied malacology. For the last 20 years, horizon of malacological research has been expanding by leaps and bound globally. The World Congress of Malacology held at Washington DC in 1998 is a major landmark initiative towards consolidation and exchange of ideas among the fellow researchers. In this book, the authors made significant contribution by addressing the frontier areas of organismal, molecular and biochemical malacology. Multiple chapters with current malacological information were collated in two separate sections. Contribution of the scientists from diverse research interests would make this edited volume a well-informative one for the concerned researchers, teachers and students.

In current years, organismal and molecular aspects of research have been identified as the two broad attributes of malacological investigation adopted in different laboratories. A paradigm shift that occurred in the trend of malacological research during the last two decades has well been reflected in these two sections. In the 'Organismal Malacology' section, a review was presented on the bioecology and aquaculture of pen shell, an economically important mollusc of the Gulf of California, Mexico. Issues related to the taxonomical problem of Conoidea had been dealt in this section with appropriate citations and research evidence. Mussels are considered as sentinel species in ecotoxicology. Functional significance of mussel as a tool to monitor the watershed quality was presented following the biomarker approach. An overview on the biology and conservation of patellid limpets, a keystone species of rocky shore, highlighted the importance of biological traits on effective management of this group in their natural habitat. In another chapter of this section, the author discussed the morphological characteristics of a gastropod *Babylonia spirata* of Tuticorin coast of India from taxonomical point of view. In the other section, entitled 'Cellular and Molecular Malacology', current trends of biochemical and molecular malacological research were highlighted in several chapters. Arsenic and sulphates are the chemical contaminants linked to coal mining and related processing. Differential patterns of gene expression of mussel were characterised under the exposure of these toxins in controlled conditions. The prime objective of the study presented in this chapter is to assess the efficacy of gene transcriptional markers in monitoring the degree of contamination. In another chapter, the author reported an elaborative study of spatiotemporal neural activity during olfactory processing in a land slug using state-of-the-art technology of imaging. Bacterial symbionts of molluscan gut bear industrial prospect and are a less explored area of research. Authors addressed an important area of applied malacology concerning the functional spectrum of heterogeneous bacterial groups inhabiting the gastrointestinal niches. Highly advanced innate immune system is a characteristic evolutionary trend as recorded in mollusc. This section included a chapter on molecular immunological adaptation of molluscs of deep sea hydrothermal vent—a less studied ecosystem of earth. In another review, authors documented their holistic view on the biochemical adaptation of lipid composition in marine bivalve in the face of various ecological factors. Evolutionary success of *Achatina fulica* was discussed with references to its innate immunological status and prospect as a source of ethnomedicine and other bioactive molecules. Molecular biology and biochemistry of C-reactive protein, glycosaminoglycan and immunoregulatory responses were explained on the basis of available scientific information.

At last, I remain thankful to Ms. Ana Pantar and Ms. Mirena Calmic of InTech for their earnest cooperation and help that I received in every step of editing of this book. My thanks go to my wife and research colleague Dr. Mitali Ray and my doctoral students of my laboratory of Aquatic Toxicology for providing me with appropriate academic inputs and technical assistance during this exercise.

**Sajal Ray**
Professor of Zoology
University of Calcutta, India

Section 1

# Organismal Malacology

Chapter 1

# Ecology, Fishery and Aquaculture in Gulf of California, Mexico: Pen Shell *Atrina maura* (Sowerby, 1835)

Ruth Escamilla-Montes, Genaro Diarte-Plata,
Antonio Luna-González,
Jesús Arturo Fierro-Coronado,
Héctor Manuel Esparza-Leal,
Salvador Granados-Alcantar and
César Arturo Ruiz-Verdugo

Additional information is available at the end of the chapter

http://dx.doi.org/10.5772/68135

**Abstract**

The pen shell *Atrina maura* bears economic importance in northwest Mexico. This chapter considers a review on diverse ecology, fishery, and aquaculture topics of this species, carried out in northwest Mexico. In ecology, biology, abundance, spatial prospecting, sex ratio, size structure, reproductive cycle, first maturity sizes, variation of gonadosomatic indexes and growth are discussed. In fishery, the information analysed corresponds to the structure of the organisms in the banks susceptible to capture, institutional and ecological interaction for fishing regulation, evaluation of fishing effort, improvement in fishing performance using the knowledge and attitudes of the fishermen on fisheries policies in the Gulf of California, resilience and collapse of artisanal fisheries and public politics. In aquaculture, they are long-line culture, bottom culture, reproductive cycle, growth, production of larvae and seeds, biochemistry of oocytes, nutritional quality of the muscle, evaluation of diets based on microalgae, immunology in larval and juvenile and probiotic use. The present work shows a status based on information published in theses and articles indexed 15 years ago to the date on the ecology, fishery and aquaculture in the pen shell *Atrina maura* carried out in the lagoon systems of northwest Mexico.

**Keywords:** growth, reproduction, culture, immunology, populations, estuaries

# 1. Introduction

## 1.1. Taxonomic position

The taxonomy of the pen shell according to Keen (1971):

Phylum: Mollusca (Linnaeus 1758)

Class: Bivalvia

Subclass: Lamellibranchia

Order: Anisomyaria

Family: Pinnidae

Genus: *Atrina maura* (Sowerby 1835)

Synonymy: *Pinna lanceolata* (G.P. Sowerby 1835)

## 1.2. Biologic

### *1.2.1. Current status*

The Pinnidae ('mussels, scallops, oysters or pen shells') are bivalve marine mollusks that during their life cycle have a planktonic stage and later will incorporate to the benthos where they live until they die. *A. maura* adults measure approximately between 170 and 350 mm, they live in muddy or sandy bottom places, where they strongly attach themselves to the substrate by using their fixation organ; also, due to the proximity of the organisms, they tend to form dense banks by living in groups with other species of lamellibranchs [1]. The shell of these organisms is triangular, reduced in the anterior part and has a pointy shape, whereas the posterior one is a long, flat and well-developed part similar to a wide fan, with a rounded truncated edge, the shape of the shell is strongly influenced by a series of adaptations related to the species life style. The reduction of the anterior part and of the anterior adductor muscle is associated to the substrate fixation by byssal threads, as it occurs with other anisomyarian mollusks [2]. The shell is colonized by a great amount of epibionts that make it look irregular; usually on the shell, there are numerous species of algae, mollusks, polychaetes, bryozoans, ascidians, etc. The outer surface is purplish-amber or dark brown in colour, while the inner surface is shiny [3] (**Figure 1**).

In the Pinnidae, the opening and closing of the valves does not occur in the normal way by the ligament, since the only function of this structure is to keep the valves together; therefore, it takes place by the flexion of the posterior part of the shell, which has high protein content. These organisms have a big posterior adductor muscle located at the central part of the shell (the organism's tissue that is put on the market), as well as a small anterior muscle located at the umbonal vertex. There are two more pairs of muscles that help the movement of the organism's foot, which are the anterior retractor pedals that together with the foot carry out the excavation and motility function [4]. These organisms can even bear the loss of the anterior

**Figure 1.** *Atrina maura* procured from the bottom culture at San Buto estuary in Baja California Sur, México [20].

adductor muscle or even the fusion of the ventral margins of the shell, as long as the posterior part can still be closed by the action of the posterior adductor muscle [5]. The byssus is kept in a chamber found at the posterior basal region of the foot; the bundle of fibrous strands can measure 25 cm in length. The byssus gland is near the area where the foot begins. The soft parts hold a series of characters that differ from other bivalves and are confined mainly in the area between both adductor muscles and overstepping this region, the mantle's lobs and the elongated gills are found, which are extended toward the posterior region, overtaking the adductor muscle [4]. The mantle is attached to the shell as it occurs with other species with similar living habits (some Mytilidae), so it is very retractable without a pallial line. The mantle's retraction confers the family great regeneration ability since all the posterior part of the shell can be rebuilt. The mantle's cavity is divided by a septum that forms the external and internal chambers [5]. The mouth is a tiny orifice located at the anterior portion of the labial lips and continues with the oesophagus, a narrow circular canal that connects with the stomach, where the intestine begins, running toward the posterior area and coming out from the digestive gland, entering the gonad until it reaches the posterior adductor muscle [4]. A very distinctive and unique organ from the Pinnidae family is the pallial organ, whose function has been controversial for a long time. Originally, it was considered as a cleansing organ, in charge of extracting broken pieces from the shell that would remain in the pallial cavity [6, 7]. The heart is located in the dorsal region at the same level as the posterior adductor muscle. The crystalline style begins in the stomach up to the posterior intestine grip, and it secretes enzymes that aid in digestion [8]. The respiratory organ is composed by gills or ctenidia that resemble light brown elongated leaf shape structures which appear in groups of four, located by pairs on each side of the organisms [4]. There are waste ducts that cover the mantle in an anterior-posterior direction, starting from the mouth's palps reaching up to the end of the posterior back part of the gills, completing the division between the inhalant and exhalant chambers.

These canals allow the disposal of pseudofeces and other type of wastes from the inhalant chamber, which can be used to clean the sandy cavity and other residues introduced by the swell before the animal shuts its valves [5]. *A. maura* is a gonochoric organism whose zygotes arise from the union of female and male gametes deriving from those organisms with separated genders. The sexual maturity state can be determined by the gonad's coloration, which sometimes can be observed macroscopically when the valves are opened. The females present a deep orange coloured gonad (like the colour of a brick), and the males have a whitish gonad [9]. The fertilization is external and the larval development is planktonic with trochophore and veliger larvae [10]. The gonads are anastomosed glandular structures that branch out invading the digestive gland. As the maturation progresses, the gonad occupies an even larger space, and it turns out to be more notorious up to the point in which it reaches the typical follicular tubular-acinar system structure that characterizes all bivalves [8, 11].

## 2. Ecology

### 2.1. Current status

The organisms of the Pinnidae family spend a short but important stage of their life cycle in the plankton and later will incorporate to the benthic zone where they will live until they die.

### 2.2. Distribution and habitat

Since the first research of [12], in the Eastern Pacific, it is known that the Pinnidae family (in Mexico, commonly known as 'Callo de Hacha') has a wide biogeographic distribution that runs from the Baja California Peninsula, Mexico, down to Panama. However, in many cases, the regional boundaries of this distribution are unknown. As part of the Pinnidae population, four species have been identified: *Pinna rugosa, Atrina tuberculosa, Atrina oldroydii* and *Atrina maura,* the latter presents a relative abundance of approximately 95% [1]. These scallops are distributed in the Indo-Pacific from the southwest of Africa to Malaysia and New Zealand and the north of Japan, they are found in Mediterranean waters and America [13–15]. In the American continent, the *A. maura* species has a wide biogeographic distribution from the Baja California Peninsula down to the south of Peru, forming not so dense banks in bays and coastal lagoons on both littorals of the California Peninsula. Nevertheless, in many cases, the regional boundaries of this distribution are unknown [1, 12]. In the Mexican Pacific, there are four species of the Pinnidae family, particularly in the Gulf of California, *Atrina maura, Atrina tuberculosa, Atrina oldroydii* and *Pinna rugosa* are distributed [9]. These are benthic organisms that live in protected intertidal zones, as well as in estuaries with slimy-clay, muddy-sandy, sandy limestone or sandy-rocky bottoms [16]. These animals can be found from the lower limit of the tide up to a maximum of 10 meters, although some have been recorded in more than 45 m of depth in some bays. *A. maura* lives semi-buried in different types of substrates such as soft substrates that can have roots and rhizomes of sea grass, small gravels, biodetritic waste and sand grains to which these animals attach to by using their byssus. *A. maura* is a halotolerant and thermotolerant species, which makes it suitable for culture in the Gulf of California for having an adequate environmental production framework [17].

## 3. Fisheries

### 3.1. Current status

*Atrina maura* also known as 'Hacha China' represents one of the most valuable fishery resources in the coasts of the Mexican Pacific; it is highly valued in the national and international market, because it has an edible adductor muscle that can be commercialized. These animals are dominant mollusks in the benthic community of the sites they live in, where they form dense size banks of variable permanence [1]. The Gulf of California fishery comprises four species: *Pinna rugosa, A. maura, A. tuberculosa* and *A. oldroydii*, which represent approximately 25% of the national scallop catch [9]. The commercialized part of the organism is the posterior adductor muscle, known as 'callo', that has a wide demand due to its nutritional quality, texture, taste and high prices in the national market, reaching up to $120 and 180 (Mexican pesos) per kilogram in the beach buying it directly from the fishermen, and if the product is sold by an intermediary, the price to the public can reach up to $300 Mexican pesos per kilogram [18, 19]. The product is sold fresh or iced, and it is mainly marketed in Nayarit, Sinaloa, Sonora, Baja California, Baja California Sur and in the most important cities of the country such as México City, Guadalajara and Monterrey [9].

During the past 30 years, the exploitation of these scallops has increased, overfishing *A. maura* in some zones of the Mexican Pacific [20]. Catch records in Baja California Sur (*A. maura, P. rugosa* and *A. oldroydii*) between 1985 and 1995 show a maximum intake of 1148 tons of fresh product (adductor muscle). Since 1991, there has been a low catch trend, reaching a minimum catch of 91 tons in 1995 [9]. According to Mexican official sources, the catch of pen shells has been included within the catch of other species of clams. Between 1966 and 2010, the pen shell represented 25% of the pen shell production as reported by the fisheries sub-delegation in Baja California Sur showing the highest catch volume in 2004 and 2008 with 280 tons, whereas the lowest value occurred in 2000 with only 20 tons, respectively [21] (**Figure 2**).

Due to the decrease in the production by fishing and the lack of culture development, the current global production levels have not been able to satisfy the national market demand, so different alternatives have been searched in order to allow a sustainable exploitation of this wild resource, such as the one that is taking place in some Seri fishing communities from Punta Chueca and Kino Viejo in Sonora, by using a population growth function which emulates how the carrying capacity affects small scale fishing communities, as well as the different institutional development levels, the opportunity to adopt new institutions and the degree in which the fishing effort is reduced, thus avoiding overfishing of the species. With these

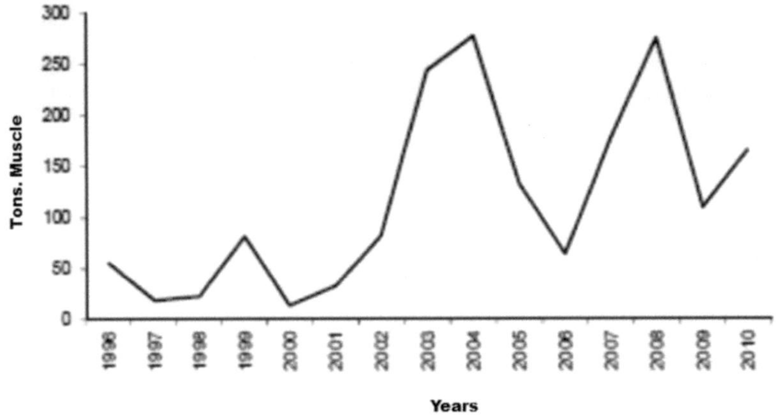


**Figure 2.** Catch volume expressed in weight tons of total live weight of pen shell *A. maura* in Baja California Sur, from 1996 to 2010 (source: Ref. [21]).

measures, the fishing Seri communities of Punta Chueca have avoided overfishing of the pen shell banks; while in their neighbour fishing community in Kino Viejo occurs otherwise, where exploitation of the resource without a proper measure occurs, in spite of the fact that these two communities are only separated by a 30 km distance.

With these two cases, there was a need to create governmental organizations because there are not enough ones to regulate and keep a sustainable fishery through the time. As soon as the communities adopt strong and solid institutions, there will be more chances of sustaining the resource. In which moment should institutions be adopted and in what measure the fishing effort must decrease? This will depend on the ecological capacity of the species in those places where the fishing takes place, as well as the fishing capacity among sites, even in those settings of similar institutional development [22]. In addition, studies have been made in order to increase the production levels through biotechnical alternatives that can be applied to aquaculture to solve the problem faced by *A. maura* in the coasts of the Mexican Pacific [23], specifically in a regional level (Gulf of California) like the realization of projects that generate information regarding the production [20, 24], larval survival up to the phase of pre-fattening and fattening [25], genetic improvement and triploid and seed production [26], economic feasibility [27], among others.

## 4. Aquaculture

### 4.1. Current status

Bivalve aquaculture in Mexico is performed almost exclusively in the coasts of the Pacific Ocean and Gulf of California, where more than 54 species of mollusks are exploited [28], as they have water bodies suitable for the development of cultures. It occupies the fourth place in Latin America after Chile, Brazil and Peru. The production data from FAO began in 1987 with 20 tons, and subsequently, the production increased to 2200 tons in 1990. In 1993, there was a decrease in the production to 1053 tons and increasing afterwards in 1995 up to 2500 tons and to 3038 tons in 1997. After this year, the production decreased again to an average of 1500 annual tons, which sustained until 2005 [29].

In the coasts of the Mexican Pacific, the pen shell *Atrina maura* is one of the most important commercialized mollusks, with a broad national and international market. The natural populations of this species are increasingly seen as decimated due to overfishing. That is why research has been directed toward the experimentation of diverse culture stages of such a valuable mollusk. Nevertheless, there have not been any successful results about its production under controlled conditions over time. The entrepreneurial and educational sector has a big interest on this resource as they have invested in lab facilities for seed production, using a developed methodology by the Company Acuacultura Robles S.A. de C.V., Centro de Reproducción de Especie Marinas del Estado de Sonora (CREMES) and el Centro de Investigaciones Biológicas del Noroeste (CIBNOR), from La Paz, Baja California Sur, as well as from Sea Farmer S.A. de C.V. in the northern part of Sinaloa. The seeds that are produced in the lab have been planted in the sea, but it has been observed that there has been a struggle with huge mortality rates; as a result, there is no culture technique that can escalade up to an industrial level. Therefore, it has been of great importance the collaboration between the academic and industrial sector with the aim of conducting studies to deal with those bottlenecks that take place during the larval phase of the species which substantially affects the seed production in an industrial level. Regarding the culture stages in the phase of pre-fattening and fattening, there is no problem whatsoever; once a certain size has been achieved, the fattening stage does not demand a higher maintenance and care because those are phases in which more level of research has been developed. On the other hand, it has been considered to contribute with cultures in a wild level and therefore promotes the occurrence of the species in its natural habitat.

### 4.2. Seed production

Bivalve seed production is practically based on the Pacific oyster *Crassostrea gigas* and in a lower degree in the Cortez oyster (*Crassostrea corteziensis*), the Mediterranean mussel (*Mytilus galloprovincialis*), the Pacific scallop (*Argopecten ventricosus)* and the rainbow-lipped pearl oyster (*Pteria sterna)*. Production of emerging species under a commercial level has been carried out such as the case of the lion's paw scallop *Nodipecten subnodosus* and the pen shell *Atrina maura*, but such production has not been maintained over time [29]. *A. maura* is a species with a highly cultured potential in the northwest of Mexico because of its nutritional content, resistance to transport and handling, as well as for its highly commercial value [20]. Within the past years, there have been huge efforts to perform diverse studies in order to know the basic biology, physiology and reproductive biology of the species [30]. For example, some research has been done to accomplish the seed production in labs to carry out cultures in the natural environment with high success probabilities [31]. Nevertheless, the species *A. maura* must be handled under physical-chemical optimum conditions as well as nutritional ones in those zones of culture, in which unfortunately there is not enough research done about the subject, which has made this situation to be considered as a big problem for seed production [32]. On the other hand, there is a need to perform studies in search of other alternatives to solve such a problem; one would be the manipulation of the species reproductive cycle under specific conditions during gametogenesis.

This is important because natural banks have been significantly diminished due to overfishing since there is no law that regulates this activity, which in turn limits the massive removal of seed from their natural environment [31, 33]. In turn, the seed production obtained from artificial collectors seems uncertain for a sustainable aquaculture activity. There are some studies regarding the recollection of juvenile pen shells in the Gulf of California [33, 34] and in the Pacific coast of the Baja California Peninsula [35, 36]. In 1988, the first juveniles of *A. maura* were produced at the Centro Reproductor de Especies Marinas del Estado de Sonora (CREMES). In 1991, a new attempt was performed with encouraging results due to the increase of produced seeds. By the end of 1994 and the beginning of 1995, the first commercial production of seeds in the lab was produced with over 800,000 organisms ranging between 15 and 25 mm in length. In 1996, the production increased to double, but there was a massive mortality of 4 or 5 days after being delivered to the producer [37]. In 2004 [19], biotechnology was developed to artificially produce larvae and diploid and triploid seeds of *A. maura* under lab conditions, establishing the proper temperature, feeding and care parameters for each of the stages such as the spawning induction, larval culture, fixation, metamorphosis and pre-fattening stage. In 2009 [38], a comparative study between the growth and survival of floating and swimming larvae of *A. maura* under intense culture lab conditions was performed, observing a higher growth in swimming larvae whereas the floating larvae showed the lowest sizes; survival was lower for both types of larvae but being higher (around 16%) for the swimming type. In 2011 [39], a lab production of larvae and seeds of *A. maura* was analysed, determining each one of the stages of the larval cycle, as well as the period of time for the growth of each stage, concluding that the pen shell larvae and seed production are different from that of other bivalves. In this species, a lower survival rate was recorded which seems to be linked to the buoyancy of the larvae within larval cultures, combined with the production facilities which must be suitable for the species culture.

### 4.3. Culture

In Mexico, there are few research advances regarding *A. maura* that can determine with certainty the establishment of a sustainable culture. The first study for this type of activity was initiated at the early 80s with the research carried out in San Blas, Nayarit by a group of workers from the former Secretaria de Pesca. The accomplished advances regarding the extraction and transport of breeders, the spawning induction and at a certain point the handling of larvae turned out to be encouraging, and therefore, some hundred larvae were produced, which were followed up to the fattening stage in the field [19].

The fattening culture could start after capturing seeds in collectors that come from natural populations or from juveniles produced under controlled lab conditions. In both cases, the culture is extended to the sea until a commercial size is reached (200 mm in valve height). This type of culture is performed in extensive systems so the nutritional or feeding studies and the carrying capacity of the ecosystems in which the fattening is carried out are essential, since the growth rate, survival, accumulation of energy reserves and biochemical composition of tissues depend on them.

This carrying capacity is defined by terms of appropriate feeding particle availability in the area as well as by physical and chemical variables that allow culture of juvenile and adult biomass which constitute the sustainability of the culture [40].

#### 4.3.1. Culture technologies

The use of two types of fattening techniques has developed based on the location of the organisms in the water column, which are the suspension and bottom cultures.

##### 4.3.1.1. Suspension cultures

In those suspension cultures, the fundamental principle requires that the organisms are confined in boxes or in any artefact (pearl baskets, 'Nestier', purse nets, Japanese lanterns, etc.), and keeping them by modules along the water column, assisted by supporting and floatation structures to keep them suspended over the surface or at a specific depth, such as the 'Long line' or stem lines, floating rafts, etc.

In 2012 [20], a suspended culture of *A. maura* was performed after produced seeds, evaluating their growth in San Buto estuary. Baja California Sur, Mexico, observing significant differences in both the size and weight growth of *A. maura* according to the length, width and thickness ($p = 0.0001$) during the experimental period. This trend was shown in the total weight of the organisms, having a final average value of 39.11 ± 3.13 g by the end of this stage, and in the case of the adductor muscle, a final average weight of 0.77 ± 0.09 g (**Figure 3**).

##### 4.3.1.2. Bottom cultures

Bottom cultures are defined as the planting of freely dispersed seeds over the ocean floor in which they are expected to develop until they reach a commercial size [20]. Both technologies have been widely used with different species and in different parts of the world, and in each species and locality obtaining different results among them [41].

The culture of *A. maura* has high potential because it can develop under temperatures of 16–30°C, and due to the fact that it presents fast growth rates. The first culture was started at the beginning of the 1990 decade. With the first group of seeds obtained in the lab of the Dirección de Fomento Pesquero del Estado de Sonora, who donated some seed batches (between 1000 and 2000 seeds per social group) that were distributed in different localities at the northwest part of the country (Gulf of California), some good results were obtained. Such results were satisfying in terms of survival and growth that even until now the seed demands from the different social and private groups continue [19]. In 1997 [32], an experimental culture of *A. maura* was carried out at Agiabampo lagoon, in Sonora, considering the seed transport, pre-growth and final growth phase. In this study, it was observed that the species adapts to artificial pre-growth conditions and stands being planted subsequently in the sea bottom after 17 months of culture; the organisms reached an average size of 208 mm, with an adductor muscle weight of 22.75 g and an accumulated population rate of 28%. Another research [31] performed an experimental culture in Bahía Magdalena Baja California Sur obtaining adductor muscles around 14 g in weight in a period of 20 months of culture, although the harvest time

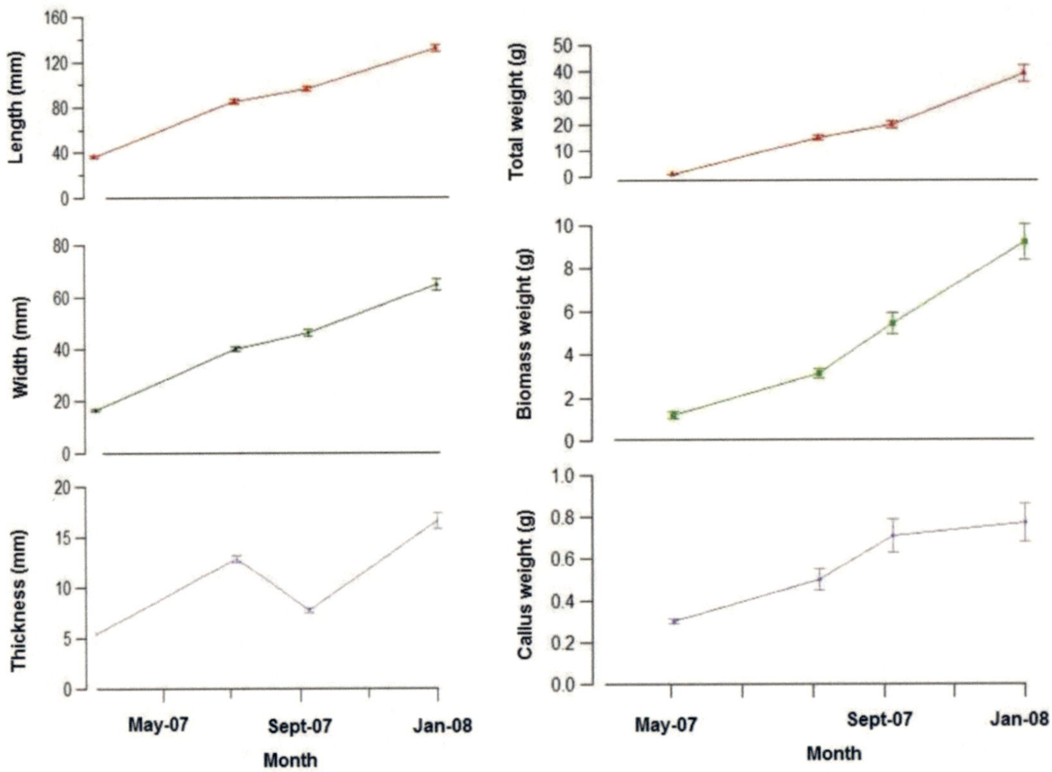


**Figure 3.** Growth in length, width, thickness, total weight, adductor muscle weight and biomass weight (media ± standard error) of the shell of *Atrina maura* during the fattening stage at San Buto estuary, Baja California Sur, Mexico.

was of approximately 2 years. In 2008 [63], the growth and survival of a culture of *A. maura* in La Palmita cove, Navolato, Sinaloa, was evaluated revealing that the organisms reached an average size of 167.81 mm in length and a total weight of 153.16 g with a survival rate of 71.57% during 19 months of culture.

In 2012 [20], a bottom culture of *A. maura* was carried out at San Buto estuary, Baja California Sur, Mexico, where a total of 584 organisms were analysed under three experimental densities of 12, 24 and 48 org/m$^2$, resulting in similar average sizes under the three densities: 243.61 ± 3.37 mm of length, 148.17 ± 3.77 mm in width and 44.86 ± 0.41 mm of valve thickness. Regarding the total weight (382.89 ± 19.08 g), biomass weight (107.2 ± 5.96 g) and adductor muscle weight (18.39 ± 1.24g), the results were similar under all three densities. The survival rate during the study was greater for those organisms kept in the density of 24 org.m$^{-2}$ (90%) rather than in the other two densities.

### 4.4. Growth

The growth in bivalves is affected by a complex combination of biological and environmental factors such as temperature, salinity, oxygen concentration, quantity and quality of food, affecting the size, reproductive condition and genetic characteristics of animals, etc. These growth aspects in mollusks are sometimes difficult or impossible to manipulate under aquaculture situations that assure an optimum growth; however, a limited control can be executed over such conditions by a careful selection of culture sites [42–44]. Different studies agree that growth rate is one of the main factors that determine the feasibility of those cultures of bivalve mollusks. The growth is inversely proportional to the population density due to the intraspecific competition over available resources in the environment [45–47], and the increase of 'fouling' can also decrease the growth rate, which also differs among size classes and change according to geographical regions [48]. Food availability is limited in reproductive periods, the gonadic growth is favoured as an evolutionary strategy in mollusks, and therefore, it is probably that the species starts forming gonads in small sizes, focusing the energy mainly for reproduction more than for somatic growth [49].

Fewer are the studies performed on organisms of the Pinnidae family related to growth. In 2011 [50], growth was analysed in a bottom culture of *A. maura* under two tide levels at San Buto estuary in Bahía Magdalena and Baja California Sur, Mexico, using the von Bertalanffy model, where it was revealed that *A. maura* reached a maximum valve height of 127 mm during the first year of life and of 207 mm in the second year; survival rate after transplanting the organisms was high (99%), and after 8 months of culture, it was of 78% for both tide levels. In 2015 [51], growth and survival were evaluated for pen shell organisms, *A. maura*, that were harvested for 15 months (May 2010–August 2011), at La Piedra estuary, Guasave, Sinaloa, Mexico. A total of 3000 seeds were obtained (61.50 ± 0.5 mm of valve height, 4.98 ± 0.2 g of total average weight). The temperature (26.15 ± 6.35°C), dissolved oxygen (6.94 ± 1.67 mg/L), pH (6.79 ± 1.5) and water salinity (32.25 ± 7.25 ups) were recorded every other week. Final growth values were of 193.17 ± 11.50 mm for shell height and of 156.54 ± 25.30 g for the weight. Significant differences were found in both, the shell height and total weight, during the culture. Growth rate was of 0.29 mm/d and 0.33 g/d for the shell height and weight, respectively. The morphometric relation of valve height with total weight was allometric (2.39), and showed a positive correlation ($r = 0.88$). The final survival rate was of 92.79%. A recommendation was given for future cultures of *A. maura* at the mentioned estuary, considering the effect of the environmental variables on the species growth.

### 4.5. Reproduction

Studies about reproduction are important because they provide necessary information for obtaining seeds in aquaculture [52]. One of the bases to know the reproductive biology of an organism is the determination of the reproductive cycle, based on the analysis of the ecology and physiology of the species. The reproductive cycle is defined as the group of events that start with the activation of the gonad, going through gametogenesis, maturity and spawning (or gamete release) and gonad recession. In every bivalve, the gonadic development consists on gonad growth drawn from an undifferentiated germinal epithelium with mesodermic origin which remains dormant within the connective tissue after each reproductive cycle. During this stage known as 'undifferentiated', no gametes are observed, and it is characterized by intense metabolic activity conducted to store reserve substances. Subsequently, the germinal epithelium proliferates and originates the gametes. Generally, during the reproductive season, the spawning consists of a series of partial releases of the gonadic content of a certain population percentage that varies within a very small percentage and 100% of the overall population [53].

Reproductive studies of bivalves indicate that gamete maturity is mainly controlled by temperature and food and secondarily by salinity and the light period. In temperate environments, temperature is the main regulatory component, but in tropical environments, emphasis has been given to food availability which in turn acts as the regulatory factor on the reproductive cycle of many species [54]. In relation with the reproductive matters of *A. maura*, some research has been done about the reproductive biology and studies on the cytology of gonads of *A. maura* [18, 22, 30], where the cytologic evolution of the ultra-structural histologic level of the gonads was determined, showing similar aspects as those described for *Pinna nobilis* [30]. In 2000 [55], the variations of protein and lipid concentration in gonads of *A. maura* were established before the maturation process. The results showed that unlike other mollusks, the energy used during the gonadic maturation process was provided by the food as no evidence was found toward the nutrient supply from other somatic compartments. Rodríguez-Jaramillo [56] compared morphometric variables and the quality of female eggs of *A. maura* by inducing gametogenesis in three temperatures (20, 25 and 30°C) under lab conditions; on the other hand, a histochemical analysis was also performed which determined the lipid concentration in oocytes, revealing that temperature does influence the size of oocytes as well as triglyceride quantity. By comparing the spawning of those females brought from their natural habitat with that of females from a lab, a significant difference was found in the lipid index of immature normal oocytes where the lipid index of oocytes coming from natural spawning environments was more heterogeneous than the one from mature lab females. In another study about the reproductive cycle of *A. maura* [23], in Bahía Magdalena, Baja California Sur, only one reproductive period was observed with synchronous spawning that took place from January to March, giving the evidence that this species habits different biogeographic regions and has a great reproductive plasticity, adapting to local environments because reproduction is an endogenous answer genetically controlled by the environment known as the reproductive latitudinal gradient. In 2007 [57], it was found that this species reproduces itself year-round with two important reproductive peaks, one from April to July and the other one from October to November and with only one resting phase from August to September. The reproductive cycle showed a direct relation with the gonadosomatic index and an inverse relation with muscle activity, as well as a spawning and post-spawning period directly related with water temperature. The influence of phenomena such as 'El Niño' and 'La Niña' over the reproductive cycle of *A. maura* at la Ensenada de La Paz, Baja California Sur, Mexico [8] is different; during 'El Niño' (2004–2005), a high proportion of mature organisms (55%), high condition index values (ICG) (33–44%) and low spawning proportions were recorded with the decrease of temperature, whereas a high proportion of organisms with gonads under a reabsorption phase (17–100%) was registered in those areas with highest temperature anomalies; therefore, assuming negative effects of 'El Niño' during spawning processes turn out to be abnormal. In contrast, during 'La Niña', lower values of ICG (29–36%) were found and a decrease in the amount of mature organisms was noticed. The latter reveals that there were massive spawning events (36–71%) when the temperature increases. In conclusion, *A. maura* at the Ensenada de La Paz Baja California Sur showed an opportunistic reproductive strategy, which allows gamete reabsorption when unfavourable environmental conditions prevail and a distribution of the exceeding energy for reproduction purposes under favourable conditions. In 2016 [58], the advantages and disadvantages of three methods used to estimate the potential fecundity in commercial valued organisms of *A. maura* were analysed in northwest Mexico. Gonadic samples were taken during the reproductive season in March of 2003 and were processed by histological combined techniques with the Cavalieri Principle, stereological analysis of the gonad with the gauge method and the theoretical radius estimate of the oocytes. Comparing with other methods, the determination of the potential fecundity by the gauge method was precise in *A. maura* ($9.8 - 15 \times 10^6$ cells $ind^{-1}$). The reproductive cycle and the growth of *A. maura* were determined at the Ensenada Pabellones lagoon system in Sinaloa [59], where the sex proportion was of 0.57 females:1.72 males within the studied population. In reference to growth, there were no valve height difference between the males and females. The average height of those sample specimens ranged between 50.99 ± 4.86 mm and 218.16 ± 8.87 mm. Histological results confirmed that *A. maura* is a gonochoric organism that presents a synchronic development of the gonads. Maturation and spawning stages were observed during the entire study period except of March and May of 2008. The frequency of the development of the gonad stages obtained on a monthly bases suggested that this species reproduces twice a year with an important reproductive period from June to September, a minor reproductive period from November to February and two resting periods from July and August and in January and February of 2009.

### 4.6. Sensory and nutritional quality of the adductor muscle ('Callo')

In 2010 [60], the effect on seasonality and tide level on the adductor muscle quality of *A. maura* was assessed at San Buto estuary Baja California Sur, Mexico, by determining the optimum harvest periods, the muscle mass index (IRM) and some physical variables such as quality, texture, colour, percentage of water release and pH. Likewise, the proximal composition and gonadic index were analysed. According to the muscle mass index, June turned out to be the best month to harvest organisms because it showed the highest IRM value (26.9%), while not enough evidences were found in order to consider that the tide level was a decisive factor that could be affecting texture quality. Furthermore, the muscle colour of *A. maura* showed high chromatic values related with a dark colour and low luminosity indexes. A proximal analysis indicated a high protein content which represents good nutrient quality muscles. As for the pH, no relation between these variations and the percentage of released water was found, as there was no relation with the season or tide level. The quality of the adductor muscle [20] was evaluated in a bottom culture at San Buto estuary, Baja California Sur, Mexico, under three different densities (12, 24 and 48 org $m^{-2}$), showing a high correlation between the pH and the ability of water retention (CRA), where low pH values correspond to high CRA values. In terms of the components of colour parameters in the adductor muscles under the different culture densities, two types of colorations appeared, and based on the obtained values, a small part was white, and a big one was darker (cream colour). The most luminous part was the smallest in the 24 org.$m^2$ density, and the less luminous was the bigger part at 12 and 48 org.$m^2$ densities. Regarding texture, the muscle with best hardness and adhesiveness was found at the 48 org.$m^2$ density, which matches with the biggest adductor muscles in size and weight, whereas the most elastic and cohesive muscles were those found in the 24 org.$m^2$ density, revealing that those pH values recorded in these densities were found within the established ranges for considering an adductor muscle as a fresh product and therefore having a better texture. In some commercial catch samples of the pen shell, *A. maura* [61] diverse features related with the dietary quality were analysed and compared. The quality of the adductor muscle was analysed by sensory (preference and hedonic test) and instrumental methods (colour analysis, capacity of water retention, pH, colour, texture and proximal chemical analysis). The results revealed that the sensorial panel used for the preference analysis and hedonic acceptability only showed a significant preference for the general appearance of *A. maura*, and no differences in the colour and texture parameters were detected by the sensorial evaluation. The quality comparison by instrumental methods indicated that the adductor muscle of *A. maura* had less luminosity. Higher pH values in *A. maura* originated less luminosity; nevertheless, it gives it an advantage for a better capacity of water retention (CRA) and could even have influence in the higher hardness and chewiness found in the species.

### 4.7. Genetics

The degree of variability and the population genetic structure, as well as the estimates of demographic and phylogenetic history patterns of *A. maura* were evaluated by the isolation and characterization of the sequences of the gene COI [62]. A total of 45 nucleotidic sequences of the gene COI with a longitude of 650 bp from Guerrero Negro, Bahía Magdalena and La Paz Baja California Sur Mexico were evaluated. High haplotypic diversity (0.7333–0.9714) and lower nucleotidic diversity (0.00185–0.002912) were found for the three studied localities. A total of seven haplotypes were identified from which three are endemic for La Paz Baja California Sur. These results indicated that this could be a consequence of the larval dynamics of the species mainly related with its prolonged larval plankton stage and to the fact that during this stage its distribution is being highly influenced by the pattern of marine currents along the Pacific and Gulf of California. On the other hand, the neutral evolutionary tests suggest a population expansion event for the species in La Paz and the presence of a bottleneck for Guerrero Negro and Bahía Magdalena.

## Author details

Ruth Escamilla-Montes[1], Genaro Diarte-Plata[1]*, Antonio Luna-González[1], Jesús Arturo Fierro-Coronado[1], Héctor Manuel Esparza-Leal[1], Salvador Granados-Alcantar[3] and César Arturo Ruiz-Verdugo[2]

*Address all correspondence to: gdiarte@ipn.mx

1 Instituto Politécnico Nacional. Centro Interdisciplinario de Investigación para el Desarrollo Integral Regional Unidad Sinaloa, Guasave, Sinaloa, México

2 Departamento de Ingeniería en Pesquerías, Universidad Autónoma de Baja California Sur, Baja California Sur, México

3 Dpto. Biología, Universidad de Occidente, Unidad Los Mochis, Los Mochis, Ahome, Sinaloa, México

## References

[1] Ahumada-Sempoal M A, Serrano-Guzman S, Ruíz-García N. Abundance, population structure, and growth at the Pen shell *Atrina maura* (Bivalvia: Pinnidae) in a coastal tropical lagoon of the Mexican Pacific. Rev. de Biol. Trop. 2002. 50, 1092-1100.

[2] García-March J R. Contribution to the knowledge of the status of Pinna nobilis (L.) 1758 in Spanish coasts. Mem. Inst. Oc. Paul Ricard. 2003. 29-41.

[3] Giacobbe S. Epibiontic mollusc communities on *Pinna nobilis* L. (Bivalvia, Mollusca). J. Nat. Hist. 2002. 36: 1385-1396.

[4] Coronel, J S. Estudio gonadal de *Pinna rugosa* (Sowerby, 1835), Pinnidae, Mollusca, en el período comprendido entre agosto de 1979 y diciembre de 1980 en la Bahía de La Paz. Tesis de Licenciatura, Universidad Autónoma de Baja California Sur. 1981. 32 p.

[5] García-March J R. Aportaciones al conocimiento de la Biología de *Pinna nobilis* Linneo, 1758 (Mollusca: Bivalvia) en el litoral mediterráneo Ibérico. Publication Service of the University of Valencia, Valencia. 2006. 100 p.

[6] Yonge C M. Form and habitat in *Pinna cornea* Gmelin. Phil. Trans. Roy. Soc. Lond. B., Biol. Sci. 1953. 648: 335-374.

[7] Liang X Y. Morton B. The palial organ of *Atrina pectinata* (Bivalvia: Pinnidae): Its structure and function. J. Zool. Soc. L. 1988. 216: 469-477.

[8] Camacho-Mondragón M. Efecto de los fenómenos de el 'Niño' y 'La Niña' en la reproducción del hacha china *Atrina maura* (Sowerby, 1835) (Bivalvia: Pinnidae) en la Ensenada de La Paz, B.C.S., México. Tesis de Maestría. Centro Interdisciplinario de Ciencias Marinas IPN. 2009. 81 p.

[9] Vélez-Barajas J A, Fajardo-León M C. Pesquería de hacha. In: Casas-Valdez M. y Ponce-Díaz G. (eds.). I. Estudio del Potencial Pesquero y Acuícola de Baja California Sur. CIBNOR, La Paz, México. 1996. pp. 101-111.

[10] Barnes R D. Zoología de los invertebrados. Ed. Interamericana, México. 1988. 1240 p.

[11] Aguilar F. Contribución al estudio histológico de las gónadas de *Atrina maura* Sowerby 1835 (Mollusca, Fam. Pinnidae). Tesis de licenciatura. Escuela Nacional de Ciencias Biológicas, IPN, México. 1964. 27 p.

[12] Keen M. Sea shells of tropical west America. Marine Mollusks from Baja California to Peru. 2a. ed. Stanford University Press. Stanford, California, USA. 1971. 1064 p.

[13] Butler A J. Ecology of *Pinna bicolor* Gmelin (Mollusca: Bivalvia) in Gulf St Vincent, South Australia: Density, reproductive cycle, recruitment, growth and mortality at three sites. Aust. J. Mar. Freshw. Res. 1987. 38: 743-769.

[14] Munguia P. Succesional patterns on pen shell communities at local and regional scales. J. Anim. Ecol. 2004. 73: 64-74.

[15] Hanafi M I, Arshad A, Sidik J B, Abd M G, Khalijah S D. Biodiversity and distribution of pen shell (Bivalvia: Pinnidae) from the seagrass beds of Sungai Pulai, Peninsular Malaysia. Res. J. Fish. Hydrob. 2008. 3: 54-62.

[16] Poutiers J M. Pinnidae, bivalvos (Acephala, Lamellibranchia, Pelecypoda) In: Guía FAO para la identificación de especies para los fines de pesca. Pacifico Centro oriental, Plantas e invertebrados. 1995. 1: 174-177.

[17] Leyva-Valencia I. Termotolerancia, halotolerancia y temperatura óptima en juveniles de callo de hacha (*Atrina maura* Sowerby, 1835). Tesis de Licenciatura Nacional Autónoma de México, Campus Iztacala, México. 1999. 54 p.

[18] González-Corona M E. Estudio de la Fisiología Reproductiva y Gametogénesis del Callo de Hacha *Atrina maura* (Sowerby, 1835). Tesis de Maestría. Universidad de Sonora, Hermosillo Sonora. 2003. 74 p.

[19] Robles-Mungaray M. Desarrollo de la biotecnología para la producción de semilla en laboratorio, diploide y triploides de callo de hacha *Atrina maura* (Sowerby, 1835). Tesis de Maestría. Universidad Autónoma de Baja California Sur. La Paz, México. 2004. 66 p.

[20] Escamilla-Montes R. Cultivo del callo de hacha *Atrina maura* (Sowerby, 1835) (Bivalvia: Pinnidae) en el estero San Buto, Baja California Sur, México. Tesis de Doctorado. CIMACO-UABCS. La Paz, Baja California Sur, México. 2012. 232 p.

[21] Carta Nacional Pesquera. Callo de Hacha *Atrina maura*. Diario Oficial de la Federación. Ciudad de México, México. 2012. 236 p.

[22] Rodríguez-Jaramillo C, Maeda-Martínez A N, Valdez M E, Reynoso-Granados T, Monsalvo-Spencer P, Prado-Ancona D, Cardoza-Velasco F, Robles-Mungaray M, Sicard M T. The effect of temperature on the reproductive maturity of penshell *Atrina maura* (Sowerby, 1835) (Bivalvia: Pinnidae). J. Shellfish Res. 2001. 20: 39-47.

[23] Ceballos-Vazquez B P, Arellano-Martinez M, García-Domínguez F A, Villalejo- Fuerte M. Reproductive cycle of the rugose pen shell, *Pinna rugosa* Sowerby, 1835 (Mollusca: Bivalvia) from Bahia Concepcion, Gulf of California and its relation to temperature and photoperiod. J. Shellfish Res. 2000. 19.

[24] Robles-Rocha E. Caracterización histoquímica y morfológica de las larvas veliger de *Atrina maura* (Sowerby, 1835) durante su cultivo larvario. Tesis de Maestría. Universidad Autónoma de Baja California Sur, Mexico. 2010. 80 p.

[25] Robles-Mungaray M, Serrano-Guzmán S J, Hoyos-Chaires F. Cultivo de larvas y semillas de callo de hacha *Atrina maura* (Sowerby, 1835) en el CREMES de Bahía Kino, Son. México. Resúmenes del XI Simposium Internacional de Biología Marina. La Paz, B.C.S. 25 al 30 de Noviembre de. 1996.

[26] Mendo T, Koch V, Wolff M, Sínsel F, C Ruiz-Verdugo. Feasibility of intertidal bottom culture of the penshell Atrina maura in Bahia Magdalena, Baja California Sur, Mexico. Aquaculture. 2011. 314: 252-260.

[27] Basurto X, E Coleman. Institutional and ecological interplay for successful self-governance of community-based fisheries. Ecol. Econ. 2010. 69: 1094-1103.

[28] Baqueiro-Cárdenas E. Status of molluscan aquaculture on the Pacific coast of México. Aquaculture. 1984. 39: 83-93.

[29] Maeda-Martínez A N. Estado actual del cultivo de bivalvos en México. En A. Lovatelli, A. Farías e I. Uriarte (eds.). Estado actual del cultivo y manejo de moluscos bivalvos y su proyección futura: factores que afectan su sustentabilidad en América Latina. Taller Técnico Regional de la FAO. 20-24 de agosto de 2007, Puerto Montt, Chile. FAO Actas de Pesca y Acuicultura. No. 12. Roma, FAO. 2008. pp. 91-100.

[30] Enriquez-Díaz M R, Cáceres-Martínez C, Chávez-Villalba J, Le Pennec G, Le Pennec M. Gametogenesis of *Atrina maura* (Bivalve: Pinnidae) under artificial conditions. Invert. Reprod. Dev. 2003. 43: 151-161.

[31] Cardoza-Velásco F, Maeda-Martínez A N. An Approach to aquacultural production of the penshell *Atrina maura* Sowerby, 1835 (Bivalvia: Pinnidae) in northwest Mexico. J. Shellfish Res. 1997. 16: 311-318.

[32] Miranda Baeza A, Balderas-Cortés J. La pesquería de Callo de hacha (*Atrina sp., Atrina maura*) (Pelecypoda: Pinnidae) en la Bahía de Yavaros, Sonora, su inicio, estado actual y tendencia. Oceanología. 1994. 1: 89-98.

[33] Cendejas J M, Carballo M G, Juárez L M. Experimental spat collection and early growth of the pen shell, *Pinna rugosa* (Pelecypoda: Pinnidae) from the Gulf of California. Aquaculture. 1985. 48: 331-336.

[34] Ruiz-Verdugo C A, Cáceres-Martínez C. Estudio preliminar de captación de juveniles de moluscos bivalvos en la Bahía de La Paz, B.C.S. México. Inv. Mar. 1990. 15: 29-38.

[35] Bojorquez-Verástica G. Reclutamiento (captación de semilla mediante colectores) de *Pinna rugosa* en Bahía de La Paz, B.C.S. Resumen de I Simposium Nacional de Acuacultura México. 1986.

[36] Singh- Cabanillas J, Vélez-Barajas A, Fajardo-León M, Masso-Rojas C, Michel-Guerrero E. Aspectos biológicos del callo de hacha *Pinna rugosa* (Sowerby, 1835) en Bahía Concepción, Baja California Sur, México. Documento Interno CRIP-La Paz INP. SEMARNAT. 1995.

[37] Robles-Mungaray, M., Reynoso-Granados T., Cardosa-Velasco F., Monsalvo-Spencer P., 1999. Efecto del tamaño del organismo y el tipo de sedimento en el enterramiento de juveniles de callo de hacha *Atrina maura*. Participación en el VI Congreso de Ciencia y Tecnología del Mar. Mazatlán, Sinaloa, México. 17 al 19 de Noviembre de 1999.

[38] Goldchain-Goldin, Y.L. Estudio comparativo de crecimiento y supervivencia de las larvas flotadoras y nadadoras del hacha china *Atrina maura* (Sowerby, 1835) bajo condiciones de cultivo intensivo. Tesis de licenciatura del área de conocimiento de Ciencias del Mar. Universidad Autónoma de Baja California Sur, México. 2010. 52 p.

[39] Melguizo-Robles A. Producción de larvas y semillas del callo de hacha *Atrina maura* (Sowerby, 1835) en el laboratorio experimental de acuacultura de Pichilingue de la universidad Autónoma de Baja California Sur. Tesis de Licenciatura. UABCS. La Paz, México. 2011. 50 p.

[40] Farías A. Nutrición y alimentación en moluscos bivalvos. En A. Lovatelli, A. Farías e I. Uriarte (eds.). Estado actual del cultivo y manejo de moluscos bivalvos y su proyección futura: factores que afectan su sustentabilidad en América Latina. Taller Técnico Regional de la FAO. 20-24 de agosto de 2007, Puerto Montt, Chile. FAO Actas de Pesca y Acuicultura. No. 12. Roma, FAO. 2008. pp. 297-308.

[41] Serrano-Casillas G. Crecimiento y Supervivencia de la almeja catarina *Argopecten ventricosus* (Sowerby II, 1842) bajo condiciones de cultivo en suspensión y fondo en Bahía Concepción. Tesis de Maestría Centro Interdisciplinario de Ciencias Marinas-Instituto Politécnico Nacional. La Paz, B.C.S. México. 2004. 76 p.

[42] Numaguchi K, Y Tanaka. Effects of salinity on mortality and growth of the spat of the Pearl oyster *Pinctada fucata martensii*. Bull. Nat. Res. Inst. Aquacult. 1986. 9: 41-44.

[43] Lodeiros C, Freites L, Núñez M, Himmelman J. Growth of the Caribbean scallop *Argopecten nucleus* (Born 1780) in suspended culture. J. Shellfish Res. 1993. 12: 291-294.

[44] Leyva-Valencia I, Maeda-Martínez A N, Sicard M T, Roldan L, Robles-Mungaray M. Halotolerance, upper thermotolerance, and optimum temperature for growth of the penshell *Atrina maura* (Sowerby, 1835) (Bivalvia: Pinnidae). J. Shellfish Res. 2001. 20: 49-54.

[45] Côté J, Himmelman J H, Claereboudt M R, Bonardelli J. Influence of density and depth on the growth of juvenile giant scallops (*Placopecten magellanicus*, Gmelin, 1791) in suspended culture in the Baie des Chaleurs. Can. J. Fish. Aquat. Sci. 1993. 50: 1857-1869.

[46] Kingzen B C N. Overwintering growth of juvenile Japanese scallops *Patinopecten yessoensis* in pearl nets effects of density and wave exposure. J. Shellfish Res. 1990. 10: 238-246.

[47] Parsons J, Dadswell M. Effect of stocking density on growth, production, and survival of the giant scallops, *Placopecten magellanicus*, held in intermediate suspension culture in Passamaquoddy Bay, New Brunswick. Aquaculture. 1992. 103: 291-309.

[48] Claereboudt M, Bureau D, Cotê J, Himmelman J. Fouling development and its effect on the growth of juvenile giant scallops (*Placopecten magellanicus*) in suspended culture. Aquaculture. 1994. 121: 327-342.

[49] Vélez A. Crecimiento, edad y madurez sexual del ostión *Crassostrea rhizophorae* de Bahía de Mochima. Bol. Inst. Oceanogr. Univ. Oriente. 1976. 15: 65-72.

[50] Mendo-Aguilar T. Suistability of bottom culture of the penshells *Atrina maura* and *Pinna rugosa* (Family Pinnidae) in Bahia Magdalena, Baja California Sur, Mexico. Thesis in International Studies in Aquatic Tropical Ecology. Universität Bremen. 2008. 44 p.

[51] Góngora-Gómez A M, Hernández-Sepúlveda J A, A L Domínguez-Orozco, B P Villanueva-Fonseca, M García-Ulloa. Crecimiento, supervivencia y condición del callo de hacha Atrina maura (Sowerby, 1835) (Bivalvia: Pinnidae) en el estero La Piedra, Sinaloa, México Avances en Investigación Agropecuaria. 2015. 19 (3): 73-86.

[52] Arsenault D J, Himmelman J H. Spawing of the Iceland Scallop (*Chlamys islandica* Müller, 1776) in the Northern Gulf of St. Lawrence and its relationship to temperature and phytoplankton abundance. The Veliger. 1998. 41: 180-185.

[53] Thompson J T, Newell R I E, Kennedy V S, Mann R. Reproductive processes and early development, In: Kennedy V.S., R.I.E. Newell & A.F. Eble (Eds.) The Eastern Oyster *Crassostrea virginica*. Maryland Sea Grant Book, Maryland. 1996. pp. 335-370.

[54] Prieto A S, Vásquez M, Ruiz L. Dinámica energética en una población del mejillón *Perna perna* (Filibranchia: Mytilidae) en el noreste del Estado Sucre, Venezuela. Rev. Biol. Trop. 1999. 47: 499-410.

[55] Enríquez-Díaz M R. Estudios citológicos y químicos en el músculo, glándula digestiva y gónada durante la gametogénesis inducida de *Atrina maura* (Sowerby, 1835). Tesis de Maestría en Ciencias en Acuacultura, Universidad Autónoma de Baja California Sur, La Paz, B.C.S., México. 2000. 63 p.

[56] Rodríguez-Jaramillo M C. Efecto de la temperatura sobre la gametogénesis en el callo de hacha *Atrina maura* (Sowerby, 1835). Tesis de Maestría. Centro Interdisciplinario de Ciencias Marinas-Instituto Politécnico Nacional. La Paz, B.C.S. México. 2004. 69 p.

[57] Ángel-Pérez C, Serrano-Guzmán S J, Ahumada-Sempoal M A. Ciclo reproductivo del molusco *Atrina maura* (Pterioidea: Pinnidae) en un sistema lagunar costero, al sur del Pacífico tropical mexicano. Rev. Biol. Trop. 2007. 55: 839-852.

[58] Cáceres-Puig J I, Saucedo P E, Chávez-Villalba J, C Cáceres-Martínez. Analysis of different methods to estimate fecundity in bivalve mollusks. J. Invert. Reprod. Dev. 2016. 60 (1): 12-16.

[59] Góngora-Gómez A, M García-Ulloa, M Arellano-Martínez, S Abad, A Domínguez, J Ponce-Palafox. Annual reproductive cycle and growth of the pen shell Atrina maura (Pterioidea: Pinnidae) on sand-bottom culture in the Ensenada Pabellones lagoon system, Gulf of California, Mexico. J. Invert. Reprod. Dev. 2016. 60(1): 12-16.

[60] García-Gutiérrez J O. Análisis de la variación estacional de la calidad del músculo aductor de los callos de hacha Pinna rugosa y Atrina maura en cultivo en fondo. Tesis de Maestría. Ciencias Marinas y Costeras. Universidad Autónoma de Baja California Sur. México. 2010. 82 p.

[61] Mora-Mayo I J. Evaluación de características químicas, físicas y sensoriales relacionadas con la calidad alimenticia de los callos de hacha Pinna rugosa (Sowerby, 1835) y *Atrina maura* (Sowerby, 1835). Tesis de Licenciatura de Ingeniero en Pesquerías. Universidad Autónoma de Baja California Sur. México. 2014. 78 p.

[62] Loeza-Quintana T R. Evaluación de la estructura Genético-Poblacional del callo de hacha *Atrina maura* en aguas de Baja California Sur mediante la caracterización de las secuencias del citocromo Oxidasa I. Tesis de Licenciatura de Biología Marina. Universidad Autónoma de Baja California Sur, México. 2009. 93 p.

[63] Almaraz-Salas J C. Primer cultivo experimental de callo de hacha *Atrina maura* (Sowerby, 1835) en la ensenada La Palmita, Navolato, Sinaloa. Tesis de Licenciatura. Universidad de Occidente Unidad Guasave, Guasave, Sinaloa, México. 2008. 74 p.

Chapter 2

# Possible Poecilogony Due to Discontinuous Multifactorial Inheritance in Some Mediterranean Species of *Raphitoma* (Mollusca, Conoidea, Raphitomidae)

Thanasis Manousis, Constantinos Kontadakis,
George Mbazios, Georgios Polyzoulis and
Sofia Galinou-Mitsoudi

Additional information is available at the end of the chapter

http://dx.doi.org/10.5772/67847

**Abstract**

At least 10 pairs of similar, most probably closely related, species of *Raphitoma* are often sampled in the same Mediterranean localities. In each pair, one member bears a planktotrophic protoconch and the other a lecithotrophic one. We propose that the phenomenon may be attributed to a simple gene that functions in conjunction with others and environmental factors to exhibit a discontinuous multifactorial inheritance leading to poecilogony. Below a threshold, the animals may produce fewer and larger germ cells, giving rise to fewer and larger eggs and large lecithotrophic embryos with large paucispiral protoconch I, while above that threshold, more and smaller germ cells leading to smaller eggs and to planktotrophic larvae with small protoconch I and large multispiral protoconch II. Preliminary measurements are in support of our hypothesis. Analysis of mitochondrial DNA markers as well as interbreeding experiments could bring an end to the existing confusion.

**Keywords:** *Raphitoma*, protoconch, Mediterranean sea

## 1. Introduction

Marine shelled gastropods follow, in general, three types of larval development: (1) direct, in which embryo development is completed in the egg capsule and the juvenile is ready to commence its benthic life immediately after hatching. (2) Lecithotrophic development in which, after hatching, the larva spends from a few minutes to several days as a veliger without feeding or growing. (3) Planktotrophic development, in which the larva hatches as a veliger and spends a few days to several months in the planktonic mode of life during which it feeds and grows [1]. In all cases, a larval shell is formed which is called protoconch.

The size of fossilized early protoconches suggests that feeding (planktotrophic) embryos first appeared at the transition from the Cambrian era to the Ordovician as Ordovician protoconches were smaller than Cambrian ones indicating smaller Ordovician eggs and offspring. It is believed that planktotrophy would prolongate their escape from benthic predators and that an increasing nutrient supply and availability of photoautotrophic plankton in the world oceans of that period have facilitated both planktotrophy and suspension feeding [2]. The eventual loss of planktotrophy in the Pliocene has been documented in some taxa of Conoidea (*Raphitoma, Bela*) and is believed to be irreversible [3]. The same phenomenon has occurred in the families such as Cerithiidae (*Bittium*), Turritellidae (*Turritella*), Rissoidae (*Rissoa, Pusillina*), and Nassariidae (*Nassarius*) [4, 5].

Planktotrophic mode of life with long-living planktonic larvae is considered advantageous in the dispersal of the larvae as they may drift considerable distances with the currents [6, 7]. In addition, it has been demonstrated that the switch in some conoidean taxa from the planktotrophic to the non-planktotrophic development has increased their adaptive radiation, especially in polar or insular region, or in groups with narrow bathymetric distribution [3, 7]. In the case of *Oenopota elongata* Bogdanov, 1989, and in the polar genus or bathyal Indo-Pacific genus *Bathytoma* Harris & Burrows, 1891, the loss of planktotrophy actually preceded adaptive radiation [8–11].

Within the Neogastropoda, Conoidea is a diverse superfamily of venomous and exclusively marine gastropods which harbors more than 300 genera, 4000 known species, and an estimated number of over 12,000 existing species [3, 12, 13]. Due to the species richness and the extensive homoplasy among shell's features and the anterior alimentary system, they have resisted repeated attempts to be permanently classified with those attempts to be hindered primarily by the absence of a stable phylogenetic framework. Rather recently, DNA analysis provided an updated classification and divided the superfamily of Conoidea into 13 families [14–16]. Among them, the position of Raphitomidae is not sufficiently secured as a clade of Conoidea and for that reason there has been an on-going attempt to clarify the phylogeny of its Mediterranean members through a greater number of taxonomic data [17, 18]. These later publications have taken into account the pioneering works of Thorson [19, 20], concerning the relationship between the morphology of the protoconch and the type of the larval development in Caenogastropoda, e.g., the dichotomy "multispiral protoconch/planktotrophic development" and "paucispiral protoconch/lecithotrophic development" that has been widely accepted [6, 21]. Although some authors have used this dichotomy to divide species into planktotrophic genera and non-planktotrophic ones [3, 22, 23], it has been clearly demonstrated that such a division based exclusively on the morphology of the protoconch produces artificially separated polyphyletic taxa of otherwise distinct genera of closely related species. A separation, for instance, between the genus *Raphitoma* (with multispiral protoconch) and the genus *Philbertia* (with a paucispiral protoconch) is inconsistent and, therefore, must be rejected [17]. Nevertheless, this planktotrophic or lecithotrophic mode of larval life as reflected in the morphology of its protoconch comprises a useful basis for distinguishing different species in several genera in Prosobranchia and Opisthobranchia Mollusca [24–29].

Under the weight of the above unsettled situation, the aim of this study was to aid the classification of the Mediterranean *Raphitoma* species by presenting new material from the Hellenic seas, demonstrating them in detailed descriptions and coming up with new ideas on the protoconch issue. More specifically, our objective was to put to test the hypothesis that protoconch poecilogony exists within a species population, with lecithotrophic larvae being produced from fewer and larger eggs, while planktotrophic ones from more and smaller eggs that result merely by additional cleavages of the maternal stem cells.

## 2. Materials and methods

Sampling of gastropods specimens was conducted from October 2008 to October 2016 in certain locations of Greece, according to Manousis and Galinou-Mitsoudi [30]. The *Raphitoma* species recognition was based on (a) systematic guides and atlases [23, 31–38], (b) faunistic and review articles [39, 40], (c) studies on the Mollusca fauna in the Hellenic seas [30, 41–47]. Information from specific websites was also taken into account. For the species nomenclature update (31 December 2016), besides the Marine Biodiversity and Ecosystem Functioning EU Network of Excellence (MarBEF, www.marbef.org) and the World Register of Marine Species (WoRMS, http://www.marinespecies.org), the Taxonomic on-line Database on European Marine Mollusca (CLEMAM, www.somali.asso.fr,) was followed. In addition, the Ellenic Network on Aquatic Invasive Species (ELNAIS, https://services.ath.hcmr.gr, 31 December 2016) and the Marine Mediterranean Invasive Alien Species database (MAMIAS, http://mamias.org/taxonomicgroup.php) were used for the status of possible alien species in the Hellenic and Mediterranean seas. Protoconch whorls were counted according to Verduin [48]. The measurement of the protoconch maximum diameter was performed at top view according to Gofas and Oliver [49]. The shell's slenderness (h/w) was estimated including the outer lip of the aperture in the shell's width. The protoconch I maximum diameter deriving after each hypothetical additional cleavage of the stem cells was calculated by multiplying measured maximum diameter of the lecithotrophic protoconch of each, so-called, sibling species by the approximate factor of 0.7937 receiving as a fact that each stem cell is a sphere that divides into two equal spheres at cleavage.

The specimens are deposited in the premises of the Alexander Technological Educational Institute of Thessaloniki and those of Dr T. Manousis, Mr Constantinos Kontadakis, Mr George Mbazios and Mr Georgios Polyzoulis. Scientists are welcome to have access to the biological material at will.

## 3. Results

Among the 570 specimens collected, three pairs of 'sibling' species of *Raphitoma* appropriate for comparison were identified, which are presented in **Figures 1–4** and described below in an alphabetical order.

*Raphitoma contigua* (Monterosato, 1884) (**Figure 1c** and **d**).

*Collection stations*: 13 shells (5.00–10.00 mm long, 2.10–4.20 mm wide) were collected from detritus material trapped in small-scale fishing nets at 10–120 m from mixed bottoms of Central Saronikos Gulf: Pantokratoras, Preveza.

*Description*: The shell is hyaline, fusiform, and 2.2 times as long as wide. Its multispiral protoconch, which is approximately 390 μm (mean) wide and 490 μm (mean) high, bears a 230 μm (mean) protoconch I of slightly more than 1 whorl that is decorated with regularly placed small tubercules. Protoconch II consists of almost 1.5 diagonally cancellated and convex whorls, the last of which bears a weak keel with erasures before the onset of the teleoconch. The teleoconch consists of five convex whorls separated by a deep suture. The body whorl occupies almost 65% of the total length and bears 15–16 orthocline axial ribs with interspaces approximately 1.5 times wider than the ribs themselves and 17–18 spiral cords slightly thinner than the ribs, six of which are situated above the aperture and the rest 12 below the extension of the suture. The spiral cords in their intersections with the axial ribs form erasures in the form of small elongated rectangular tubercules. The tubercles on the first two adapical cords are spiny and close to each other forming a subsutural ramp. The shell's inner wall viewed through the aperture exhibits a transparency. The aperture occupies approximately 45% of the shells length and exhibits a smooth and sigmoid columella, angled at its upper part. The anterior siphonal canal is short and wide, while the posterior one is deep and narrow. The color is uniformly light yellowish tan and in certain shells darker with irregularly situated areas of cream-white and a white subsutural band by the ramp and one more as an extension of the suture on the body whorl. Protoconch I is white and the rest is light yellowish tan with white erasures.

*Similar species*: *R. contigua* superficially resembles a number of congeneric Mediterranean *Raphitoma* species, but it is different from: *R. alternans* (Monterosato, 1884), in which, the latter has a paucispiral protoconch, a more elongated shell, and a different color pattern; *R. atropurpurea* in its color, which is honey-red instead of purple-brown in *R. atropurpurea* and in its less slender spire than *R. atropurpurea*; *R. densa* (Monterosato, 1884) in the color and the paucispiral protoconch of the late; *Raphitoma lineolata* (Bucquoy, Dautzenberg and Dollfus, 1883) in its more inflated profile, its more robust shell, in the presence of a narrow subsutural ramp in *R. contigua,* in its wider aperture, and in the absence in *R. contigua* of a subsutural white cord present in *R. lineolata*; *R. oblonga* (Jeffreys, 1867) in its wider aperture and the different color pattern; *R. spadiana* mainly because the latter bears a paucispiral protoconch [17].

*Habitat and distribution*: Infralittoral, on sandy and muddy bottoms, also under stones and in the holes of the rocks [38]. Central Mediterranean Sea [17, 36, 38].

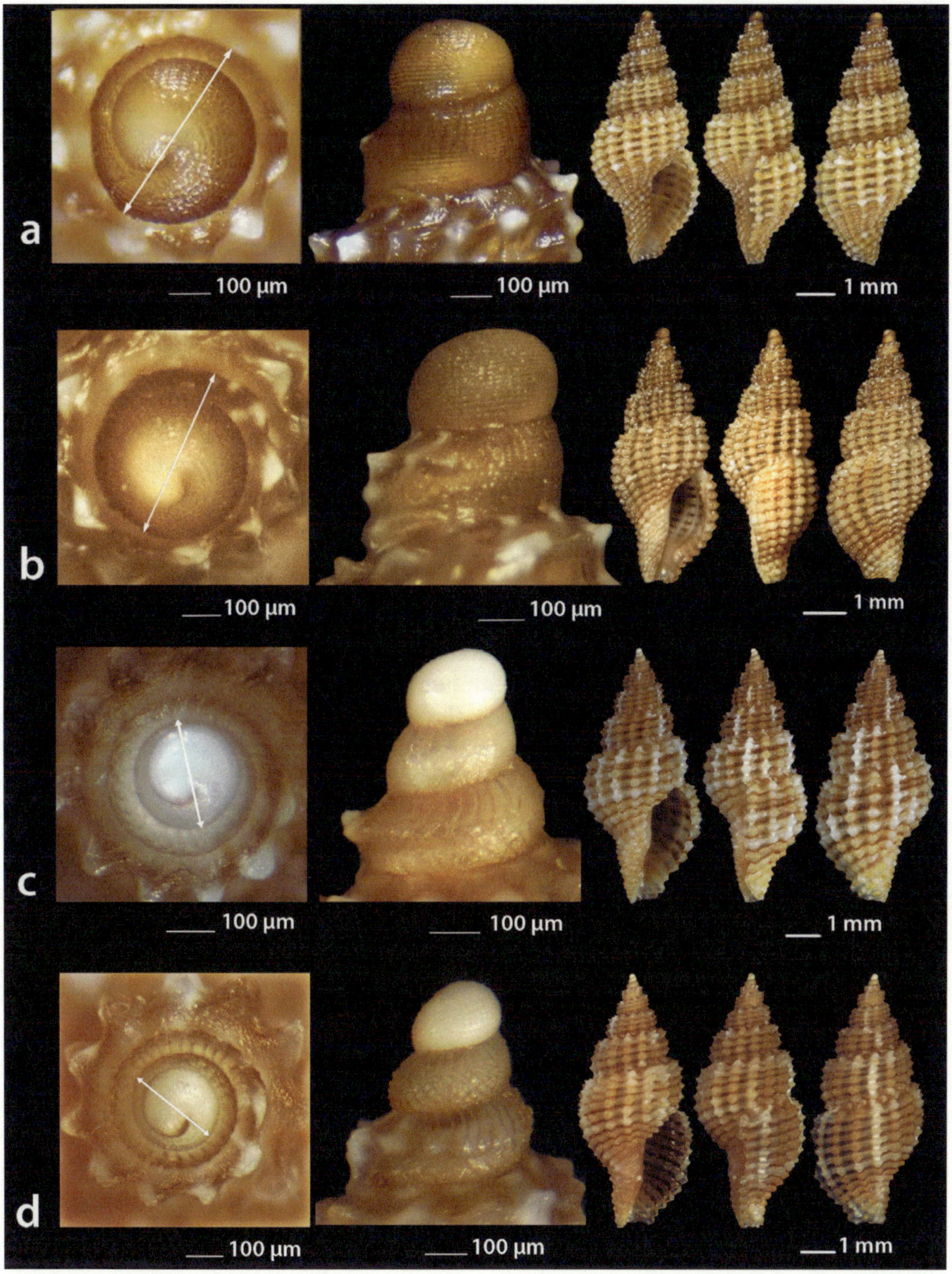


**Figure 1.** (a and b): *R. spadiana*, (c and d): *R. contigua*.

*Status*: Uncommon [34]. First record for the Greek seas.

*Raphitoma lineolata* (Bucquoy, Dollfus & Dautzenberg, 1883) (**Figure 2c** and **d**).

*Collection stations*: 44 shells (4.00–8.00 mm long, 1.40–2.80 mm wide) were collected from detritus material trapped in small-scale fishing nets at 10–40 m depth from mixed bottoms of Palioura, Epanomi; Pyrgadikia, Chalkidiki; Psakoudia, Chalkidiki; Anavyssos, Attiki; Legrena, Attiki; Central Saronikos Gulf; Syros Island; and Pantokratoras, Preveza.

*Description*: The slim, small, thin but solid and fusiform shell is almost 2.5 times as long as wide. Its multispiral protoconch, which is 410 μm (mean) wide and 510 μm (mean) high, bears a protoconch I of 226 μm (mean) and consists of 2.7 convex whorls with a white nucleus decorated with diagonally cancellated striae and the last whorl with a weak keel before the onset of the teleoconch. The later consists of five moderately convex whorls separated by a deep and canaliculated suture with a narrow ramp. The body whorl occupies in certain shells almost 75% and in others almost 60% of the total length, bearing 17–18 orthocline axial ribs with interspaces approximately as wide as the ribs and 16–17 spiral cords slightly thinner than the ribs, six of which are situated above the aperture and the rest bellow. The spiral cords in their intersections with the axial ribs exhibit mammiliform tubercules. The tubercules on the first two adapical cords are spiny and close to each other. The shell's inner wall viewed through the aperture exhibits a transparency. The aperture occupies something more than 40% of the shells length and exhibits a smooth and S-shaped columella angled at its upper part. The anterior siphonal canal is short and wide, while the posterior one is conspicuous. The outer lip bears 9–10 strong teeth with the first one delimiting the posterior canal and the last the anterior. The whole shell exhibits either a lemon-yellow or a honey-red background color with beige or honey-yellow the tubercules, respectively, and irregularly placed white highlights all over and an interrupted white band as a prolongation of the suture on the body whorl.

*Similar species*: *R. lineolata* superficially resembles *R. contigua* (Monterosato, 1884), from which it differs in that *R. lineolata* is slender, less robust, bears a narrow subsutural ramp, and has a smaller aperture and only a narrow white band on its body whorl [17, 18].

*Habitat and distribution*: Infralittoral, on sandy and muddy bottoms. During the day, also under stones and crevices. It can be found live among the rhizomes of *Posidonia oceanica* [38]. Mediterranean Sea [[18, 34, 38], http://www.conchigliedelmediterraneo.it].

*Status*: Uncommon [34, http://www.conchigliedelmediterraneo.it]. First record for the Greek seas.

*Raphitoma locardi* Pusateri & Giannuzzi 2013 (**Figures 3c** and **d**, and **4**).

*Collection stations:* Two shells (7.90 and 8.00 mm long, 2.90 and 2.95 mm wide, respectively) were collected from detritus material trapped in small-scale fishing nets at about 60 m depth from mixed bottoms of Central Saronikos Gulf.

*Description:* The solid and fusiform shell is almost 2.3 times as long as wide. Its multispiral protoconch is 400 μm (mean) wide and 440 μm (mean) high, bears a protoconch I of 221 μm (mean)

http://dx.doi.org/10.5772/67847

**Figure 2.** (a and b): *R. smriglioi*, (c and d): *R. lineolata*.

and consists of 3.0 convex whorls with nucleus decorated with diagonally cancellated striae and the last whorl with a weak keel before the onset of the teleoconch. The latter consists of five moderately convex whorls separated by a deep and canaliculated suture with a ramp. The inflated body whorl occupies almost 65% of the total length, is rounded at its base, and bears 17–18 opisthocline axial ribs with slightly wider interspaces and 19–20 spiral cords slightly thinner than the ribs, seven of which are situated above the aperture and the rest below. The spiral cords in their intersections with the axial ribs exhibit pointed tubercules that lend a rough appearance. The tubercules on the first two adapical cords are particularly spiky. The shell's inner wall viewed through the aperture exhibits a transparency. The aperture occupies some 44% of the shells length and exhibits a smooth and S-shaped columella, angled at its upper part. The anterior siphonal canal is rather long and wide, while the posterior one is narrow and shallow. The outer lip bears 10–11 strong teeth with the first one delimiting the posterior canal and the last the anterior. The whole shell exhibits a chestnut-purple background color with beige or honey-yellow irregularly placed highlights all over and an interrupted white band as a prolongation of the suture on the body whorl, while the aperture is lilac and the apex dark chestnut-purple with irregularly placed yellow spotlets. The animal's body is gray turning lighter towards the foot, the foot and the antennas are off white with white speckles and there is a white area around the eyes. The siphon is dark gray with irregularly placed white speckles but its white edge.

*Similar species: R. locardi* superficially resembles *R. contigua* (Monterosato, 1884), from which it differs in that *R. locardi* is slender, less robust, bearing a narrow subsutural ramp, and has a smaller aperture and only a narrow white band on its body whorl [17].

*Habitat and distribution:* Mediterranean Sea [[18, 34], http://www.conchigliedelmediterraneo.it].

*Status:* Uncommon ([34], http://www.conchigliedelmediterraneo.it). First record for the Greek seas.

*Raphitoma philberti* (Michaud, 1829) (**Figure 3a** and **b**).

*Collection stations*: 35 shells (8.90–9.15 mm long, 3.70–3.80 mm wide) were collected from *Zostera* bed, 0.2 m, Cape, Epanomi; several depths of Psakoudia, Chalkidiki; Nea Roda, Chalkidiki; Alonissos Island; Amorgos Island; Elaphonisos, Lakonia; Siros Island; Elaphonisi, Crete; Karpathos Island.

*Description:* Shell solid, fusiform, and almost 2.4 times as long as wide. Its paucispiral protoconch is 358 μm (mean) wide and 420 μm (mean) high and consists of approximately 1.25 convex whorls, a white nucleus decorated with diagonally cancellated striae and bears a weak keel before the onset of the teleoconch. The later consists of five moderately convex whorls separated by a deep and canaliculated suture with a ramp. The inflated body whorl occupies almost 65% of the total length, is rounded at its base, and bears 17–18 slightly opisthocline axial ribs with slightly wider interspaces and 17–19 spiral cords slightly thinner than the ribs, 6–7 of which are situated above the aperture and the rest below. The spiral cords in their intersections with the axial ribs exhibit pointed tubercules lending a rough appearance. The tubercules

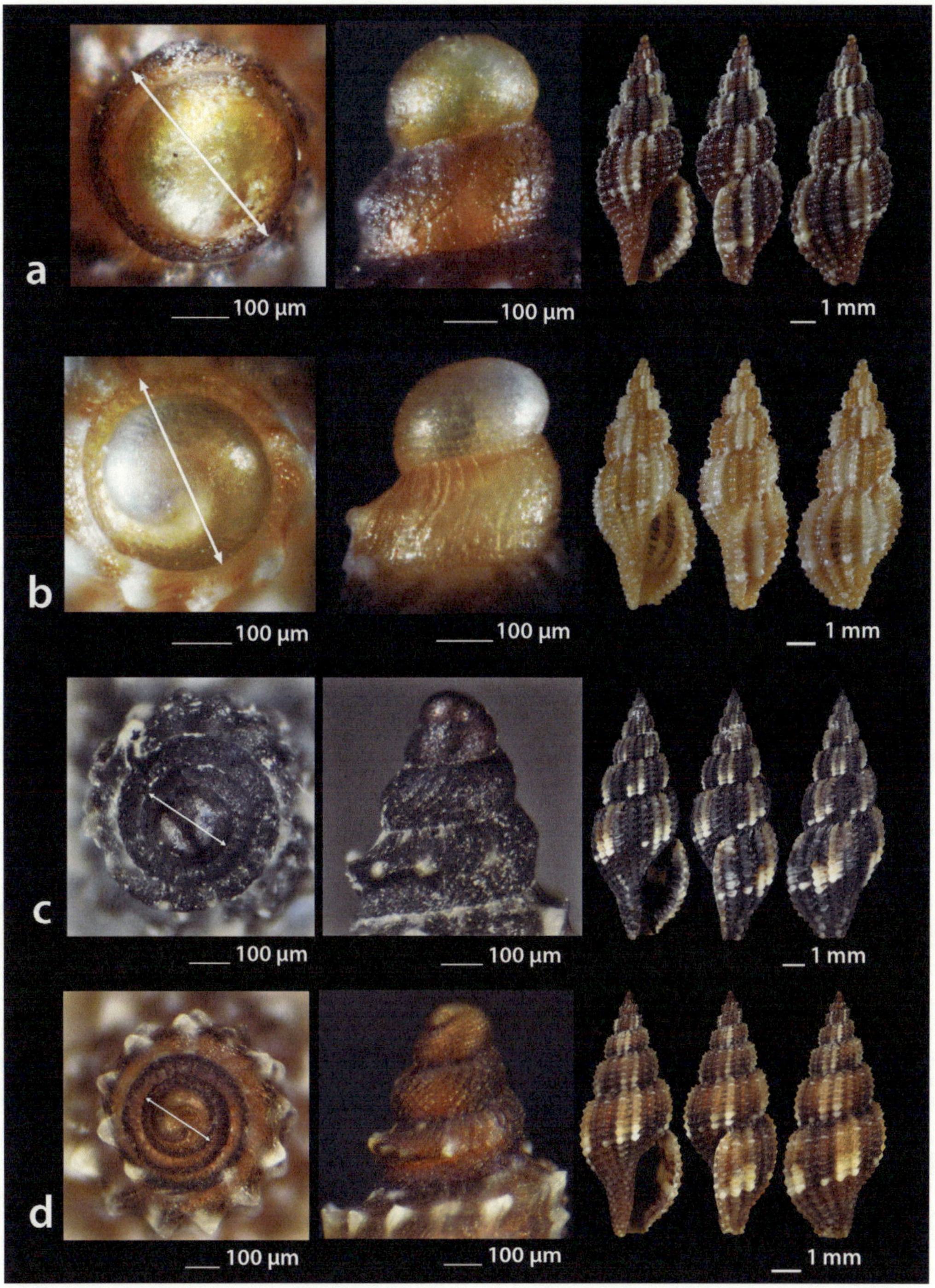


**Figure 3.** (a and b): *R. philberti,* (c and d) *R. locardi.*

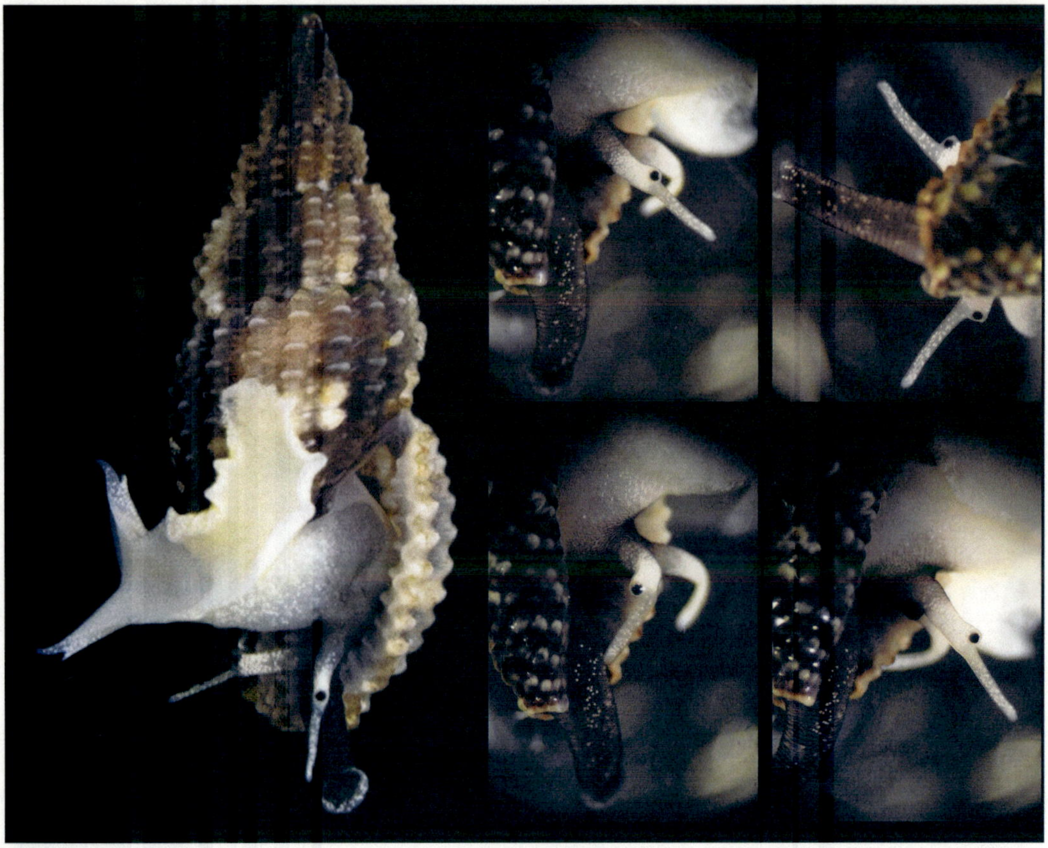

**Figure 4.** *R. locardi* live individual.

on the first two adapical cords are particularly spiky. The shell's inner wall viewed through the aperture exhibits a transparency. The aperture occupies some 40% of the shells length and exhibits a smooth and S-shaped columella angled at its upper part. The anterior siphonal canal is rather long and wide while the posterior one is narrow and shallow. The outer lip bears 9–11 strong teeth with the first one delimiting the posterior canal and the last the anterior. The whole shell exhibits a chestnut-purple or, rarely, a honey-yellow background color with white or light beige irregularly placed highlights all over and an interrupted white band as a prolongation of the suture on the body whorl, while the aperture is lilac and the apex dark chestnut-purple with irregularly placed yellow spotlets.

*Similar species*: *R. philberti* superficially resembles *R. contigua* (Monterosato, 1884), from which it differs in that *R. philberti* is slender, less robust, and has a smaller aperture. It differs from *R. locardi*, in which it bears a lecithotrophic protoconch while *R. locardi* bears a planktotrophic one [17].

*Habitat and distribution*: Mediterranean Sea [18, 34, http://www.conchigliedelmediterraneo.it], Central and Eastern Mediterranean Sea [18].

*Status*: Uncommon [34, http://www.conchigliedelmediterraneo.it). Species already reported for the Greek seas.

*Raphitoma smriglioi* Pusateri & Giannuzzi-Savelli, 2013 (**Figure 2a** and **b**).

*Collection stations*: Nine shells (3.75–9.70 mm long, 2.15–3.80 mm wide) were collected from detritus material trapped in small-scale fishing nets at about 50 m depth from mixed bottoms of Central Saronikos Gulf; found beached in Legrena, Attiki; mixed bottom of Pantokratoras, Preveza; *Zoster*a bed, 0.2 m, Cape, Epanomi.

*Description*: The slim, small, thin but solid and fusiform shell is almost 2.5 times as long as wide. Its paucispiral protoconch, which is 375 μm (mean) wide and 390 μm (mean) high, consists of approximately 1.25 convex whorls, bears a white nucleus, and is decorated with irregularly cancellated fine striae. The mature teleoconch consists of five moderately convex whorls separated by a deep and canaliculated suture. The body whorl occupies some 70% of the total length and bears 17–18 orthocline axial rips with interspaces almost twice as wide as the ribs. Spiral decoration of 17–18 spiral cords is slightly thinner than the ribs, six or seven of which are situated above the aperture and the rest below. The spiral cords in their intersections with the axial ribs exhibit small and elongated tubercles. The first two adapical cords are vestigial, spiky and close to each other. The shell's inner wall viewed through the aperture exhibits a transparency. The aperture occupies some 44 % of the shell length and exhibits a smooth and S-shaped columella angled at its upper part. The anterior siphonal canal is short and wide while the posterior one (anal) is conspicuous. The whole shell exhibits a variety of shades from lemon-yellow to honey-red background color with slightly lighter the cords, and irregularly placed white highlights all over.

*Similar species*: *R. smriglioi* differs from *R. contigua* (Monterosato, 1884) in its paucispiral protoconch and the lack of a subsutural ramp; *R. lineolata* (Bucquoy, Dollfus & Dautzenberg, 1883) in its paucispiral protoconch; from *R. philberti* (Michaud, 1829) in its white and smaller nucleus of the protoconch and its relatively smaller body whorl; from *R. spadiana* Pusareti & Giannuzzi-Savelli, 2012 in the lack of a subsutural ramp and in the protoconch which is slightly smaller and slender in *smriglioi* (395 μm × 400 μm) while almost 10% larger in *R. spadiana* (425 μm × 450 μm) [17, 18].

*Habitat and distribution*: Mediterranean Sea [18, 34, http://www.conchigliedelmediterraneo.it], Central and Eastern Mediterranean Sea [18].

*Status*: Uncommon ([35]; http://www.conchigliedelmediterraneo.it). First record for the Greek seas.

*Raphitoma spadiana* Pusareti & Giannuzzi-Savelli, 2012 (**Figure 1a** and **b**).

*Collection stations*: Two live specimens and three shells (2.00–9.80 mm long, 1.05–4.00 mm wide) were collected: detritus material trapped in small-scale fishing nets at 60–80 m depth from mixed bottoms of Kardamili, Messinia and South Saronic Gulf.

*Description*: The shell is hyaline, fusiform and almost 2.3 times as long as wide. Its paucispiral protoconch, which is approximately 457 μm (mean) wide and 470 μm (mean) high, consists of almost 1.35 irregularly cancellated and convex whorl, the first of which is decorated only with fine spiral striae. The teleoconch consists of five convex whorls separated by a deep suture. The body whorl occupies almost 65% of the total length and bears 16–17 orthocline axial ribs with interspaces approximately two times wider than the ribs themselves and 19–20 spiral cords thinner than the ribs, six of which are situated above the aperture and the rest 14 below the aperture. The spiral cords in their intersections with the axial ribs form erasures in the form of small elongated rectangular tubercles. The tubercles on the first two adapical cords are spiky and close to each other. The shell's inner wall viewed through the aperture exhibits a transparency. A narrow ramp is evident immediately below the suture, formed by the vestigial first two spiral cords and the much prominent and spiky third cord. The aperture occupies approximately 45% of the shells length and exhibits a smooth and slightly sinuous columella in its lower part, angled at its upper part. The anterior siphonal canal is short and wide, while the posterior one is deep and narrow. The outer lip bears 11 strong teeth with the first one delimiting the posterior canal and the last the anterior. The shells are of yellow-beige background color, while the tubercles and some irregularly situated areas or isolated tubercles are of lighter color. The body whorl usually bears at its middle a lighter color spiral band as a prolongation of the suture.

*Similar species*: *R. spadiana* is different from: *R. alternans* (Monterosato, 1884), in which the latter is slender and with a different color pattern; *R. atropurpurea* in its color which is light red-brown instead of purple-brown in *R. atropurpurea* and in its more inflated spire; *R. contigua*, in which the latter bears a multispiral protoconch, is larger, of lighter color, and more robust; *R. densa* (Monterosato, 1884) in the color and the less dense sculpture; *R. lineolata* (Bucquoy, Dautzenberg and Dollfus, 1883), in which *R. lineolata* has a less robust shell, a more narrow aperture, and a more narrow subsutural ramp; *R. oblonga* (Jeffreys, 1867) in its wider aperture and the different color pattern [17, 18].

*Habitat and distribution*: Whole Mediterranean Sea [17].

*Status*: Uncommon [17]. First record for the Greek seas.

## 4. Discussion

It is generally accepted that the morphology of the gastropod protoconch determines the mode of development and the duration of larval stage in the ontogenesis [3, 6, 11]. We are of the opinion that such a change in the larval morphology in members of different genetically isolated molluscan lower taxa must have taken place independently and must be due to a simple genetic change involved in the cell's contingency. Otherwise, it would be rather improbable for the same, yet complex, genetic changes to have taken place simultaneously in different individuals belonging to different taxa. Point mutations or even reverse mutations could easily occur leading also to switches. Such multiple switches in the mode of protoconch development are shown to have occurred in the evolutionary history of the Indo-Pacific *Kermia–Pseudodaphnella* complex and the diversity of protoconch morphologies exhibited in this group points to a high developmental and evolutionary plasticity [10].

http://dx.doi.org/10.5772/67847

Both, planktotrophy and lecithotrophy commence with an initial short lecithotrophic stage equipped with the larval shell I. The initial difference between the two types of larval shells I lies in their size which, in turn, is associated with the size of the egg. Larval shell I, leading to multispiral protoconch II, is smaller in width and accounts for the planktotrophic developmental mode, whereas larval shell I, directly linked to lecithotrophy (paucispiral), is larger. In the case of the Mediterranean pair of the so-called sibling species, *R. spadiana* and *R. contigua* (**Figure 1**) were examined in this work; the first bears a larger lecithotrophic larval shell (protoconch I), while the second bears a smaller protoconch I (of its planktotrophic larval shell) of a diameter that corresponds to 3 additional cell cycles (**Table 1**) of the stem cells. In the case of *R. smriglioi* and *R. lineolata* (**Figure 2**), the size of the protoconch I of the latter corresponds to 2 additional cell cycles (**Table 1**), while in that of *R. philberti* and *R. locardi* (**Figure 3**) also to 2 (**Table 1**). This kind of difference in the size of protoconch I could be attributed simply to a single gene intervening in the control of the germ line cell cycle in the gonads, functioning either in favor of a few additional mitoses and thus to a larger number and smaller germ cells and eggs (reduced parental investment per offspring) and eventually smaller in diameter (by a factor of approximately 0.7937 per cell cycle) stage I embryos, or in favor of no further mitoses and thus to a smaller number of larger germ cells, eggs and embryos (increased parental investment per offspring). At the same time, the reproductive output per adult biomass (total energy expenditure or reproductive effort) in those alternative gene actions would not constrain any of the two types of larval development; otherwise, all mollusks should tend to have the same mode of development if that mode gave higher returns per effort [50].

Larger embryos (lecithotrophic) leave the water column early, while smaller ones (planktotrophic) later and thus continue with the development of protoconch II. In spite that time latency, both types of larvae eventually lose their buoyancy and sink. There seems to be no

| Pair of 'sibling' species | Measured lecithotrophic protoconch I maximum diameter (μm) (mean) | Expected planktotrophic protoconch I maximum diameter (μm) after one additional cell cycle | Expected planktotrophic protoconch I maximum diameter (μm) after two additional cell cycles | Expected planktrotophic protoconch I maximum diameter (μm) after three additional cell cycles | Measured planktotrophic protoconch I maximum diameter (μm) (mean) | Corresponding additional cell cycles |
|---|---|---|---|---|---|---|
| *R. spadiana – R. contigua* | *R. spadiana* 457 | *R. contigua* 363 | *R. contigua* 88 | *R. contigua* **228** | *R. contigua* **230** | 3 |
| *R. smriglioi – R. lineolata* | *R. smriglioi* 375 | *R. lineolata* 297 | *R. lineolata* **235** | *R. lineolata* 186 | *R. lineolata* **226** | 2 |
| *R. philberti – R. locardi* | *R. philberti* 358 | *R. locardi* 284 | *R. locardi* **225** | *R. locardi* 179 | *R. locardi* **221** | 2 |

**Table 1.** Expected and measured maximum diameter of protoconch I in planktotrophic sibling species after hypothetical additional cell cycles. Bold characters indicate concurrence between expected and measured protoconch I maximum diameter.

reason why, at least some of those two different types of larvae, not to find themselves in the same locality and, as the initial mutation responsible for the differentiation of the larval mode of life has not led to the establishment of a genetic barrier, when maturation is reached, to interbreed. There are no publications on interbreeding of Mediterranean 'sibling' species so that one can draw conclusions on the existence of a genetic barrier and thus to a confirmation that loss of either planktotrophy or lecithotrophy in the past has eventually led to speciation.

If there is no genetic barrier, then a rising question is associated with the type of inheritance imposed by the initial mutation on that gene controlling the germ cell cycle prior to meiosis. If it displayed a Mendelian inheritance, we would expect also the production of heterozygotes exhibiting a kind of semi-planktotrophic mode of life of shorter duration and presumably protoconches with fewer whorls. In conclusion, someone would expect to find in the same morphological species all three types of protoconches, e.g., paucispiral, multispiral, and intermediate. As this is not the case, at least in Mediterranean *Raphitoma* species [17, 18], we are inclined to propose that the mutated gene cooperates in conjunction with other genes and environmental factors in a discontinuous multifactorial inheritance in which environmental or even population factors also effect a threshold. *Ceteris paribus*, below that threshold, the animals would produce fewer and larger germ cells giving rise through meiosis to fewer and larger eggs that after fertilization produce large lecithotrophic embryos with large paucispiral protoconch I, while, above that threshold, more and smaller germ cells would be produced leading after meiosis to smaller eggs which eventually give rise to planktotrophic larvae with small protoconch I and large multispiral protoconch II. Those two possibilities would jointly constitute protoconch-related poecilogony, a phenomenon already known in some sacoglossan mollusks [51].

In support to the above hypothesis, there are at least 10 pairs of similar, most probably closely related species of *Raphitoma*, one of which bears planktotrophic protoconch and the other lecithotrophic, often sampled in the same localities [17, 18]. That implies the same mode of adulthood life supported by a common gene pool maintained by free gene exchange. Populations of such pairs would employ simultaneously (under different environmental conditions) different dispersal strategies that might reduce interspecific competition. Apparently, long-living planktotrophic larvae maintain a wide geographic range of a species and high genetic integrity between distant subpopulations [11]. Comprehensive accounts on the benefits of these strategies are already given [10, 50]. At the same time, it is generally accepted that, in shelled molluscs, the presence or the absence of any nutritional resource during development affects egg size, which, in turn, affects the size, the number of whorls, and the morphology of the protoconch [1, 52].

Apart from the loss of planktotrophy (in our view, in some members of the same population, as mutations are random phenomena) in Raphitomidae, there is also a well-documented tendency for repeated loss of other conoidean important foregut structures such as radula, proboscis, and venom gland without alteration of the teleoconch morphology [12, 53–55]. Nevertheless, it is worth noting that the loss of planktotrophy in some turrids, like *Raphitoma*, is not necessarily related to simplification in shell morphology [56] which means that the teleoconch morphology could remain unaltered in a species population consisted of individuals with either lecithotrophic or planktotrophic protoconch. We are of the opinion that at least

http://dx.doi.org/10.5772/67847

some of the Mediterranean *Haedropleura, Mangelia,* and *Bela* species also fall in the same case as *Raphitoma,* but we conservatively refrain from formally proposing that possibility pending the mitochondrial DNA markers analysis and employment of interbreeding experiments that could answer the questions raised and solve existing paradoxes.

Finally, through the present work the Raphitomid fauna of the Hellenic sea has been enriched by five new members as the search targeted appropriate environments such as biogenic backgrounds, maerl beds, and deeper waters.

## Author details

Thanasis Manousis[1]*, Constantinos Kontadakis[2], George Mbazios[3], Georgios Polyzoulis[4] and Sofia Galinou-Mitsoudi[5]

*Address all correspondence to: athanasiosmanousis@gmail.com

1 Epanomi, Greece

2 Aristotelous, Hellinicon, Athens, Greece

3 Orfanidou, Athens, Greece

4 N. Redestos, Greece

5 Alexander Technological Educational Institute of Thessaloniki (A.T.E.I.Th.), School of Technological Applications, Department of Civil Engineering, Thessaloniki, Greece

## References

[1] Bouchet P, Warèn A. Planktotrophic larval development in deep-water gastropods. Sarsia. 1979; **64:** 37-40.

[2] Nützel A, Lehnert O, Frýda J. Origin of planktotrophy - evidence from early molluscs. Evolution and Development. 2006; **8:** 325-330.

[3] Bouchet P. Turrid genera and mode of development: The use and abuse of protoconch morphology. Malacologia. 1990; **32:** 69-77.

[4] Oliverio M, Tringali L. Two sibling species of Nassariinae in the Mediterranean Sea (Prosobranchia, Muricidae, Nassariinae). Bollettino Malacologico. 1992; **28:** 157-160.

[5] Lieberman BS, Allmon WD, Eldredge N. Levels of selection and macroevolutionary patterns in the turritellid gastropods. Paleobiology. 1993; **19**: 205-215.

[6] Jablonski D, Lutz RA. Larval ecology of marine benthic invertebrates: paleobiological implications. Wiley Online Library; 1983; **58**: 21-89.

[7] Leal JH, Bouchet P. Distribution patterns and dispersal of prosobranch gastropods along a seamount chain in the Atlantic Ocean. Journal of Marine Biological Association of the. 1991; **71:** 11-25.

[8] Ruppert EE, Fox RS, Barnes RD. Invertebrate Zoology. A functional evolutionary approach. 7th Ed. Brooks/Cole, Thomson Learning, Inc., Belmont, CA. 2004; 990 pp.

[9] Puillandre N, Sysoev AV, Olivera B, Couloux A, Bouchet P. Loss of planktotrophy and speciation: geographical fragmentation in the deep-water gastropod *Bathytoma* (Gastropoda, Conoidea) in the western Pacific. Systematics and Biodiversity. 2010; **8:** 1-24.

[10] Fedosov AE, Puillandre N. Phylogeny and taxonomy of the *Kermia–Pseudodaphnella* (Mollusca: Gastropoda: Raphitomidae) genus complex: a remarkable radiation via diversification of larval development. Systematics and Biodiversity. 2012; **10:** 447-477.

[11] Castelin M, Lambourdiere J, Boisselier MC, Lozouet P, Couloux A, Cruaud C, Samadi S. Hidden diversity and endemism on seamounts: focus on poorly dispersive neogastropods. Biological Journal of the Linnean Society. 2010; **100:** 420-438.

[12] Taylor JD, Kantor YI, Sysoev AV. Foregut anatomy, feedings mechanisms and classification of the Conoidea (Toxoglossa) (Gastropoda). Bulletin of the Natural History Museum of London (Zoology). 1993; **59:** 125-170.

[13] Tucker JK. Catalogue of recent and fossil turrids (Mollusca: Gastropoda). Zootaxa. 2004; **682**: 1-1295.

[14] Puillandre N, Samadi S, Boisselier MC, Sysoev AV, Kantor YI, Cruaud C, Couloux A, Bouchet P. Starting to unravel the taxoglossan knot: Molecular phylogeny of the "turrids" (Neogastropoda: Conoidea). Phylogenetics and Evolution. 2008; **47:** 1122-1134.

[15] Puillandre N, Kantor YI, Sysoev AV, Couloux A, Meyer C, Rawlings T, Todd J, Bouchet P. The dragon tamed? A molecular phylogeny of the Conoidea (Mollusca Gastropoda). Journal of Molluscan Studies. 2011; **77:** 259-272.

[16] Bouchet P, Kantor YI, Sysoev AS, Puillandre N. A new operational classification of the Conoidea (Gastropoda). Journal of Molluscan Studies. 2011; **77:** 273-308.

[17] Pusateri F, Giannuzzi-Savelli R, Oliverio M. A revision of the Mediterranean Raphitomidae 1: on the sibling species *Raphitoma contigua* Monterosato, 1884 and *Raphitoma spadiana* n. sp. (Gastropoda, Conoidea). Iberus. 2012; **30:** 41-52.

[18] Pusateri F, Giannuzzi-Savelli R, Oliverio M. A revision of the Mediterranean Raphitomidae 2: On the sibling species *Raphitoma lineolata* (B.D.D., 1883) and *Raphitoma smriglioi* n. sp. Iberus. 2013; **31**: 11-20.

[19] Thorsom G. Reproduction and larval development of Danish marine bottom invertebrates, with special reference to the planktonic larvae in the Sound (Øresund). Meddelelser fra Kommissionen for Danmarks Fiskeri Og Havundersögelser, Serie. Plankton 1. 1946; **4**: 1-523.

[20] Thorson G. Reproductive and larval ecology of marine bottom invertebrates. Biological Review. 1950; **25**: 1-45.

[21] Rex MA, Warėn A. Planktotrophic development in deep-sea prosobranch snails from the western North Atlantic. Deep-Sea Res. 1982; **29**: 171-184.

[22] Powell AWB. The molluscan families Speightiidae and Turridae an evaluation of the valid taxa, both recent and fossil, with lists of characteristic species. Bulletin of the Auckland Institute and Museum. 1966; **5**: 1-184.

[23] Cachia C, Mifsud C, Sammut PM. The Mollusca of the Maltese Islands (part III: sub-class Prosobranchia to sub-class Pulmonata, order Basommatophora). Backhuys Publishers, Leiden. 2001.

[24] Bouchet P. A review on poecilogony in gastropods. Journal of Molluscan Studies. 1989; **55**: 67-78.

[25] Gibson GD. Why be choosy? Temporal changes in larval sensitivity to several naturally-occurring metamorphic inducers in the opisthobranch *Haminaea callidegenita*. Journal of Experimental Marine Biology and Ecology. 1995; **194**: 9-24.

[26] Gould AA. Report on the Invertebrata of Massachusetts. Second edition, comprising the Mollusca. W. G. Binney. Wright and Potter, Boston. 1870; 524 p.

[27] West HH, Harrigan JF, Pierce SK. Hybridization of two populations of a marine opisthobranch with different developmental patterns. Veliger. 1984; **26**: 199-206.

[28] Gibson, GD. Chia FS. Development variability in the poecilogonous opisthobranch *Haminoea calledegenita*: life history traits and effects of environmental parameters. Marine Ecology - Progress Series. 1995; **121**: 139-155.

[29] Krug PJ, Ellington RA, Burton R, Valdes A. A new poecilogonous species of sea slug (Opisthobranchia: Sacoglossa) from California: Comparison with the planktotrophic congener *Alderia modesta* (Lovėn, 1844). Journal of Molluscan Studies. 2007; **73**: 29-38.

[30] Manousis T, Galinou-Mitsoudi S. New and Uncommon Bivalvia Mollusca of Thermaikos Gulf (Northern Aegean Sea). Journal of Biological Research –Thessaloniki. 2013b; **20**: 339-366.

[31] Oliver PG. Bivalved seashells of the Red Sea. Verlag Christa Hemmen and National Museum of Wales, Cardiff. 1992.

[32] Poppe GT, Goto Y. European seashells, Vol. II. ConchBooks, Hachenheim. 1993.

[33] Doneddu M, Trainito E. Conchiglie del Mediterraneo: guida ai molluschi conchigliati. Castello, Trezzano sul Naviglio. 2005.

[34] Repetto G, Orlando F, Arduino G. Conchiglie del Mediterraneo. Amici del Museo "Federico Eucebio", Alba. 2005; 392 p.

[35] Delamotte M, Vardala-Theodorou E. Sea Shells from the Greek Seas. Goulandris Museum of Natural History, Kifisia. 2008.

[36] Cossignani T, Ardovini R. Malacologia Mediterranea. Atlante delle conchiglie del Mediterraneo. L' Informatore Piceno, Ancona, Italia. 2011.

[37] Gofas S, Moreno D, Salas C. Clase Gastropoda (Heterobranchia), clase Bivalvia, clase Scaphopoda, clase Cephalopoda, glosario e indices. Vol. II. Universidad de Málaga. 2011.

[38] Scaperrotta M, Bartolini S, Bogi C. Stadi di accrescimento dei Molluschi marini del Mediterraneo, Volume VI. L'Informatore Piceno, Ancona, Italy. 2014.

[39] Demir M. Shells of Mollusca collected from the Seas of Turkey. Turkish Journal of Zoology. 2003; **27**: 101-140.

[40] Aslan-Cihangir H, Ovalis P. Seasonal Variations and Structure of the Molluscan Assemblage in the Canakkale Strait (Turkey). Acta Zoologica Bulgarica. 2013; **65**: 233-250.

[41] Zenetos A, Gofas S, Russo G, Templado J. Atlas of exotic species in the Mediterranean. Vol. 3. CIESM Publishers, Monaco. 2003.

[42] Zenetos A, Vassilopoulou V, Salomidi M, Poursanidis D. Additions to the marine alien fauna of Greek waters (2007 update). JMBA2-Biodiversity records. 2007. 5928.

[43] Zenetos A, Ovalis P, Houart R. Eastward spread of *Ergalatax junionae* Houart, 2008 (Gastropoda, Mollusca), a recent alien species in the Mediterranean. Journal of Biological Research-Thessaloniki. 2008; **10**: 221-223.

[44] Zenetos A, Ovalis P, Kalogirou S. Closing the gap: *Cerithium scabridum* Philippi, 1848 found in the South Aegean (Greece, Mediterranean Sea). Journal of Biological Research-Thessaloniki. 2009a; **11**: 107-110.

[45] Zenetos A, Pancucci-Papadopoulou MA, Zogaris S, Papastergiadou E, Vardakas L, Aligizaki K, Economou AN. Aquatic alien species in Greece (2009): tracking sources, patterns and effects on the ecosystem. Journal of Biological Research-Thessaloniki, 2009b; **12**: 135-172.

[46] Manousis T, Mpardakis G, Zamora Silva A, Paraskevopoulos C, Manios D, Galinou-Mitsoudi S. New findings of Gastropods in the Hellenic Seas with emphasis on their origin and distribution status. Journal of Biological Research. 2012; **18**: 249-264.

[47] Manousis T, Galinou-Mitsoudi S. New findings of Gastropods for the Hellenic Seas. Preliminary presentation. 15th Pan Hellenic Symposium of Ichthyologists, Thessaloniki. 2013a.

[48] Verduin A. On the systematics of recent *Rissoa* of the subgenus *Turboella* Gray, 1847, from the Mediterranean and European Atlantic coasts. Basteria. 1976; **40**: 21-73.

[49] Gofas S, Oliver JD. The species of the genus *Chauvetia* (Gastropoda, Neogastropoda, Buccinidae) in the Ibero-moroccan area, with the description of four new species. Iberus. 2010; **28**: 23-60.

http://dx.doi.org/10.5772/67847

[50] Strathmann, RR. Feeding and nonfeeding larval development and life-history evolution in marine invertebrates. Annu. Rev. Ecol. Syst. 1985; **16:** 339-361.

[51] Levin LA. Multiple patterns of development in *Streblospio benedicti* Webster (Spionidae) from three coasts of North America. Biological Bulletin. 1984; **166**: 494-508.

[52] Lima and Lutz. The relationship of larval shell morphology to mode of development in marine prosobranch gastropods. Journal of the Marine Biological Association of the UK. 1990; **70:** 611-637.

[53] Kantor YI, Sysoev AS. The morphology of toxoglossan gastropods lacking a radula, with a description of a new species and genus of Turridae. Journal of Molluscan Studies. 1989; **55:** 537-550.

[54] Kantor YI, Taylor JD. Foregut anatomy and relationships of raphitomine gastropods (Gastropoda: Conoidea: Raphitominae). In: Oliverio M, Chemello R, editors. Systematics, Phylogeny and Biology of the Neogastropoda. Bollettino Malacologico, Supplement 2002; 161-174.

[55] Fedosov AE. Anatomy of accessory rhynchodeal organs of *Veprecula vepratica* and *Tritonoturris subrissoides*: new types of foregut morphology in Raphitominae (Conoidea). Ruthenica. 2007; **17:** 33-41.

[56] Fedosov AE. Five new species of the genus *Lienardia* (Conidae: Gastropoda) from the shallow waters of central Philippines. Ruthenica. 2011; **21:** 123-135.

Chapter 3

# Mussel as a Tool to Define Continental Watershed Quality

Mélissa Palos Ladeiro, Iris Barjhoux, Aurélie Bigot-Clivot, Marc Bonnard, Elise David, Odile Dedourge-Geffard, Elodie Geba, Emilie Lance, Maxime Lepretre, Gabrielle Magniez, Damien Rioult, Dominique Aubert, Isabelle Villena, Gaëlle Daniele, Arnaud Salvador, Emmanuelle Vulliet, Jean Armengaud and Alain Geffard

Additional information is available at the end of the chapter

http://dx.doi.org/10.5772/67995

**Abstract**

Bivalves appear as relevant sentinel species in aquatic ecotoxicology and water quality assessment. This is particularly true in marine ecosystems. In fact, several biomonitoring frameworks in the world used mollusks since several decades on the base of contaminant accumulation (Mussel Watch, ROCCH) and/or biological responses called biomarker (OSPAR) measurements. In freshwater systems, zebra and quagga mussels could represent alternative sentinels, which could be seen as the counterparts of mussel marine species. This chapter presents original studies and projects underlying the interest of these freshwater mussels for water quality monitoring based on contaminant accumulation and biomarker development measurements. These sentinel species could be used as a tool for chemical/biological monitoring of biota under the European water framework directive and for the development of effect-based monitoring tools.

**Keywords:** *Dreissena*, biomonitoring, bioaccumulation, biomarkers

## 1. Introduction

The continental ecosystems are subjected to many stresses related to the human activities through the emission of a large number of molecules whose toxicological and ecological effects are currently poorly studied. There needs to be further development of tools not only for highlighting the exposure of nontarget organisms to these molecules but also for assessing possible risk to the stability of their population and, by domino effect, to the biodiversity and ecological functioning of the ecosystems. Water quality evaluation, management, and protection represent a strong challenge of our society to maintain the sustainability of freshwater ecosystems and protect the biodiversity and ecosystem services for human benefits such as the potabilization or recreational uses. This challenge results in particular in a better management of all the new chemical products likely to be introduced into the environment (program recording of chemical substances REACH) as well as the evaluation of its ecological state (European framework directive on water 2000/60/CE). Currently, the texts of the European Community dated October 23, 2000, set a number of environmental objectives in order to reach the "chemical and ecological good state" of freshwaters. Although environmental quality standards appear in European texts, this monitoring program is primarily based on the evaluation of the level of contaminants in water and on the biological diversity of the aquatic communities. In addition to the biocenotic approaches, many researches focus on biological responses evaluated at sub-individual and individual levels, gathered under the generic term of "biomarkers." Among these biomarkers, a very close attention is paid to the processes whose disturbances may induce a chain reaction of damage on the population and community levels. These biomarkers include physiological functions such as the energy metabolism, the reproduction, or the immune system. The development of biomarkers is conducted on a large number of species known as "environmental sentinels" with respect to criteria related to their sedentary lifestyle, their large distribution, knowledge on their biology and ecology, and the feasibility of using them for experimental exposures. Among these species, bivalves appear to be a genuine tool for the detection of the chemical and biological contamination of the environment [1]. Indeed, the filter feeders have the ability to accumulate and concentrate the contamination in a time- and dose-dependent manner that allows to recall events of pollution prior to the sampling point. The studies are mainly based on the marine species of economic interest as the mussels (*Mytilus* sp.), the clams (*Ruditapes* sp.), and the oysters (*Crassostrea* sp.). However, at the continental level, a model organism, zebra mussel *Dreissena polymorpha* (Pallas, 1771), proves its interest and could represent the freshwater counterpart of *Mytilus* sp. [2]. Zebra mussel is an invasive species originated from Ponto-Caspian region in central Europe. It has spread around the world and became invasive in Europe and North-America freshwater ecosystem, where it has been observed in the great lakes at the end of the 1980s. *D. polymorpha* is characterized by a large abundance and widespread distribution, great filtration capacities leading to high levels of xenobiotic accumulation, and a good tolerance of environmental stressors. Being sedentary, their individual responses may be correlated to the quality of the site. Moreover, dreissenids present a lifetime of several years, can be sampled all year around, and their biology and ecology are well known [3]. Therefore, zebra mussel constitutes a valuable bioindicator species, largely used as freshwater biomonitoring tool [4–7].

## 2. Interest for bioaccumulation capacities

Bioaccumulation is defined as the accumulation of a target compound in an organism relative to its concentration in the surrounding environment. It depends on several intrinsic (e.g., detoxification abilities, lipid content, maturity, or sex of the organism) and environmental (e.g., water temperature or pH) factors that have to be considered in bioaccumulation assessment. The ability of species to bioaccumulate contaminants could represent an indirect evaluation tool of water quality allowing going beyond the limitations associated both to the sporadic nature of water sampling and to potent high dilution of contaminant in the water matrix. In fact, contrary to water matrix, using attached filter-feeding organisms makes contamination measurements representative of their living environment. Measurements in appropriate biological matrices such as mussels will (i) limit the variability (temporal integration) of measurements as compared to measurements on water samples, (ii) convey the degree of contamination of water bodies more reliably, and thus facilitate comparisons. Zebra mussel, *D. polymorpha*, has a high filtration activity and filters between 0.018 and 0.402 L/mussel/h [8] that may give them a capacity to accumulate environmental contaminants.

### 2.1. Pharmaceuticals

There are only a few studies dealing with the uptake and bioaccumulation of pharmaceuticals in freshwater mussels; more studies are available for fish. Quantification of pharmaceuticals at trace levels in mollusks is often analytically challenging, due to the complexity of the biological matrix, requiring suitable extraction and purification steps combined with selective and sensitive detection techniques. Various methods are employed for the extraction of pharmaceuticals in bivalves: pressurized liquid extraction [9, 10], microwave-assisted extraction [11, 12], sonication [13], or QuEChERS (Quick, Easy, Cheap, Effective, Rugged, and Safe) extraction [14, 15]. The last one has the advantages of a low solvent consumption combined with rapid extraction and clean-up steps. Most current analytical methods for the separation and detection of pharmaceuticals are based on gas chromatography-mass spectrometry (GC-MS) or more frequently on liquid chromatography (LC) coupled to ultraviolet (UV), diode array (DAD), fluorescence (FL) detectors, or mass spectrometry (MS) and tandem mass spectrometry (MS/MS). Due to large differences in their physicochemical properties (e.g., polarity and solubility), efficient extractions of all the targeted pharmaceuticals are difficult to obtain, especially in multiresidue methods. As a consequence, the few studies related to the bioaccumulation of pharmaceuticals in freshwater mussels generally focused on a very small number of target compounds [15–17]. Li et al. [18] studied the distribution of 22 antibiotics in 11 mollusk species sampled from the Bohai Sea of China, with concentrations up to 1575.10 μg/kg for quinolones, 76.75 μg/kg for sulfonamides, and 36.21 μg/kg for macrolides. Du et al. [19] studied the bioaccumulation of several selected pharmaceuticals in two mussel species (*Pondhorn mussel* and *Paper pondshell* mussel) and fish, among other matrices, collected from a lower-order effluent-dependent stream in Texas, USA. They concluded that pharmaceuticals accumulated to higher concentrations in invertebrates in comparison to fish. Fluoxetine, carbamazepine, and diphenhydramine were quantified at concentrations inferior to 22 μg/kg, whereas elevated concentrations of the antidepressant sertraline and its primary metabolite desmethylsertraline were observed in both mussel species at levels higher than 130 μg/kg. Indeed, as they could exhibit equal or higher bioaccumulation capacities, it is relevant to include metabolites and transformation products in bioaccumulation studies [15, 19]. Bioconcentration factors (BCFs) have been calculated for the anti-inflammatory drug diclofenac in different studies. BCF of 10 have been displayed for blue mussels exposed during 8 days in aquarium to diclofenac at 1 μg/L [20], and comprised between 4 and 13 in another study where diclofenac was introduced in mesocosms at concentrations between 0.05 and 5 μg/L [15]. This study shows the accumulation in *D. polymorpha* as a function of exposure time, and highlights the presence of one of its transformation products in this bivalve species. The bioaccumulation of the contraceptive hormone levonorgestrel was also examined in *D. polymorpha* [21]. The lowest concentration (0.312 μg/L) was 100-fold bioconcentrated within 4 days. A decrease of the BCF was observed within 1 week for the highest tested concentrations (3.12 and 6.24 μg/L).

### 2.2. Protozoa pathogens

Protozoa such as *Toxoplasma gondii, Cryptosporidium parvum,* and *Giardia duodenalis* represent a threat to human health because they are infectious at low doses and their (oo)cysts can remain infectious in the environment for months due to a robust wall [22, 23]. *T. gondii* is the agent of toxoplasmosis and infection is mainly acquired by the ingestion of food or water that is contaminated with oocysts. *C. parvum* and *G. duodenalis* are intestinal protozoa found in mammals, fish, and birds. (Oo)cysts can be transmitted via drinking and recreational waters contaminated by agricultural and urban runoff [24]. Protozoa are difficult to detect in environmental samples. Actually, no standardized method has been developed for the detection of *Toxoplasma* oocysts in water samples. On the contrary, *Cryptosporidium* oocysts and *Giardia* cysts are currently detected in samples from 100 L of filtered water, using indirect immunofluorescence after immunomagnetic separation (ISO 15553:2006). Limitations have been identified in this method: it requires large volumes of filtered water and high parasite concentrations, it is costly and time-consuming, and thus does not allow for rapid routine detection. Moreover, filtration and purification techniques from water supplies can yield variable results depending on water quality, sampling period, place, and quantity [24].

Several studies have demonstrated the interest of using aquatic organisms to assess and monitor the quality of water bodies [25] and *D. polymorpha* can represent a tool to reveal protozoan contamination. Indeed, during a laboratory experiment, zebra mussels were exposed to various *C. parvum, G. duodenalis,* and *T. gondii* (oo)cyst concentrations for 1 week, and bioaccumulation by mussels proportionally to ambient contamination was observed. This study showed for the first time *T. gondii* accumulation by zebra mussel [26]. In a second in vivo experiment, *D. polymorpha* were exposed to 1000 *T. gondii* oocysts per individual and per day for 21 consecutive days, followed by 14 days of depuration time in protozoa-free water. *T. gondii* was detected in all organs in a time-dependent manner, but oocysts were found in greater amount in the hemolymph and mantle tissues compared to the other organs (gills, gonads, and foot).

This study also shows that mussels accumulate protozoan proportionally to water contamination. Moreover, *T. gondii* was still present in the mussel tissues for up to 14 days postinoculation, reflecting the integrative character of *D. polymorpha* [27]. A field study was also carried out in order to test protozoa detection under environmental conditions. *D. polymorpha* were caged for two seasons (autumn and spring) for 1 month upstream and downstream of wastewater treatment plants (WWTPs). Concerning the study realized in autumn, *T. gondii* was detected in mussels caged downstream of two WWTPs. In spring, *T. gondii* was detected upstream of one WWTP. These results highlight the interest to use *D. polymorpha* caged as a new effective tool in sanitary biomonitoring of water bodies [28].

### 2.3. Cyanobacteria

The eutrophication of aquatic ecosystems, associated to the climate change, enhances the frequency and the severity of cyanobacterial proliferations. Cyanobacteria are photosynthetic organisms producing endotoxins such as neurotoxins, hepatotoxins, dermatotoxins, and cytotoxins, threatening target organisms and humans [29]. The World Health Organization derived a guideline value of 1 µg/L in tap water for the hepatotoxins microcystins (MCs), the most reported cyanotoxins in fresh waters worldwide. In addition to being exposed to cyanotoxins through drinking, humans can also be exposed by ingestion or inhalation of contaminated water during recreational activities [30, 31], or by trophic transfer of MCs as demonstrated between invertebrates and vertebrates [32]. The bioaccumulation of MCs and their effects on organisms are overall quite well documented. However, some cyanobacteria and diatom strains may also produce the neurotoxin BMAA (β-N-methylamino-L-alanine) [33–35], a nonproteinogenic amino acid that has been associated with neurodegenerative diseases like amyotrophic lateral sclerosis/Parkinsonism dementia complex (ALS-PDC) that occurred on the Guam Island [36]. The bioaccumulation of BMAA in marine organisms is little documented [37–39]. However, BMAA accumulation and impact on freshwater organisms during laboratory exposure or field investigations remains poorly investigated [35, 37].

Even though monitoring programs are in place to evaluate the concentration of cyanotoxins in fresh waters, prompt management of ecological and public health risk remains difficult due to the high spatiotemporal variability of the cyanobacterial proliferations. Freshwater mussels, and particularly *D. polymorpha*, are known to ingest phytoplankton and cyanobacteria and accumulate cyanotoxins at the laboratory [40, 41] and in the field [42–44]. Therefore, the use of sentinel species to integrate and reveal early MC and BMAA contamination at low cyanobacterial or phytoplankton densities may allow a preventative management of contaminated sites. As MC accumulation by *D. polymorpha* has been largely demonstrated, we primarily focused on the capacity of the mussel to accumulate BMAA using a laboratory approach. The mussel *D. polymorpha* was exposed to exogenous BMAA at 500 µg/L for 3 weeks, and we followed up both free and protein-bound BMAA in tissues. The free amino acid BMAA was detected from day 2 to day 21 and really high concentrations (reaching 283 µg/g dry weight) were quantified in mussels at the end of the contamination experiment. BMAA was therefore taken up by *D. polymorpha* and the internalized concentration of the toxin, detected as its free form, increased linearly over time ($r^2 = 0.9635$) while no steady state was reached (**Figure 1**).

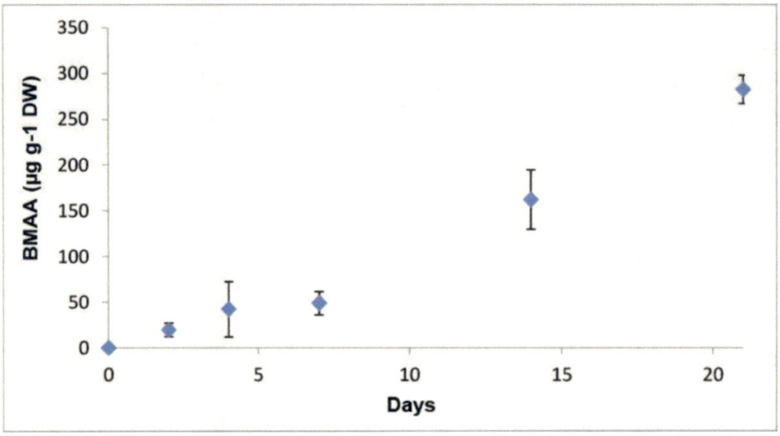


**Figure 1.** Concentration of free BMAA in *D. polymorpha* with fitted linear regression after free exogenous exposure for 21 days at 500 µg/L. Error bars denote standard deviation of $n = 5$ (except for D21, $n = 2$).

It is conceivable that this freshwater species would have continued to accumulate BMAA from the medium after a longer exposure time. Concentrations of total BMAA in mussels of day 2 and day 21 were not significantly different from the concentrations of free BMAA in the same individuals (data not shown). Thus, the incorporation into or in association of BMAA to proteins was not evident in this study.

Overall current projects also include other freshwater bivalves as potent predictive sentinel species of the presence of MCs and BMAA in freshwater environments through laboratory and *in situ* approaches. The results that will be achieved in these projects will facilitate the long-term tracking of the contamination of ecosystems by cyanotoxins, which will provide an advance in the knowledge about the ecodynamic of cyanotoxins and the main conditions of human exposure. This step forward could constitute a decision-making tool for public health authority in charge with the management of food safety risks.

## 3. Interest for biomarker measurement

Different biological responses have been developed as biomarkers on freshwater mussels. Biomarkers related to energy metabolism could be an interesting prognostic tool bearing in mind the close relationships existing between the energy balance and health of individuals. As well, the reproduction function is a key element for the preservation of the species in their environment and can be disrupted by numerous stresses of anthropogenic origin. Furthermore, organisms are subjected to various pathogens. Thus, interaction with hemocytes, a key component of the immune system, undoubtedly occurs. Exploring these three relevant functions will permit to develop early warning and sensitive diagnosis tools with an ecological relevance which seems to be promising for the early assessment of the ecosystem degradation resulting from damaging contaminants.

### 3.1. Energy and metabolism

All the biochemical and physiological processes involved in the vital cycle of the organisms depend strictly on the energy metabolism. Besides, within the framework of an evaluation of the quality of the environment, the variations at the molecular level (e.g., enzymatic activities) are one of often first answers to the stress; they thus represent useful points in the confirmation of the toxic effects before these last ones are perceptible at superior biological levels of organization (cellular, tissular, and physiological). Biomarkers related to energy metabolism could then constitute interesting prognostic tools considering the casual relationships linking the energy balance and the health status of individuals. A very large number of biological (adenylate energy charge, energy reserves, cellular energy allocation, scope for growth, etc.) responses to stress linked to the metabolism and energy of living organisms may be investigated. While the concept is clear, linking cellular or biochemical responses to the individual and population level remains relatively scarce. However, several studies underline the relevance of cellular responses to extrapolate effects at higher levels of biological organization in *D. polymorpha* under laboratory [45] and environmental conditions [46]. Zebra mussel and all other animals obtain the energy necessary for life from an external source. So, in heterotrophic organisms, the acquisition of energy is controlled by feeding and the subsequent breakdown of food to release the energy contained (assimilation). Feeding rate/clearance was studied in different species of mussels (particularly blue mussel) exposed to various contaminants [47–49]. Generally, feeding appears unaffected and/or less sensitive compared to biochemical responses. Inversely, during the study of different zebra mussel populations, among several genes involved in cellular metabolic activities, detoxification process, oxidative stress, and digestive functions, digestive enzyme genes expression appears particularly relevant to discriminate the sites according to the contamination level [50]. This study has also pointed out different sources of variability in gene expression (individual size, season, trophic resources, and origin of mussels) which are inevitable in natural fluctuant environment. Accordingly, in *D. polymorpha,* we decided to investigate the breakdown of food's mechanisms and particularly the digestive capacity. Digestion in bivalves occurs through a biphasic process involving an extracellular digestion phase of ingested food particles followed by an intracellular digestion phase. The first step occurs within the stomach under the mechanical and chemical actions of the crystalline style (CS), a gelatinous rod saturated with digestive enzymes. The CS protrudes from the style sac into the stomach and revolves against a cuticular structure, the gastric shield, promoting the grinding of food materials and their mixing with the enzymes released by its former dissolution. This extracellular digestion phase allows a preliminary breakdown of food particles into molecules small enough to enter the digestive diverticulae where digestion is completed intracellularly within phagocytic digestive cells. In this way, amylolytic and cellulolytic activities in the crystalline style and the digestive diverticulae of the freshwater bivalve *D. polymorpha* were characterized [51]. In the same time, to avoid the influence of biotic parameters (age, size, etc.), we proposed to use individual from the same population transplanted to the study sites (see part 5). Thus, during active biomonitoring carried out with *D. polymorpha,* in the catchment of the River Vesle, upregulation of digestive activities in digestive gland was observed in mussels exposed to various chemical contaminants and thus potentially facing an increase in their energy expenditure (general adaptation syndrome) [52]. However, this upregulation has not been recorded at all contaminated sites, lower levels of pollution or higher levels of food availability probably explaining the low need for organisms to "invest" in their digestive activities (least energy requirement). These results highlight adjustment of digestive enzymatic activities according to intake and energy requirements of the organism, factors which, if not taken into account, constitute potential confounding sources in the interpretation of response measured. According to literature data, digestive carbohydrate activity of bivalve species shows a great adaptability to variations in food availability, quantity, and quality. However, if food availability exerts a primary control on digestive activities, the latter could also vary in response to other factors, both external (e.g., temperature) and internal (e.g., reproductive state). These "regulation" factors exhibit seasonal variations, which, logically, should be reflected on digestive enzyme activities [52]. Nevertheless, if seasonal amylase and cellulase activities were well recorded in the digestive gland of zebra mussels, no seasonal activities were recorded in their crystalline style. The absence of seasonality in the crystalline-style activities could be related to the innate nature of the style, for example, a secretion organ whose weight does not vary according to the seasons. Inversely to digestive gland, the crystalline-style activity (downregulated) appeared to be more associated with the location and metal exposure level, allowing the discrimination of the most chemically impacted site from the others. Thus, the high discrimination potential of crystalline-style enzymes activities, in association with their low seasonal dependency, makes these parameters very promising biomarkers to develop in biomonitoring studies using *D. polymorpha*. In this way, zebra mussels were transplanted during 2 months along a metal and organic pollution gradient in spring 2008 (Seine River, France; PIREN-Seine program) [53]. Amylolytic and cellulolytic activities measured in the CS displayed similar patterns of response with lower activities at the downstream site compared to the upstream site (20–30% lower). By contrast, digestive enzyme activities in the digestive gland displayed only slight variations during the whole period of exposure, confirming the interest of the crystalline style as matrix in biomonitoring program.

### 3.2. Reproductive process

There is a large body of literature on ecophysiology and reproductive biology of dreissenid species [54]. The reproductive function is a key element for the preservation of species, implying a lot of responses which may be studied in reproductive ecotoxicology at different levels of biological organization, for example, from toxic effects on biological macromolecules to the consequences on the reproductive capacity of organisms and the maintenance of their population in the environment. Until today, literature in dreissenids mainly focused on disturbances of gametogenesis and on biomarkers related to the integrity and functionality of gametes. Numerous stressors of anthropogenic origin are susceptible to modify the reproductive physiology of mussels, either by direct or by indirect effects. In dreissenid species, existing data on the disturbances of the reproductive cycle by environmental factors, such as water temperature, food availability, or chemical contamination, highlighted early or delayed sexual maturation as well as asynchronous development between congeners in exposed mussels [55–58]. Many studies also revealed that gonads were the primary organs infected by trematodes of the family *Bucephalidae* in dreissenid mussels, known as the first intermediate host. The development of cercariae in sporocysts coincides with the host sexual maturation and causes in infected mussels an energetic depletion allowed to the reproductive tissue, leading to disturbances of their reproductive cycle as well as in the most severe cases to complete castration and host sterility [59, 60].

The gametogenesis and reproductive effort in dreissenid species has mainly been investigated by either nonhistological examinations, with the measure of the reproductive tissue biomass (gonad) compared to the rest of soft tissues (somatic) and the definition of a *gonado-somatic index*: GSI (the difficulty for discriminating the reproductive tissue and the digestive gland renders this methodology quite imprecise); either histological examination of gonads with the definition of various reproductive stages (*resting, development, prespawning, or postspawning stages*) and the possible measurements of morphological parameters by computer-assisted image analysis systems (e.g., oocyte number in follicles, oocyte diameter, etc.). Although informative, histological examination of the gonads is both time-consuming and experimenter-dependent. Recently, a novel strategy has been developed for a rapid and sensitive determination of an *index of sexual maturity* in male zebra mussels. This index considers the proportion of germ cells in male gonad according to their DNA content ((sub)-haploid, diploid, or tetraploid cells) measured by flow cytometry. This *index of sexual maturity* was used to describe the reproductive cycle in a control population of zebra mussels over a year. This biomarker also highlighted the reprotoxicity of carbamazepine, an anti-epileptic compound which was experimented in a mesocosm study in caged zebra mussels [61]. Initiation of gametogenesis, proliferation, and differentiation of immature germ cells determines both the quantity and quality of mature germ cells or gametes implied in fertilization and *de facto* in the reproductive success of zebra mussels [62]. Gametes represent "cells of choice" for the development of biomarkers in ecotoxicology, by relating the exposure of adult mussels with the long-term consequences in offsprings [63]. Until today, spermatozoa of dreissenid mussels have mainly been studied in comparison with oocytes for the development of biomarkers like sperm mortality (or membrane permeability), sperm motility (duration, velocity, etc.) with computer-assisted sperm analysis (CASA) system or fecundity [64]. Other authors investigated the integrity or the functionality of specific organites, like acrosomal integrity or mitochondrial membrane potential, and demonstrated the sensitivity of zebra mussel's spermatozoa to the toxicity of commercial formulations of pesticides (Bayluscide, Round-Up Ready-to Use Plus®) [65]. In their study, Seaver *et al.* [66] observed in spermatozoa of zebra mussel that *in vitro* irradiated with a physical agent (UV-B) decreased acrosomal integrity, incorporation of sperm into egg cytoplasm, and first zygotic cleavage. These defects in first zygotic cleavages, as also observed by these authors with the irradiation of eggs to UV-B, could be related to genomic damage in gametes. To our knowledge, the use of gametes (spermatozoa and oocyte) for evaluating the genotoxicity of water contaminants has not yet been tested in dreissenid mussels. Other authors also highlighted the sensitivity of early-life stages in zebra mussels to the toxicity of various water contaminants and proposed their use for the development of biomarkers [67].

### 3.3. Hemocytes

To be relevant, biomarkers must gather several properties such as being flexible according to the various pressures exerted by the environment and associated with a specific mode of action which is measurable by significant and reproducible techniques. Moreover, the early analyses of these biomarkers have to predict the effects at higher levels of organization, that is, the population. In this context, the hemocytes responses show a particular interest since they are key cells involved in the major physiological functions in invertebrates. Indeed, the hemolymph is a remarkable mussel tissue. With an opened circulatory system, hemolymph circulates through the bodies allowing to bring the nutrients to the various organs. Drawn from the adductor muscle of the bivalves after having carried out a little breach on the shell, hemolymph sampling is not lethal for the individuals and represents a clear advantage for biomonitoring. Circulating cells, the hemocytes, are present in the cellular fraction of the hemolymph and have been recently characterized [68]. These cells, mainly granulocytes and hyalinocytes, provide various functions involved in the essential physiological activities such as the transport and the digestion of the nutrients, breathing, excretion, repairs as well of the tissues and of the shell and, of course, immune defenses [69]. Contrary to vertebrates, which have an *acquired immunity* (or specific), the immunity of the invertebrates is only composed of the *innate immunity* (or not specific) ensured by the hemolymphatic compartment (hemocytes and plasma). In a global context, hemocytes allow the measurement of various markers involved in the physiological processes usually studied in ecotoxicology. Current studies are based mainly on the evaluation of the biomarkers of genotoxicity (measurements of DNA damage in hemocytes [70]), of cellular parameters (oxidative activity, mortality [71]), and immune defense (capacity of phagocytosis [72]). However, all these responses suggest the implication of another component of the immune system and complementary to the cellular responses of the organism, the humoral-mediated responses in plasma [73]. Humoral responses are regrouped in different effectors such as the lysosomal enzymes, lectins, protease inhibitors, or antimicrobial peptides [74]. The phenoloxidase cascade [75] and the fast production of reactive oxygen species (ROS) also take part in organism homeostasis. Both systems interact to give an effective answer relative to the various types of aggressions which the organism undergoes.

A recent study by Juhel et al. [45] highlights the effect of cyanotoxins on the immune functions of *D. polymorpha*. The authors underline a reduction in the full number of hemocytes per milliliter of hemolymph as well as a modulation of the subpopulations of hemocytes, an inhibition of the phagocytosis capacity and an increase in the concentration of lysozyme after 21 days of *in vivo* exposition to various strains of cyanobacteria. However, this study raises the problematic of the complexity to define the real immune effects based only on cellular approaches. Currently, very few studies take into account the entire hemolymphatic compartment, for example, cells and plasma, in the analysis of the biomarkers. However, the evaluation of the interactions and the relations between these two types of responses would permit to understand the functioning of this complex tissue which maintains actively the homeostasis of organism (for review see [76]). Indeed, considering that a panel of enzymes and hydrolytic molecules can influence the cellular responses, it seems paramount to study the humoral compounds in parallel. Considering the central role of hemolymph in the physiology of the organisms, the comprehension of the different functions it provides in relation to other major biological functions is also necessary to ensure relevant conclusions in ecotoxicological studies. However, to date, very little information is available on the hemocyte implications on the physiological responses other than those of immune functions and defenses. This lack of knowledge is certainly due to difficulties in studying simultaneously both cellular and acellular compartment, because it necessarily implies a global and multi-markers approach which is difficult to perform with the techniques commonly used.

## 4. Interest of *D. polymorpha* in aquatic biomonitoring framework

For being usable as early warning system, a sentinel species should be able to accumulate contaminants at much higher levels than those in the environment and/or to show variations in biological parameters at one or several levels of organization. In fact, even if analyses of bioaccumulated pollutants have the great benefit of traducing the bioavailable fraction of xenobiotic, they give no information about the potential biological effects. Therefore, the relationship between exposure to pollutants and associated adverse effects is of growing importance in environmental-risk assessment (ERA) and management. In this way, a sentinel species could be designated as a necessary tool of choice for monitoring both the contamination levels and the associated risks for organisms. The zebra mussel was extensively used as model organism for the quality assessment and the biomonitoring of freshwater systems for large amounts of well-known mineral and organic trace pollutants (for review, see [2]). However, more recently zebra mussel appeared also particularly relevant for the assessment of chemical (pharmaceutical, part 2.1) and biological (protozoa and algae toxins, parts 2.2 and 2.3) emergent contaminants as underlined in this chapter. Similarly, an increasing development of biological responses, called as biomarkers, under controlled laboratory conditions (*in vivo* and *in vitro* approaches considering particularly hemocytes and gill cells) and field studies ([2] and part 3 of this chapter) is observed. Regarding field studies, several passive biomonitoring approaches based on autochthonous populations underline a significant modulation of biological responses not only by environmental parameters (e.g., temperature, conductivity, etc.) but also by intrinsic parameters (e.g., size, age, genetic background, life history according to contamination, and potential adaptive responses) [50, 7]. These confounding factors could induce a misunderstanding in the data interpretation according to the toxicity of pollutants. Also, as discussed in section 4, zebra mussel seems to coexist with another *Dreissena* species and it remains difficult to distinguish between the two species. These potent physiological discrepancies could put into questions the biomarker response interpretation.

An active biomonitoring approach appears as an interesting strategy with the view to improving the usefulness of biomarkers and particularly defining reference and threshold values allowing (i) a qualification of the toxicological effects of water quality on organisms and (ii) a suitable comparison between the monitored sites. In fact, the caging of individuals from the same population (and *a fortiori* the same species) in different sites allows (i) to limit or avoid influence of intrinsic parameters and (ii) to expose organisms for a time-limited period, so recent contamination can be detected. As for contamination level and/or biological effects associated, several studies had underlined the interest of the active biomonitoring approach using the zebra mussel, *D. polymorpha* [28, 52, 53, 77, 78]. In the purpose of improving the ecological relevance of biomarkers as early-warning responses, many studies proposed to apply a multi-biomarkers approach considering physiological processes particularly those implied to maintain an individual and/or its population (e.g., reproduction, energy, and immunity) as indicated in part 3 of this chapter. Therefore, the proposition of an integrated tool is mandatory to ensure the use of zebra mussel (together with developed biological responses) in environmental-risk assessment and management in order to establish environmental quality standards and to favor ecosystem protection. Through the process, several developments and approaches were proposed including "Integrated Biological Response" (IBR) to summarize biomarker responses [79]. This approach takes into account "only" biomarker measurements that may be limiting for an environment quality assessment requiring data from a variety of multidisciplinary sources. Then, the weight-of-evidence (WOE) approach including multiple parameters such as for instance bioaccumulation, biomarkers, bioassays, or life history traits appears particularly relevant (see part 6.2 of this chapter).

## 5. The usefulness of another dreissenid species, *D. rostriformis bugensis*

Despite its high invasive potential, zebra mussel populations have been recently reported to decline such as in the Rhine River [80] or in the Seine River where some populations seem to have disappeared since a dozen years (personal observations). In the meantime, the closely related quagga mussel *D. rostriformis bugensis* [81] became in turn invasive in Europe and North America. In Western Europe, the first observation of quagga mussels was made in 2006 in the Hollands Diep [82, 83] and since 2008 in the Meuse River [81, 84]. In some sites, *D. rostriformis* seems to have become gradually the dominant species [85], as it has been observed in some German and Dutch rivers [86]. Some researchers tried to explain the dominance shift from zebra to quagga mussels through biological, ecological, and ecophysiological studies. The two species are indeed characterized by differential physiological performances. They disclose different temperature, salinity and low oxygen tolerance levels, byssal thread attachment, growth, respiration rates, assimilation efficiency, or reproduction [87–89]. Quagga mussels reveal a lower tolerance to high temperature, although there is a better adaptability to lower temperature, allowing a development in deeper waters. A higher physiological adaptability of quagga mussel than zebra mussel is also highlighted with higher filtration and lower respiration rates, decreasing the energetic cost in quagga [88, 90]. A higher assimilation efficiency allows quagga mussel to maintain higher growth and fecundity rates even at low food levels [8]. Both species also reveal different reproductive strategies. Quagga mussel spawn earlier in the season and in deeper waters, but zebra mussel releases more eggs than quagga [87, 88, 90]. Quagga mussel seems to invest less energy into reproduction than zebra mussel: its lower fecundity is offset by an earlier maturity [90]. Quagga larvae settle at a larger size as they grow more rapidly. Quagga higher somatic growth rate and size [91] are supposed to contribute to a better stress survivorship. A differential sensitivity to chemicals could also be involved in the replacement of a species by the other. Some studies focused on differential xenobiotics accumulation between the two species, mostly on metals and given scarce or inconclusive information [92–95]. Schäfer et al. [80] also showed differences between *D. polymorpha* and *D. rostriformis* in the bioaccumulation potential of pesticides from resuspended sediments. They showed that quagga mussel accumulates more organochlorine pesticides, with greater DNA damage and lower stress protein hsp70 content. These results suggest a difference of sensitivity to genotoxic stress between the two species that may be related to "threshold" levels of DNA damage or to differential capacity of DNA repair [80]. More recently, Potet et al. [96] exposed zebra and quagga mussels to nickel and chromium at two temperatures in laboratory conditions. They measured a set of 14 biomarkers that revealed differential bioaccumulation patterns, filtration activity, and cellular antioxidant and detoxification responses between the two species, with more marked effect of metals in *D. polymorpha*.

In most of the comparative studies, the morphological criterion is chosen to identify the two species. However, the two species show high morphological similarities [97, 98] that prevent from morphological consistent species identification [99]. Mussel's "angularity" has been proposed anyway to improve morphometric identification with discriminating thresholds for zebra and quagga mussels [100]. Molecular genetic markers have then been developed to differentiate *Dreissena* species [97, 98, 101], but most of them are based on restriction fragment length polymorphism (RFLP) of cytochrome-C oxidase subunit I (COI gene). Such mitochondrial genetic tool does not allow identifying potential interspecific hybrids [97]. If natural hybridization has been reported only once by Voroshilova et al. [97] using allozyme patterns, no evidence exists that the two species are unable to hybridize in natural environment. In ecotoxicological studies, a reliable identification of *Dreissena* species is a key step to ensure the efficiency of the use of such freshwater sentinel species and their associated biomarkers to reflect exposure and toxic effects of xenobiotics. It is currently essential to deepen our knowledge on the levels of responses and sensitivity of these two species with a view to their use in biomonitoring projects.

## 6. Perspective: toward new markers and an integrative approach

Contamination of aquatic environment is multifaceted as it combines chemical, physical, and biological pollutions, which could be influenced by a wide range of (a)biotic factors. Both short- and long-term cascading negative effects of such pressures on ecosystems still remain difficult to evaluate considering the complexity and the diversity of exogenous inputs in the aquatic environment. While the contamination level in a particular site can be quite readily determined through chemical analysis as defined by the presence of "substances that would not normally occur or at concentrations above the natural background," the assessment of the "pollution status" also integrates the notion of chemicals bioavailability and biological impacts induced by contaminants within the considered environment [102]. Assessing the ecological health of an ecosystem should thus address the following questions, initially enunciated for metallic contaminants by Chapman et al. [103], but which can be generalized to all environmental contaminants: (1) do chemicals/biologicals accumulate in biota above background (or reference) concentrations? (2) once accumulated, are they bioreactive? and if so, (3) what are the incidence and severity of the induced effects (acute, long-term sublethal, individual, population effects, etc.)? Thus, a relevant and successful water quality assessment has to be based on combined analyses integrating the greatest possible number of these various parameters. This also includes the sanitary status of aquatic organisms which are chronically exposed to a cocktail of pollutants in their environment. Despite a large number of publications on invertebrate biomarkers, little of them reached a level of validation allowing to be recommended as efficient tools for an accurate evaluation of the quality of the aquatic compartment. Nonetheless, new integrative approaches could permit to remove scientific (define the reference levels) and technical (integration of several responses) obstacles encountered in environmental-risk assessment strategies. The use of OMICS tools in field studies has gained great interest for some years as it gives broad information on pollutant modes of actions, considerably increasing knowledge and an understanding of such mechanisms. Finally, clustering data in a weight-of-evidence approach represents a powerful and practical tool to facilitate the decision-making processes of environment managers within the framework strategies.

### 6.1. Genomics and proteomics in ecotoxicology

Although next-generation sequencing has made great progress over the last decade, genome sequencing of animal models is still far to be a trivial task. This is due to their large genome and the difficulty for assembling too numerous short reads without any scaffold from a closely related species. Furthermore, the delineation of coding open-reading frames and their functional annotation is difficult and subjected to an important bias toward easy annotation of conserved genes but detrimental to species-specific genes. Till now, no complete genome is available for any *Dreissena* and genetic markers are relatively scarce [104–106]. The genomes of marine mussels such as *M. galloprovincialis* are only partially known [107, 108]. Transcriptomics is fully complementary to genomics and adds value in highlighting genes of interest if their expression is regulated. However, this approach has till now been really poorly used for characterizing *Dreissena*. Very few studies based on cDNA microarray have been published, specifying the molecular mechanisms at the early stage of underwater adhesion of the zebra mussel [109], or the effect of seasonal and environmental variations on the physiology and metabolism of *D. polymorpha* [110]. Today, cDNA microarray has been advantageously replaced by RNAseq. High-throughput sequencing of cDNA allows comparing the transcriptome from several conditions and highlighting the most modulated genes in terms of expression levels. In addition, RNAseq data may be used to construct a six-reading frame RNA-translated database that can be used for discovery proteomics. This concept is the basis of numerous proteogenomics analysis of various animal models or even nonmodels [111]. For example, the amphipod *Gammarus fossarum* has been scrutinized in detail to document the reproductive system of amphipods [112] or to understand the response to endocrine disruptors [113].

Till now, *Dreissena* has been worked out in terms of proteomics only with 2D-PAGE traditional approach. For example, [114] analyzed *D. polymorpha* exposed to benzo($\alpha$)pyrene and focused on 28 proteins. A set of 16 proteins were found more abundant and 12 were noted as less abundant in exposed mussels. They could be identified after MALDI-TOF/TOF mass spectrometry measurements on their tryptic peptides. Such methodology is known to focus only on the most abundant and soluble proteins [115], thus explaining the relatively low number of protein hits identified by this approach. For sure, high-throughput shotgun proteomics based on next-generation mass spectrometers should allow today documenting thousands of proteins and delineating candidate biomarkers after the analysis of specific exposure conditions. Importantly, these candidate biomarkers require a strict validation consisting in monitoring the candidate biomarkers for a large cohort of animals and various conditions [116]. Such monitoring is ideally carried out with targeted proteomics based on the selected reaction monitoring (SRM) quantitative approach. SRM mass spectrometry assay allows high-throughput multiplex analysis, with fewer samples required. It is efficient in terms of time and cost and is able to reliably detect different proteins across a broad dynamic range of concentrations as recently reported for *G. fossarum* [117]. Thus, in our opinion, OMICS tools are today pertinent for obtaining major insights into the most important molecular mechanisms of mussels.

### 6.2. Weight of evidence

The WOE approach is based on the packaging of a wide variety of data within several lines of evidence (LOEs) in which the contamination level—assessed through chemical analyses—is combined to bioavailability (bioaccumulation) analysis, and biological responses (biomarkers) on key species and/or model organisms (bioassays) at different levels of biological organization, from the molecular to the community level [118, 119]. The resulting environmental diagnosis is then based on the calculation of a hazard index for each LOE, which is next set out on an evaluation grid allowing a clear and rapid classification of hazard [118, 120, 121]. A global hazard evaluation is also proposed through the compilation of all calculated LOE indexes within a single one that is also finally assigned to a hazard class. The WOE approach is applicable to various matrices such as effluent, water, and soil, as well as for more global environmental diagnosis like aquatic and terrestrial ERA. The WOE model developed by Piva et al. [121] was mainly applied to the quality assessment of harbor area using fish species (European eel, *Anguilla anguilla*) and/or mussels (Mediterranean mussels, *M. galloprovincialis*) as bioindicator organisms. These studies undoubtedly demonstrated the relevance and the performance of the procedure to diagnose ecosystem health status in chronically impacted area (e.g., industrial harbors, natural crude oil, and gas seepage) [122, 123] as well as in accidental pollution events as demonstrated by its use in the Mussel Watch program following the Costa Concordia wreck [124]. Mussels were relevantly used in these studies for their suitability in translocation (caging) procedures and their ability to reflect environmental pollution levels through bioaccumulation measurements (see part 2) and biomarkers analysis (see part 3). The elaboration of each hazard index within the WOE approach relies on the initial calculation of ratio-to-reference (RTR) values. It thus supposes that reference levels are available for every end point integrated in the model. In the abovementioned studies, the reference levels were determined examining the biological responses in control organisms maintained in clean water under laboratory conditions or transplanted at a reference site. However, the laboratory (controlled) conditions are far removed from those during *in situ* exposure as control organisms are not submitted to any variations of their environment that naturally occur in field (e.g., temperature variations, general physico-chemistry of the water column, etc.) and which could modulate the biological responses of the organisms with no link with the contamination status of the environment. Transplantation of organisms at a reference site is also commonly used to avoid such bias. However, a "perfect" reference station would assume that (i) the site is geographically close enough to reflect the natural state of the studied environment; (ii) the exposure conditions are exactly similar to those at the other sites (in terms of temperature, physicochemistry, etc.); and (iii) that there is absolutely no anthropogenic contamination which could induce any modulations in the biological responses, even in a limited way. There is no doubt that the determination of such station in each studied area is not realistic—if not utopic—and that the use of control organisms (at a reference station or maintained in laboratory conditions) to set the reference levels integrated in the WOE approach generates a bias in the ecological health status assessment.

An alternative was proposed by Barjhoux et al. [125] to address these concerns. Briefly, the study proposes an application of the WOE strategy to a freshwater system: the Seine River (France), well known to be submitted to heavy anthropogenic pressures through important industrial, agricultural, and urban activities. The three studied sites were located upstream (Marnay, in a non-urbanized area) to downstream from Paris conurbation (Bougival and Triel, respectively, situated at 40 and 80 km from Paris). The dataset selected for WOE integration included (i) chemical contamination levels, (ii) bioavailability (bioaccumulation) measurements, (iii) biological effects in field-transplanted organisms (biomarkers in gammarids), and (iv) (eco)toxicological responses assessed using laboratory bioassays. The strength of the quality assessment proposed in this study lies on the use of the same population of gammarids for bioaccumulation and biomarkers measurements. Reference and threshold values were established using modeling developments quantifying the natural variability of the studied markers in relation to identified confounding factors. These reference/threshold levels were integrated in the WOE approach as they clearly enhance the reliability of *in situ* methodology and allow its implementation at a large spatial and temporal scales [126, 127]. The calculated WOE indexes clearly reflected the anthropogenic gradient along the Seine River,

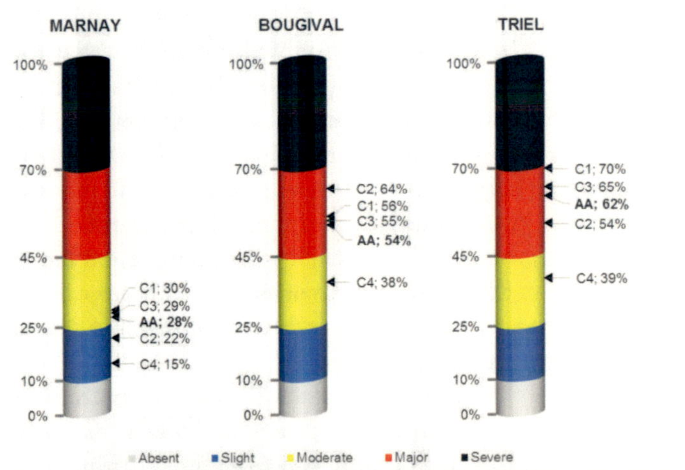


| | Summary of the hazard classes attributed to the HQs values within each LOE | | | | | | | | | | | | | | |
|---|---|---|---|---|---|---|---|---|---|---|---|---|---|---|---|
| HQs | MAR_C1 | MAR_C2 | MAR_C3 | MAR_C4 | **MAR_AA** | BOU_C1 | BOU_C2 | BOU_C3 | BOU_C4 | **BOU_AA** | TRI_C1 | TRI_C2 | TRI_C3 | TRI_C4 | **TRI_AA** |
| *ChemHQ$_{water}$* | Moderate | Moderate | Moderate | Moderate | Moderate | Severe | Severe | Severe | Major | Severe | Severe | Severe | Severe | Major | Severe |
| *ChemHQ$_{sed}$* | Severe | *N/A* | Severe | Moderate | Severe | Severe | *N/A* | Severe | Severe | Severe | Severe | *N/A* | Severe | Severe | Severe |
| *BioavHQ* | Slight | Slight | Slight | Slight | Slight | Major | Major | Major | Slight | Major | Major | Moderate | Moderate | Slight | Moderate |
| *BiomHQ* | Moderate | *N/A* | Moderate | Slight | Moderate | Severe | *N/A* | Major | Major | Major | Severe | *N/A* | Severe | Major | Severe |
| *ToxHQ$_{water}$* | Absent | *N/A* | Absent | Absent | Absent | Moderate | *N/A* | Absent | Absent | Slight | Major | *N/A* | Slight | Absent | Slight |
| *ToxHQ$_{sed}$* | Slight | *N/A* | Slight | Slight | Slight | Slight | *N/A* | Moderate | Slight | Moderate | Severe | *N/A* | Severe | Slight | Major |

**Figure 2.** WOE indices and associated hazard classes integrating the results of each LOE calculated for the three stations during four sampling campaigns (C1–C4) and annual average (AA) values. The hazard class attributed to each LOE hazard quotient (HQ) is summarized in the table below. ChemHQ$_{water/sed}$, water/sediment contamination HQ; BioavHQ, bioavailability (bioaccumulation) HQ; BiomHQ, biomarker HQ; ToxHQ$_{water/sed}$, bioassay-based HQ on water/sediment samples. Note that in the C2 campaign, only data on water contamination and bioavailability are evaluated. The integral version of the article is available at: http://link.springer.com/article/10.1007/s11356-016-6993-6. The figure was reproduced with permission of SpringerNature publisher.

with values increasing from upstream to downstream of Paris (**Figure 2**: [125]). The results also highlighted some seasonal variations in the hazard class attributed to each site with the winter campaign showing lower level of perturbation than the three other campaigns.

Accordingly, the use of external reference values and thresholds eliminated the need for a reference site in the study area, which could be very problematic in large rivers subjected to multiple and diffuse pressures. The results of this study also reveal that at the upstream site, generally used as a relative reference or control site in previous investigations in this area, the low contamination levels nonetheless resulted in low but significant biological effects. The WOE approach applied in this study proved to be efficient and relevant in terms of both global environmental hazard diagnosis and seasonality analysis. The in-depth characterization of the baseline levels and relevant effect thresholds for environmentally relevant end points is thus a challenge and might be vigorously pursued and developed further over the coming years to lead to homogenized ERA procedures between the various environmental institutions. In particular, several research programs are in progress to define basal reference levels, effect thresholds, and confounding factors of the biological responses in *D. polymorpha* (from molecular to population) in order to routinely include this promising species in WOE approaches dedicated to freshwater environment quality diagnosis.

## Author details

Mélissa Palos Ladeiro[1]*, Iris Barjhoux[1], Aurélie Bigot-Clivot[1], Marc Bonnard[1], Elise David[1], Odile Dedourge-Geffard[1], Elodie Geba[1], Emilie Lance[1], Maxime Lepretre[1], Gabrielle Magniez[1], Damien Rioult[1], Dominique Aubert[2], Isabelle Villena[2], Gaëlle Daniele[3], Arnaud Salvador[3], Emmanuelle Vulliet[3], Jean Armengaud[4] and Alain Geffard[1]

*Address all correspondence to: melissa.palos@univ-reims.fr

1 UMR-I 02 SEBIO, Environmental Stress and Water BIOmonitoring Unit, Faculty of Natural Sciences, University of Reims Champagne-Ardenne, Moulin de la Housse, France

2 EA 3800, SFR CAP-Santé, Faculty of Medicine, University of Reims Champagne-Ardenne, Reims, France

3 Institute of Analytical Sciences, University of Lyon, CNRS, UMR, Villeurbanne, France

4 CEA DRF-Li2D, PRAE Marcel Boiteux—Départementale 765, Bagnols-sur-Cèze Cedex, France

## References

[1] Zuykov M, Pelletier E, Harper D. Bivalve mollusks in metal pollution studies: From bioaccumulation to biomonitoring. Chemosphere. 2013;93:201-8.

[2] Binelli A, Della Torre C, Magni S, Parolini M. Does zebra mussel (*Dreissena polymorpha*) represent the freshwater counterpart of *Mytilus* in ecotoxicological studies? A critical review. Environ Pollut. 2015;196:386-403.

[3] Nalepa T, Schloesser D. Quagga and zebra mussels, biology, impacts and control. 2014. 775 p.

[4] Borcherding J. Steps from ecological and ecotoxicological research to the monitoring for water quality using the zebra mussel in a biological warning system. In: The Zebra Mussel in Europe. 2010. p. 279-83.

[5] Faria M, Huertas D, Soto D, Grimalt J, Catalan J, Riva M, et al. Contaminant accumulation and multi-biomarker responses in field collected zebra mussels (*Dreissena polymorpha*) and crayfish (*Procambarus clarkii*), to evaluate toxicological effects of industrial hazardous dumps in the Ebro river (NE Spain). Chemosphere. 2010;78:232-40.

[6] Voets J, Bervoets L, Smolders R, Covaci A, De Coen W, Blust R. Biomonitoring environmental pollution in freshwater ecosystems using Dreissena polymorpha. In: The Zebra Mussel in Europe. 2010. p. 301-21.

[7] Pain-Devin S, Cossu-Leguille C, Geffard A, Giamberini L, Jouenne T, Minguez L, et al. Towards a better understanding of biomarker response in field populations: a case study in eight populations of zebra mussels. Aquat Toxicol. 2014;155:52-61.

[8] Baldwin B, Mayer M, Dayton J, Pau N, Mendilla J, Sullivan M, et al. Comparative growth and feeding in zebra and quagga mussels (*Dreissena polymorpha* and *Dreissena bugensis*): implications for North American lakes. Can J Fish Aquat Sci. 2002;59:680-94.

[9] Alvarez-Muñoz D, Huerta B, Fernandez-Tejedor M, Rodríguez-Mozaz S, Barceló D. Multi-residue method for the analysis of pharmaceuticals and some of their metabolites in bivalves. Talanta. 2015;136:174-82.

[10] McEneff G, Barron L, Kelleher B, Paull B, Quinn B. The determination of pharmaceutical residues in cooked and uncooked marine bivalves using pressurised liquid extraction, solid-phase extraction and liquid chromatography-tandem mass spectrometry. Anal Bioanal Chem. 2013;405:9509-21.

[11] Cueva-Mestanza R, Torres-Padrón M, Sosa-Ferrera Z, Santana-Rodríguez J. Microwave-assisted micellar extraction coupled with solid-phase extraction for preconcentration of pharmaceuticals in molluscs prior to determination by HPLC. Biomed Chromatogr. 2008;22:1115-22.

[12] Contardo-Jara V, Lorenz C, Pflugmacher S, Nutzmann G, Kloas W, Wiegand C. Exposure to human pharmaceuticals Carbamazepine, Ibuprofen and Bezafibrate causes molecular effects in *Dreissena polymorpha*. Aquat Toxicol. 2011;105:428-37.

[13] Martínez Bueno M, Boillot C, Fenet H, Chiron S, Casellas C, Gómez E. Fast and easy extraction combined with high resolution-mass spectrometry for residue analysis of two anticonvulsants and their transformation products in marine mussels. J Chromatogr A. 2013;1305:27-34.

[14] Núñez M, Borrull F, Fontanals N, Pocurull E. Determination of pharmaceuticals in bivalves using QuEChERS extraction and liquid chromatography-tandem mass spectrometry. Anal Bioanal Chem. 2015;407:3841-9.

[15] Daniele G, Fieu M, Joachim S, James-Casas A, Andres S, Baudoin P, et al. Development of a multi-residue analysis of diclofenac and some transformation products in bivalves using QuEChERS extraction and liquid chromatography-tandem. Application to samples from mesocosm studies mass spectrometry. Talanta. 2016;155:1-7.

[16] Bringolf R, Heltsley R, Newton T, Eads C, Fraley S, Shea D, et al. Environmental occurrence and reproductive effects of the pharmaceutical fluoxetine in native freshwater mussels. Environ Toxicol Chem. 2010;29:1311-8.

[17] Gilroy E, Klinck J, Campbell S, McInnis R, Gillis P, de Solla S. Toxicity and bioconcentration of the pharmaceuticals moxifloxacin, rosuvastatin, and drospirenone to the unionid mussel *Lampsilis siliquoidea*. Sci Total Environ. 2014;487:537-44.

[18] Li W, Shi Y, Gao L, Liu J, Cai Y. Investigation of antibiotics in mollusks from coastal waters in the Bohai Sea of China. Environ Pollut. 2012;162:56-62.

[19] Du O, Haddad S, Luek A, Scott W, Saari G, Kristofco L, et al. Bioaccumulation and trophic dilution of human pharmaceuticals across trophic positions of an effluent-dependent wadeable stream. Philos Trans R Soc Lond B Biol Sci. 2014;369:20140058.

[20] Ericson H, Thorsén G, Kumblad L. Physiological effects of diclofenac, ibuprofen and propranolol on Baltic Sea blue mussels. Aquat Toxicol. 2010;99:223-31.

[21] Contardo-Jara V, Lorenz C, Pflugmacher S, Nutzmann G, Kloas W, Wiegand C. Molecular effects and bioaccumulation of levonorgestrel in the non-target organism *Dreissena polymorpha*. Environ Pollut. 2011;152:38-44.

[22] Belli S, Smith N, Ferguson D. The coccidian oocyst: a tough nut to crack! Trends Parasitol. 2006 Sep;22(9):416-23.

[23] Dumètre A, Dubey J, Ferguson D, Bongrand P, Azas N, Puech P. Mechanics of the *Toxoplasma gondii* oocyst wall. PNAS. 2013 Jul 9;110(28):11535-40.

[24] Gallas-Lindemann C, Sotiriadou I, Plutzer J, Karanis P. Prevalence and distribution of Cryptosporidium and Giardia in wastewater and the surface, drinking and ground waters in the Lower Rhine, Germany. Epidemiol Infect. 2013 Jan;141(1):9-21.

[25] Palos Ladeiro M, Bigot A, Aubert D, Hohweyer J, Favennec L, Villena I, et al. Protozoa interaction with aquatic invertebrate: interest for watercourses biomonitoring. Environ Sci Pollut Res. 2013;20:778-89.

[26] Palos Ladeiro M, Aubert D, Villena I, Geffard A, Bigot A. Bioaccumulation of human waterborne protozoa by zebra mussel (*Dreissena polymorpha*): Interest for water biomonitoring. Water Res. Elsevier Ltd; 2014;48:148-55.

[27] Palos Ladeiro M, Bigot-Clivot A, Aubert D, Villena I, Geffard A. Assessment of *Toxoplasma gondii* levels in zebra mussel (*Dreissena polymorpha*) by real-time PCR: an organotropism study. Environ Sci Pollut Res. 2015;22:13693-701.

[28] Kerambrun E, Palos Ladeiro M, Bigot-Clivot A, Dedourge-Geffard O, Dupuis E, Villena I, et al. Zebra mussel as a new tool to show evidence of freshwater contamination by waterborne *Toxoplasma gondii*. J Appl Microbiol. 2016;120:498-508.

[29] Paerl H, Huisman J. Blooms like it hot. Science (80- ). 2008;32:57-8.

[30] Wiegand C, Pflugmacher S. Ecotoxicological effects of selected cyanobacterial secondary metabolites a short review. Toxicol Appl Pharmacol. 2005;203:201-18.

[31] Ibelings B, Chorus I. Accumulation of cyanobacterial toxins in freshwater "seafood" and its consequences for public health: a review. Environ Pollut. 2007;150:177-92.

[32] Lance E, Petit A, Sanchez W, Paty C, Gérard C, Bormans M. Evidence of trophic transfer of microcystins from the gastropod Lymnaea stagnalis to the fish Gasterosteus aculeatus. Harmful Algae. 2014;31:9-17.

[33] Jiang L, Johnston E, Åberg K, Nilsson U, Ilag L. Strategy for quantifying trace levels of BMAA in cyanobacteria by LC/MS/MS. Anal Bioanal Chem. 2013;405:1283-92.

[34] Réveillon D, Amzil Z. Beta-N-methylamino-l-alanine: LC-MS/MS optimization, screening of cyanobacterial strains and occurrence in shellfish from Thau, a French Mediterranean Lagoon. Mar Drugs. 2014;12:5441-67.

[35] Lage S, Annadotter H, Rasmussen U, Rydberg S. Biotransfer of beta-N-methylamino-l-alanine (BMAA) in a eutrophicated freshwater lake. Mar Drugs. 2015;13:1185-201.

[36] Spencer P, Nunn P, Hugon J, Ludolph A, Ross S, Roy D, et al. Guam amyotrophic lateral sclerosis-parkinsonism-dementia linked to a plant excitant neurotoxin. Science (80- ). 1987;237:517-22.

[37] Faasen E. Presence of the neurotoxin BMAA in aquatic ecosystems: what do we really know? Toxins (Basel). 2014;6:1109-38.

[38] Salomonsson M, Fredriksson E, Alfjorden A, Hedeland M, Bondesson U. Seafood sold in Sweden contains BMAA: a study of free and total concentrations with UHPLC–MS/MS and dansyl chloride derivatization. Toxicol Rep. 2015;2:1473-81.

[39] Réveillon D, Séchet V, Hess P, Amzil Z. Systematic detection of BMAA (β-N-methylamino-l-alanine) and DAB (2,4-diaminobutyric acid) in mollusks collected in shellfish production areas along the French coasts. Toxicon. 2016;110:35-46.

[40] Dionisio Pires L, Karlsson K, Meriluoto J, Visser P, Siewertsen K, Van Donk E, et al. Assimilation and depuration of microcystin-LR by the zebra mussel, *Dreissena polymorpha*. Aquat Toxicol. 2004;69:385-96.

[41] Juhel G, Davenport J, O'Halloran J, Culloty S, O'Riordan R, James K, et al. Impacts of microcystins on the feeding behaviour and energy balance of zebra mussels, *Dreissena polymorpha*: a bioenergetics approach. Aquat Toxicol. 2006;79:391-400.

[42] Ibelings B, Bruning K, De Jonge J, Wolfstein K, Dionisio Pires L, Postma J, et al. Distribution of microcystins in a lake foodweb: no evidence for biomagnification. Microb Ecol. 2005;49:487-500.

[43] Paldaviciene A, Zaiko A, Mazur-Marzec H, Razinkovas-Baziukas A. Bioaccumulation of microcystins in invasive bivalves: a case study from the boreal lagoon ecosystem. Oceanologia. 2015;57:93-101.

[44] Barda L, Kankaanpää H, Purina I, Balode M, Sjövall O, Meriluoto J. Bioaccumulation of hepatotoxins—a considerable risk in the Latvian environment. Environ Pollut. 2015;196:313-20.

[45] Juhel G, Ramsay RM, Davenport J, O'Halloran J, Culloty SC. Effect of the microcystin-producing Cyanobacterium, *Microcystis aeruginosa*, on immune functions of the zebra mussel *Dreissena polymorpha*. J Shellfish Res [Internet]. 2015;34(2):433-42. Available from: http://www.bioone.org/doi/10.2983/035.034.0227

[46] Smolders R, Bervoets L, de Coen W, Blust R. Cellular energy allocation in zebra mussels exposed along a pollution gradient: linking cellular effects to higher levels of biological organization. Environ Pollut. 2004;129:99-112.

[47] Canty M, Hagger J, Moore R, Cooper L, Galloway T. Sublethal impact of short term exposure to the organophosphate pesticide azamethiphos in the marine mollusc *Mytilus edulis*. Mar Pollut Bull. 2007;54:396-402.

[48] Halldórsson H, De Pirro M, Romano C, Svavarsson J, Sarà G. Immediate biomarker responses to benzo[a]pyrene in polluted and unpolluted populations of the blue mussel (*Mytilus edulis* L.) at high-latitudes. Environ Int. 2008;34:483-9.

[49] Solé M, Shaw J, Frickers P, Readman J, Hutchinson T. Effects on feeding rate and biomarker responses of marine mussels experimentally exposed to propranolol and acetaminophen. Anal Bioanal Chem. 2010;396:649-56.

[50] Kerambrun E, Rioult D, Delahaut L, Evariste L, Pain-Devin S, Auffret M, et al. Variations in gene expression levels in four European zebra mussel, *Dreissena polymorpha*, populations in relation to metal bioaccumulation: a field study. Ecotoxicol Environ Saf. 2016;134:53-63.

[51] Palais F, Jubeaux G, Dedourge-Geffard O, Biagianti-Risbourg, S Geffard A. Studies of amylolytic and cellulolytic activities in the crystalline style and the digestive diverticule of the freshwater bivalve Dreissena polymorpha (Pallas, 1771). Molluscan Res. 2010;30:29-36.

[52] Palais F, Dedourge-Geffard O, Beaudon A, Pain-Devin S, Trapp J, Geffard O, et al. One-year monitoring of core biomarker and digestive enzyme responses in transplanted zebra mussels (*Dreissena polymorpha*). Ecotoxicology. 2012;21:888-905.

[53] Bourgeault A, Gourlay-France C, Vincent-Hubert F, Palais F, Geffard A, Biagianti-Risbourg S, et al. Lessons from a transplantation of zebra mussels into a small urban river: an integrated ecotoxicological assessment. Environ Toxicol. 2010;25:468-78.

[54] Ram J, Fong P, Garton D. Physiological aspects of zebra mussel reproduction: maturation, spawning, and fertilization. Am Zool. 1996;36:326-38.

[55] Wacker A, Von Elert E. Food quality controls reproduction of the zebra mussel (*Dreissena polymorpha*). Oecologia. 2003;135:332-8.

[56] Mantecca P, Vailati G, Bacchetta R. Histological changes and micronucleus induction in the zebra mussel *Dreissena polymorpha* after paraquat exposure. Histol Histopathol. 2006;21:829-40.

[57] Griebeler E, Seitz A. Effects of increasing temperatures on population dynamics of the zebra mussel Dreissena polymorpha: implications from an individual-based model. Oecologia. 2007;151:530-43.

[58] Bacchetta R, Mantecca P. DDT polluted meltwater affects reproduction in the mussel *Dreissena polymorpha*. Chemosphere. 2009;76:1380-5.

[59] Lajtner J. Presence of *Bucephalus polymorphus, Echinoparyphium recurvatum* and *Aspidogaster limacoides* (Platodes, Trematoda) in the visceral mass of *Dreissena polymorpha* (Mollusca, Bivalvia). Helminthologia. 2012;53:85-92.

[60] Minguez L, Buronfosse T, Giamberini L. Different host exploitation strategies in two zebra mussel-trematode systems: adjustments of host life history traits. PLoS One. 2012;7:e34029.

[61] Magniez G, Bonnard M, Franco A, Rioult D, Joachim S, Daniele G, et al. Development of a reproductive toxicity biomarker by flow cytometry. Application during exposure to carbamazepine under controlled and semi-ecosystemic conditions. In: 1st International Conference on Risk Assesment of Pharmaceuticals in the Environment. Paris, France; 2016. p. 8-9 September.

[62] Franco A, Kellner K, Mathieu M, Lelong C, Goux D, Heude-Berthelin C. Male germ cells of the Pacific oyster Crassostrea gigas: flow cytometry analysis, cell sorting and molecular expression. Aquat Living Resour. 2011;24:237-45.

[63] Lewis C, Galloway T. Sperm toxicity and the reproductive ecology of marine invertebrates. Integr Environ Assess Manag. 2010;6:188-90.

[64] Bialkowska J, Dietrich G, Demaniowicz W, Glogowski J. Evaluation of CASA system for predicting the motility of zebra mussel (*Dreissena polymorpha*) spermatozoa. Electron J Polish Agric Univ. 2004;7.

[65] Favret K, Lynn J. Flow-cytometric analyses of viability biomarkers in pesticide-exposed sperm of three aquatic invertebrates. Arch Environ Contam Toxicol. 2010;58:973-84.

[66] Seaver R, Ferguson G, Gehrmann W, Misamore M. Effects of ultraviolet radiation on gametic function during fertilization in zebra mussels (*Dreissena polymorpha).* J Shellfish Res. 2009;28:625-33.

[67] Faria M, Lopez M, Fernandez-Sanjuan M, Lacorte S, Barata C. Comparative toxicity of single and combined mixtures of selected pollutants among larval stages of the native freshwater mussels (*Unio elongatulus*) and the invasive zebra mussel (*Dreissena polymorpha).* Sci Total Environ. 2010;408:2452-8.

[68] Evariste E, Auffret M, Audonnet S, Geffard A, David E, Brousseau P, et al. Functional features of hemocyte subpopulations of the invasive mollusk species *Dreissena polymorpha.* Fish Shellfish Immunol. 2016;56:144-54.

[69] Cheng TC. Bivalves. In: Invertebrate blood cells 1. Academic P. London, UK; 1981. p. 233-300.

[70] Binelli A, Parolini M, Cogni D, Pedriali A, Provini A. A multi-biomarker assessment of the impact of the antibacterial trimethoprim on the non-target organism Zebra mussel (*Dreissena polymorpha)*. Comp Biochem Physiol - C Toxicol Pharmacol [Internet]. 2009;150:329-36. Available from: http://dx.doi.org/10.1016/j.cbpc.2009.05.011

[71] Parolini M, Magni S, Castiglioni S, Zuccato E, A B. Realistic mixture of illicit drugs impaired the oxidative status of the zebra mussel (*Dreissena polymorpha)*. Chemosphere. 2015;128:96-102.

[72] Brousseau P, Pellerin J, Morin Y, Cyr D, Blakley B, Boermans H, et al. Flow cytometry as a tool to monitor the disturbance of phagocytosis in the clam Mya arenaria hemocytes following in vitro exposure to heavy metals. Toxicology. 2000 Jan 3;142(2):145-56.

[73] Perez D, Fontanetti C. Hemocitical responses to environmental stress in invertebrates: a review. Environ Monit Assess. 2011 Jun;177:437-47.

[74] Tiscar P, Mosca F. Defense mechanisms in farmed marine molluscs. Vet Res Commun. 2004;28:57-62.

[75] Renwrantz L, Schmalmack W, Redel R, Friebel B, Schneeweib H. Conversion of phenoloxidase and peroxidase indicators in individual haemocytes of Mytilus edulis specimens and isolation of phenoloxidase from haemocyte extract. J Comp Physiol B. 1996;165:647-58.

[76] Allam B, Raftos D. Immune responses to infectious diseases in bivalves. J Invertebr Pathol. 2015;131:121-36.

[77] Camusso M, Balestrini R, Muriano F, Mariani M. Use of freshwater mussel Dreissena polymorpha to assess trace metal pollution in the lower river Po (Italy). Chemosphere. 1994;29:729-45.

[78] Bervoets L, Voets J, Chu S, Covaci A, Schepens P, Blust R. Comparison of accumulation of micropollutants between indigenous and transplanted zebra mussels (*Dreissena polymorpha)*. Environ Toxicol Chem. 2004;23(8):1973-83.

[79] Sanchez W, Burgeot T, Porcher J. A novel "Integrated Biomarker Response" calculation based on reference deviation concept. Environ Sci Pollut Res. 2013;20:2721-5.

[80] Schäfer S, Hamer B, Treursić B, Möhlenkamp C, Spira D, Korlević M, et al. Comparison of bioaccumulation and biomarker responses in *Dreissena polymorpha* and *D. bugensis* after exposure to resuspended sediments. Arch Environ Contam Toxicol. 2012;62:614-27.

[81] Marescaux J, Molloy D, Giamberini L, Albrecht C, Van Doninck K. First records of the quagga mussel, *Dreissena rostriformis bugensis* (Andrusov, 1897), in the Meuse River within France. BioInvasions Rec. 2012;1:273-6.

[82] Molloy D, Bij De Vaate A, Wilke T, Giamberini L. Discovery of *Dreissena rostriformis bugensis* (Andrusov 1897) in Western Europe. Biol Invasions. 2007;9:871-4.

[83] Matthews J, Van der Velde G, Bij de Vaate A, Collas F, Koopman K, Leuven R. Rapid range expansion of the invasive quagga mussel in relation to zebra mussel presence in The Netherlands and Western Europe. Biol Invasions. 2014;16:23-42.

[84] Marescaux J, Bij de Vaate A, Van Doninck K. First records of *Dreissena rostriformis bugensis* (Andrusov, 1897) in the Meuse River. BioInvasions Rec. 2012;1:109-14.

[85] Marescaux J, Boets P, Lorquet J, Sablon R, Van Doninck K, Beisel J. Sympatric *Dreissena* species in the Meuse River: towards a dominance shift from zebra to quagga mussels. Aquat Invasions. 2015;10:287-98.

[86] Bij de Vaate A, Van der Velde G, Leuven R, Heiler K. Spread of the quagga mussel, *Dreissena rostriformis* bugensis in Western Europe. In: Nalepa T, Schloesser D, editors. Quagga and Zebra Mussels: Biology, Impacts, and Control (second ed). CRC Press, Florida; 2014. p. 83-92.

[87] Nalepa T, Fanslow D, Pothoven S. Recent changes in density, biomass, recruitment, size structure, and nutritional state of Dreissena populations in southern Lake Michigan. J Great Lakes Res. 2010;36:5-19.

[88] Ram J, Karim A, Banno F, Kashian D. Invading the invaders: Reproductive and other mechanisms mediating the displacement of zebra mussels by quagga mussels. Invertebr Reprod Dev. 2012;56:21-32.

[89] De Ventura L, Sarpe D, Kopp K, Jokela J. Variability in phenotypic tolerance to low oxygen in invasive populations of quagga and zebra mussels. Aquat Invasions. 2016;11:267-76.

[90] Stoeckmann A. Physiological energetics of Lake Erie dreissenid mussels: A basis for the displacement of *Dreissena polymorpha* by *Dreissena bugensis*. Can J Fish Aquat Sci. 2003;60:126-34.

[91] Karatayev A, Mastitsky S, Padilla D, Burlakova L, Hajduk M. Differences in growth and survivorship of zebra and quagga mussels: size matters. Hydrobiologia. 2011;668:183-94.

[92] Mills E, Roseman E, Rutzke M, Gutenmann W, Lisk D. Contaminant and nutrient element levels in soft tissues of zebra and quagga mussels from waters of southern Lake Ontario. Chemosphere. 1993;27:1465-73.

[93] Johns C, Timmerman B. Total cadmium, copper, and zinc in two *Dreissenid mussels, Dreissena polymorpha* and *Dreissena bugensis,* at the outflow of Lake Ontario. J Great Lakes Res. 1998;24:55-64.

[94] Richman L, Somers K. Can we use zebra and quagga mussels for biomonitoring contaminants in the Niagara River? Water Air Soil Pollut. 2005;167:155-78.

[95] Matthews J, Schipper A, Hendriks A, Yen Le T, Bij De Vaate A, Van Der Velde G, et al. A dominance shift from the zebra mussel to the invasive quagga mussel may alter the trophic transfer of metals. Environ Pollut. 2015;203:183-90.

[96] Potet M, Devin S, Pain-Devin S, Rousselle P, Giambérini L. Integrated multi-biomarker responses in two dreissenid species following metal and thermal cross-stress. Environ Pollut. 2016;218:39-49.

[97] Voroshilova I, Artamonova V, Makhrov A, Slyn'ko Y. Natural hybridization of two mussel species *Dreissena polymorpha* (Pallas, 1771) and *Dreissena bugensis* (Andrusov, 1897). Biol Bull. 2010;37:542-7.

[98] Marescaux J, Van Doninck K. Using DNA barcoding to differentiate invasive *Dreissena* species (Mollusca, Bivalvia). Zookeys. 2013;365:235-44.

[99] Beggel S, Cerwenka A, Brandner J, Geist J. Open accesshell morphological versus genetic identification of quagga mussel (*Dreissena bugensis*) and zebra mussel (*Dreissena polymorpha)*. Aquat Invasions. 2015;10:93-9.

[100] Teubner D, Wesslein A, Rønne P, Veith M, Frings C, Paulus M. Is a visuo-haptic differentiation of zebra mussel and quagga mussel based on a single external morphometric shell character possible? Aquat Invasions. 2016;11:145-54.

[101] Hoy M, Kelly K, Rodriguez R. Development of a molecular diagnostic system to discriminate *Dreissena polymorpha* (zebra mussel) and *Dreissena bugensis* (quagga mussel). Mol Ecol Resour. 2010;10:190-2.

[102] Chapman P. Determining when contamination is pollution—weight of evidence determinations for sediments and effluents. Environ Int. 2007;33:492-501.

[103] Chapman P, Wang F, Janssen C, Goulet R, Kamunde C. Conducting ecological risk assessments of inorganic metals and metalloids: current status. Hum Ecol Risk Assess. 2003;9:641-97.

[104] Lamers A, Heiney J, Ram J. Cloning and sequence analysis of two cDNAs encoding cyclin A and cyclin B in the zebra mussel *Dreissena polymorpha*. Biochim Biophys Acta. 1999;1448:519-24.

[105] Wilson A, Boulding E, Naish K. Characterization of tri- and tetranucleotide microsatellite loci in the invasive mollusc *Dreissena bugensis*. Mol Ecol. 1999;8:692-703.

[106] Peñarrubia L, Sanz N, Pla C, Vidal O, Viñas J. Using massive parallel sequencing for the development, validation, and application of population genetics markers in the invasive bivalve zebra mussel (*Dreissena polymorpha)*. PLoS One. 2015;10:e0120732.

[107] Cao L, Kenchington E, Zouros E, Rodakis G. Evidence that the large noncoding sequence is the main control region of maternally and paternally transmitted mitochondrial genomes of the marine mussel (*Mytilus* spp.). Genetics. 2004;167:835-50.

[108] Mizi A, Zouros E, Moschonas N, Rodakis G. The complete maternal and paternal mitochondrial genomes of the Mediterranean mussel *Mytilus galloprovincialis*: implications for the doubly uniparental inheritance mode of mtDNA. Mol Biol Evol. 2005; 22:952-67.

[109] Xu W, Faisal M. Development of a cDNA microarray of zebra mussel (*Dreissena polymorpha*) foot and its use in understanding the early stage of underwater adhesion. Gene. 2009;436:71-80.

[110] Navarro A, Campos B, Barata C, Piña B. Transcriptomic seasonal variations in a natural population of zebra mussel (*Dreissena polymorpha*). Sci Total Environ. 2013;454:482-9.

[111] Armengaud J, Trapp J, Pible O, Geffard O, Chaumot A, Hartmann E. Non-model organisms, a species endangered by proteogenomics. J Proteomics. 2014;105:5-18.

[112] Trapp J, Geffard O, Imbert G, Gaillard J, Davin A, Chaumot A, et al. Proteogenomics of *Gammarus fossarum* to document the reproductive system of amphipods. Mol Cell Proteomics. 2014;13:3612-25.

[113] Trapp J, Armengaud J, Pible O, Gaillard J, Abbaci K, Habtoul Y, et al. Proteomic investigation of male *Gammarus fossarum*, a freshwater crustacean, in response to endocrine disruptors. J Proteome Res. 2015;14:292-303.

[114] Riva C, Binelli A, Rusconi F, Colombo G, Pedriali A, Zippel R, et al. A proteomic study using zebra mussels (*D. polymorpha*) exposed to benzo($\alpha$)pyrene: the role of gender and exposure concentrations. Aquat Toxicol. 2011;104:14-22.

[115] Armengaud J. Microbiology and proteomics, getting the best of both worlds! Environ Microbiol. 2013;15:12-23.

[116] Trapp J, Armengaud J, Salvador A, Chaumot A, Geffard O. Next-generation proteomics: toward customized biomarkers for environmental biomonitoring. Environ Sci Technol. 2014;48:13560-72.

[117] Charnot A, Gouveia D, Armengaud J, Almunia C, Chaumot A, Lemoine J, et al. Liquid chromatography coupled to tandem mass spectrometry offers new perspectives in environmental science: multiplexed assay for protein quantitation in the invertebrate *Gammarus fossarum*. Anal Bioanal Chem. in press.

[118] Chapman P, McDonald B, Lawrence G. Weight-of-evidence issues and frameworks for sediment quality (and other) assessments. Hum Ecol Risk Assess. 2002;8:1489-515.

[119] Chapman P, Hollert H. Should the sediment quality triad become a tetrad, a pentad, or possibly even a hexad? J Soils Sediment. 2006;6:4-8.

[120] Dagnino A, Sforzini S, Dondero F, Fenoglio S, Bona E, Jensen J, et al. A "Weight-of-Evidence" approach for the integration of environmental "Triad" data to assess ecological risk and biological vulnerability. Integr Environ Assess Manag. 2008;4:314-26.

[121] Piva F, Ciaprini F, Onorati F, Benedetti M, Fattorini D, Ausili A, et al. Assessing sediment hazard through a weight of evidence approach with bioindicator organisms: A practical model to elaborate data from sediment chemistry, bioavailability, biomarkers and ecotoxicological bioassays. Chemosphere. 2011;83:475-85.

[122] Benedetti M, Gorbi S, Fattorini D, D'Errico G, Piva F, Pacitti D, et al. Environmental hazards from natural hydrocarbons seepage: Integrated classification of risk from sediment

chemistry, bioavailability and biomarkers responses in sentinel species. Environ Pollut. 2014;185:116-26.

[123] Bebianno M, Pereira C, Rey F, Cravo A, Duarte D, D'Errico G, et al. Integrated approach to assess ecosystem health in harbor areas. Sci Total Environ. 2015;514:92-107.

[124] Regoli F. A multidisciplinary weight of evidence approach for environmental risk assessment at the Costa Concordia wreck: Integrative indices from Mussel Watch. Mar Environ Res. 2014;96:92-104.

[125] Barjhoux I, Fechner L, Lebrun J, Anzil A, Ayrault S, Budzinski H, et al. Application of a multidisciplinary and integrative weight-of-evidence approach to a 1-year monitoring survey of the Seine River. Environ Sci Pollut Res. 2017; in press.

[126] Coulaud R, Geffard O, Coquillat A, Quéau H, Charles S, Chaumot A. Ecological modeling for the extrapolation of ecotoxicological effects measured during in situ assays in Gammarus. Environ Sci Technol 2014;48:6428-36.

[127] Chaumot A, Geffard O, Armengaud J and Maltby L. Gammarid as reference species for freshwater monitoring. In Amiard-Triquet C, Amiard JC and Mouneyrac C, Aquatic Ecotoxicology: Advancing tools and dealing with emerging risks. London, United Kingdom: Academic Press, Elsevier. 2015. p. 253-280.

Chapter 4

# Patellid Limpets: An Overview of the Biology and Conservation of Keystone Species of the Rocky Shores

Paulo Henriques, João Delgado and Ricardo Sousa

Additional information is available at the end of the chapter

http://dx.doi.org/10.5772/67862

**Abstract**

This work reviews a broad spectrum of subjects associated to Patellid limpets' biology such as growth, reproduction, and recruitment, also the consequences of commercial exploitation on the stocks and the effects of marine protected areas (MPAs) in the biology and populational dynamics of these intertidal grazers. Knowledge of limpets' biological traits plays an important role in providing proper background for their effective management. This chapter focuses on determining the effect of biotic and abiotic factors that influence these biological characteristics and associated geographical patterns. Human exploitation of limpets is one of the main causes of disturbance in the intertidal ecosystem and has occurred since prehistorical times resulting in direct and indirect alterations in the abundance and size structure of the target populations. The implementation of MPAs has been shown to result in greater biomass, abundance, and size of limpets and to counter other negative anthropogenic effects. However, inefficient planning and lack of surveillance hinder the accomplishment of the conservation purpose of MPAs. Inclusive conservation approaches involving all the stakeholders could guarantee future success of conservation strategies and sustainable exploitation. This review also aims to establish how beneficial MPAs are in enhancing recruitment and yield of adjacent exploited populations.

**Keywords:** Patellidae, limpets, fisheries, MPAs, conservation

## 1. Introduction

The Patellidae are one of the most successful families of gastropods that inhabit the rocky shores from the supratidal to the subtidal, a marine habitat subject to some of the most variable and unpredictable environmental conditions. Therefore, many of their peculiar morphological and biological characteristics can be understood as adaptations to this environment. The biological traits of limpets vary inter- and intraspecifically as a result of genetic differences and environmental influences [1]. Parameters such as growth, reproduction, and mortality are dependent on a complex array of selective forces and are important in understanding the distribution and abundance of a species [2, 3]. Differences in limpet populations from distinct geographic areas are most probably explained by specific environmental and anthropogenic conditions, essentially oligotrophy, sea water temperature, and fishing pressure. Thus, for some of the biological traits, it is expected to find patterns, like temperature which changes somewhat consistently with latitude and has a profound effect on the growth of limpet species, with species inhabiting higher latitudes growing more slowly and achieving larger maximum sizes, therefore having a longer lifespan than limpets from lower latitudes.

Patellid limpets are also subjected to anthropogenic impacts on the coastal ecosystems such as, pollution, habitat removal, and harvest which in some cases has led to the reduction of abundance or even the disappearance of limpets from large areas. The decline of these species, which may have been further accelerated by the progressive deterioration of the coastline, continues at an alarming rate and many of their stocks are on the verge of disappearance. To avert this situation, regulators have established several measures including the implementation of closed seasons and areas where limpet harvest is interdicted, minimum size of capture, and catch limits. Limpet populations seem to respond, in general, in a positive way to these measures; however, the response is closely linked to the ability of the regulators to enforce said measures.

Another popular strategy adopted in the protection of the rocky shores and limpets is the implementation of marine protected areas (MPAs). The effectiveness of MPAs in protecting exploited populations of limpets and underlying their overall success in increasing density and abundance as well as promoting healthy size composition with impact on the reproductive output of these species is well known. Nonetheless, several limitations are recognized that can negatively affect the protective role of MPAs such as, naturally occurring variations of the species biology and ecology as well as limitations regarding the management of MPAs, for instance, the lack of surveillance and enforcement of protection regulations.

The aim of this work is to review a broad spectrum of subjects associated to Patellid limpets' biology such as growth and reproduction, also the consequences of commercial exploitation on the stocks of these species and the effects of marine protected areas in the biology and populational dynamics of these intertidal grazers. The focus is on determining the effect of identified biotic and abiotic factors that influence these biological characteristics and geographical patterns recognized to be closely connected to growth and reproduction, such as latitude. Regarding conservation of Patellidae, the authors aim to elucidate how beneficial MPAs are in their role of protection of exploited populations and in enhancing recruitment and yield of adjacent exploited populations.

http://dx.doi.org/10.5772/67862

## 2. Biology and ecology of Patellid limpets

### 2.1. Taxonomy and distribution

Patellid limpets are marine gastropod grazers belonging to the family Patellidae Rafinesque, 1815 that comprises the genera *Patella* Linnaeus, 1758, *Cymbula*, H. Adams & A. Adams, 1854, *Helcion* Montfort, 1810, and *Scutellastra* H. Adams & A. Adams, 1854. The worldwide distribution of Patellidae species is anti-tropical with half of the known species restricted to southern Africa and the North-Eastern Atlantic where a high diversity of species is found, while relatively few species are present in the Indian and Pacific Oceans [4, 5]. The Patellidae family is currently represented by, at least, 49 recognized species [6]. The genus *Patella* is comprised of 14 recognized species with a geographical distribution restricted to the North-Eastern Atlantic and the Mediterranean Sea; the genus *Cymbula* includes 10 species found in Southern Africa, South-Eastern Atlantic, and Mediterranean; the genus *Helcion* is represented by four species restricted to Southern Africa, while *Scutellastra* encompasses 21 species with a wide distribution ranging from Southern Africa to the Indo-West and Eastern Pacific [4, 7–10]. Limpets are subject to an array of environmental stresses as a result of their extended vertical distribution, which ranges from the upper to the lower shore levels. Thus, limpets can exhibit varying degrees of structural adaptations since their position relative to the shore influences their exposure to desiccation, hydrodynamic action of the waves, temperature variation, and tidal width [11–14]. This impressive phenotypic plasticity allied to the relatively simple shell geometry, convergent shell shape, and sculpturing results in an unclear Patellid limpet's taxonomy, in such a way that the initial generic names, with broad geographical range, had to be re-evaluated based on superficial similarities [15].

### 2.2. Feeding habits and ecological importance

Limpets are grazing herbivores that feed, by scraping the rocky substrate with the radula, on microbial biofilms which are primarily composed of cyanobacteria and microalgae, including diatoms, spores, and other propagules of macroalgae and invertebrates [16, 17]. Limpets' feeding habits are essential in structuring intertidal communities [16–19] since limpet grazing is a key process in rocky shores involved in determining macroalgal abundance and in modifying ecosystem stability, indirectly enhancing or inhibiting the establishment of other organisms [17]. The decline of population density of limpets might result in an abnormal development of algae diversity as reported by Boaventura et al. [20] or in the occupation of their ecological niche by competing organisms such as barnacles or sea urchins [21–23]. However, the effect of these grazers is not limited to the removal of algae, and very often they can affect other animal species through competitive interactions [24] and by providing secondary habitats for other invertebrates that settle either on top of, or beneath, their shells [19, 25, 26]. Grazers may also affect the rate of succession [27] or cause different assemblages to develop [28]. Thus, limpets are rightfully considered to be keystone species in intertidal communities [29].

### 2.3. Movement and homing

Patellid limpets are considered to some extent semi-sessile organisms; nonetheless, they perform small movements in the area surrounding their usual fixation site. This behavior is designated as homing and can often be observed through the scar that remains in the rocky substrate where the limpet settles. Limpet movement patterns and homing behavior have been extensively studied for *Patella vulgata* Linnaeus, 1758 [30], *Patella depressa* Pennant, 1777 [31], *Patella rustica* Linnaeus, 1758 [32], *Patella ferruginea* Gmelin, 1791 [33], *Scutellastra flexuosa* (Quoy & Gaimard, 1834), and *Scutellastra argenvillei* (Krauss, 1848) [34]. This homing behavior has different functions in different species such as avoiding desiccation [35, 36], reducing predation and intraspecific competition [37–40], responding to wave action [41, 42] and defending territory or asserting dominance [43, 44]. The mechanism that is most widely accepted as being responsible for the homing behavior reports limpets following chemical trails, laid down on the outward trip, on their way back to the fixation site [31, 45, 46].

### 2.4. Growth

Biological parameters such as growth rate, asymptotic length, and age structure reflect the overall state of health of a population and are commonly used as stock assessment tools of exploited marine organisms. Growth, reproductive strategy, and mortality are dependent on a complex array of selective forces [2] and are important in understanding the distribution and abundance of a species [3]. To determine these parameters, most studies usually resort to the capture-recapture method [22, 47–50] or length-frequency distribution analysis [51–53]. Over the past decades, intensive research has focused on the biology of limpets, due to their diversity and ecological significance; however, there remain gaps in the knowledge concerning these species' age structure and growth patterns.

Patellid limpets, like many marine gastropods, exhibit both intra- and interspecific seasonal variation in growth rates [54]. Although some intraspecific variation may be genetically controlled [55], external factors such as changes in food availability [56, 57], wave action [58–60], and vertical distribution on the shore [61] are thought to influence growth rates. Other factors such as population density, available grazing area, predation, and competition are indicated as influencing growth rates of mollusks supporting the idea that the strategy of diverting energy to reproduction and vice versa, according to the organisms' needs, influences growth rates [24, 49, 62, 63]. It has been suggested that limpets with greater growth rates have smaller lifespan while limpets with slow growth are generally long-lived [46]. As such, rapidly growing limpets are usually associated with early maturation, high mortality, and a short lifespan [46, 64].

Clarke et al. [49] observed a latitudinal cline in annual shell growth of the polar limpet *Nacella concinna* (Strebel, 1908). This latitudinal pattern could nevertheless be masked by inter-annual variability. The authors suggest that the observed variation could be the result of a simultaneous change in both growth rate and the duration of the growth period. This change would result from the shorter duration of the seasonal blooming of epiphytic microalgal and microbial biomass at higher latitudes. Another factor influencing growth rates in *N. concinna* is seawater temperature, with warmer temperatures that last longer producing higher growth rates.

*Scutellastra* and *Cymbula* species that occur at similar latitudes present variations in terms of growth, namely in maximum size and growth rates. When compared to tropical limpets belonging to the genus *Cellana* H. Adams, 1869, limpets from temperate regions are generally larger, with wider lifespan and slower growth rates. Additionally, limpets inhabiting the artic regions such as *N. conccina* achieve larger sizes, even wider lifespans, and slower growth rates. This latitudinal pattern has been usually associated with the latitudinal variation of temperature, photoperiod, and insolation [49]. Even though it is consensual that species from lower latitudes grow more rapidly than species from higher latitudes [49, 52], it is not yet clarified whether physiological constraints, a reduced or prolonged growing season, or combination of both might be the cause of dissimilar growth rates at differing latitudes [49].

Nevertheless, due to Patellids' anti-tropical distribution, growth patterns are difficult to observe, particularly when considering latitude. Within this family, variations in growth are mostly derived from prevalent local environmental factors. Nonetheless, when comparing to other Patellogastropoda, a latitudinal pattern becomes apparent, in which at lower latitudes limpets grow at faster rates and achieve smaller sizes, while at higher latitudes, they grow at slower rates and achieve larger sizes. For instance, for the polar limpet *N. concinna* reported growth rates range between 0.059 and 0.323 year$^{-1}$, while the highest growth rate is exceptionally high for a limpet inhabiting the polar regions, probably due to specific characteristics of the habitat in Signy Island [49]. The overall growth rates are inferior to those reported for limpets of the genus *Cellana* that inhabit lower latitudes in temperate and tropical regions with growth rates ranging from 0.400 to 1.661 year$^{-1}$. Patellid limpets exhibit intermediate growth rates ranging from 0.117 year$^{-1}$ in *Scutellastra choclear* (Born, 1778) and 1.020 year$^{-1}$ in *Cymbula oculus* (Born, 1778) reflecting their anti-tropical distribution.

However, the nonlinearity of growth of marine organisms renders the direct comparison of growth parameters impossible [65]. As such, determination and comparison of the overall growth performance of different marine species is achieved using the growth performance index (GPI) of Pauly and Munro [66], which relates the asymptotic length and growth rate [66]. Nonetheless, the growth performance index in Patellogastropod limpets exhibits the same pattern as growth rates with decreasing GPI as latitude increases and ranging from 1.942 in *N. concinna* to 3.653 in *Cymbula granatina* (Linnaeus, 1758), suggesting that growth performance of limpets varies with latitude. Within the Patellidae family the variation of GPI is reduced with values ranging from 2.42 for *S. cochlear* to 3.65 for *C. granatina* from South Africa [62], which is in agreement with Sparre et al. [67] who claim that the growth performance index remains relatively constant at similar rates between related taxa. The variability results therefore due to abiotic and biotic factors that different species are subject to, such as greater or lesser extent of hydrodynamics, desiccation, predation, competition, and temperature.

### 2.5. Reproduction

Patellid limpets have a simple reproductive system, consisting of a simple gonad inserted in the visceral mass and a reduced gonoduct leading to the right nephridium [68, 69]. These species are not externally sexually dimorphic, and sex determination is only possible through macroscopic observation of the gonads. Spawning results in the release of oocytes and sperm directly in the ocean where fecundation occurs. According to Orton et al. [68], spawning is stimulated by environmental triggers, such as high wind speed and wave action. An increase in phytoplankton concentration may also stimulate spawning as suggested by Underwood [24] who observed that gastropod species with planktotrophic larvae spawn when phytoplankton concentration is high.

Most limpet species have a reproductive cycle with a gonadal development stage culminating in a spawning period followed by a resting phase. The spawning period varies inter- and intraspecifically; it may also vary from year to year and is supposed to be triggered by temperature variations, increased wave action, and onshore winds [70]. In regions with higher temperatures, spawning occurs in a short period contrary to what happens in regions with colder waters, where the development of the gonads requires a longer time period [71]. *P. vulgata* is believed to be a winter breeder, with spawning occurring from October to March; however, in colder localities, sexual maturation occurs earlier [68]. On the other hand, in south-west England, *P. depressa* is considered a summer breeder [72] with spawning occurring between late July and early September and without a resting phase unlike *P. vulgata*. The same authors suggested that an increase in temperature associated with wave action stimulates spawning in this species. *Patella ulyssiponensis* Gmelin, 1791 has a spawning period that lasts from October to December, being also considered a winter breeder in south-west England [59, 68, 73]. Orton et al. [68] and Orton and Southward [72] suggested that although the development of the gonad in *P. vulgata* and *P. depressa*, respectively, is well related with temperature, the act of spawning is triggered by violent onshore storms. Thompson [59] also found *P. ulyssiponensis* spawning during the autumn storms. Hence, it seems likely that spawning cannot take place until a population is sufficiently mature, but after that stage is reached, the first strong windstorm will trigger spawning [59]. Another factor that potentially affects the timing of spawning in limpets is food availability; Underwood [24] reported that species with planktotrophic larval stage time spawning with periods when phytoplankton concentrations are high. One such case is that of the closely related species of *P. ulyssiponensis* from the Portuguese mainland and *Patella aspera* Röding, 1798 from Madeira Island. *P. ulyssiponensis* is reported to be a summer breeder while *P. aspera* was reported to be a winter breeder with spawning occurring when the phytoplankton concentration is higher (P. Henriques, pers. comm.). Similarly, it has been reported that in limpets with restricted geographic distribution, the reproductive cycle is influenced by geographic locality, namely in the timing of gametogenesis and spawning [62, 74]. For limpets with broader geographic distribution, it is possible that the reproductive cycle is adjusted to regional environmental conditions [74].

Limpets, like many sessile or sedentary marine invertebrates, have life cycles that include a prolonged pelagic larval phase that can last up to 2 weeks as reported by Hawkins et al. [75] for *Patella* species. Veliger larvae remain in the water column as plankton until eventually fixating in the rocky substrate on the inferior level of the coast. As the juveniles grow, they begin a slow vertical migration, colonizing different levels of the rocky shores [76], leading to variability in patterns of recruitment [77]. Moreover, larvae in the water column are subject to processes of physical transport that can disperse them from the site of reproduction [78]. Thus, the number of recruits on a specific location may be independent of the local larvae production [16, 79] and influenced by current regimes. Nonetheless, limpet populations cannot be considered fully open or fully closed, since some local larval retention is likely to occur despite larval dispersal [80, 81].

Orton [82] suggested the existence of the phenomenon of protandrous hermaphroditism in limpets of the genus *Patella* based on sexual dimorphism in size-frequency of *P. vulgata*; subsequently Thompson [59], Branch [46], and Le Quesne [83] observed that some individuals reach maturity as males and become females in the more advanced stages of their life cycle. This phenomenon of sequential hermaphroditism is also suggested to occur in species of the genera *Cymbula* [46], *Helcion* [74], and *Scutellastra* [62, 84, 85]. Not all male limpets change sex, since a considerable proportion of males can be found in the larger size-groups, these individuals might eventually change sex or remain as males if the signals that lead to sex change are not present [86]. Also, some limpet species are sequential hermaphrodites in which the sex change can be reverted as reported for *P. ferruginea* by Guallart et al. [87].

Sex change in limpet species is thought to be genetically controlled. However, high variability in the timing or on the limpet size at which the change occurs suggests that environmental factors may influence the process. Species such as *C. oculus* have a relatively fixed timing of sex change [88], while in other species, the sex change occurs at sizes that are highly variable. These differences in size and age at which the sex change occurs are often mediated by environmental factors [46, 89–92]. For instance, sex change in mollusks can be delayed in populations where large females are present [89, 90]. Additionally, in populations subjected to higher mortality rates or slower growth rates, sex change seems to occur earlier [93]. Also, it has been reported that social control of sex change occurs in Patellogastropod limpets [91, 92]. In this case, several possible cues for sex change have been suggested such as, contact frequency between individuals, available movement area, food availability, growth rate, pheromonal information, and communication by mucus traces left by individuals during foraging excursions [91].

Hermaphroditism is an evolutionarily advantageous strategy for species with low population densities or low motility such as limpets, since under such conditions, hermaphroditism is supposed to increase the likelihood of successful fertilization [87]. Reproductive success in broadcast spawners, such as limpets, is correlated to the quantity of gametes released into the water column. It is believed that larger limpets produce more gametes than smaller individuals. Additionally, sex change in protandrous hermaphrodite species results in an increase of female individuals in the larger size classes. Thus, the sex distribution through sizes in protandrous hermaphrodite limpets makes these species extremely vulnerable to harvest [33], since the depletion of larger and more fecund individuals and females in a higher percentage may potentially alter the sex ratio and reduce the reproductive output of populations [86].

## 3. Anthropogenic impact on Patellid limpets

Patellid limpets are common gastropods of intertidal rocky shores; however, some species are in serious decline mainly as a consequence of overexploitation [94]. These intertidal and shallow-water grazers are highly vulnerable because of their restricted habitat and its accessibility to human activity [26]. Worldwide, shellfish exploitation has often been shown to lead to decreased biomass and species richness and cause shifts in community composition [95–98]. These effects are driven by the increase of human population density along the coast, the replacement of subsistence by commercial exploitation, and technological advances in methods of collection, processing, storage, and transport [99, 100]. As a result, the effects of human exploitation add to those of natural processes that influence population size of exploited limpets and are a concern in conservation biology [101]. Limpets have been exploited by human populations since the Palaeolithic period [102] at a subsistence level and used as food and bait in several parts of the world, including Mexico, the United States of America [101], Hawaii [103], Australia [104], South Africa [105], Chile [106], and Macaronesia [53, 107, 108]. More recently, this subsistence activity has been replaced, in many parts of the world, by heavy and highly profitable commercial exploitation, increasing the pressure on these species' stocks. Limpet harvest results in reductions in density and shifts toward smaller individuals and can decrease reproductive output since individual fecundity is greater in larger individuals [44, 109, 110]. Thus, harvesting has both direct and indirect effects on these species. There are also effects on the overall community composition as removal of grazing limpets facilitates the growth of algae [20, 111, 112], leading to further changes within the rocky shore communities [16, 17].

The direct effects of limpet exploitation are the decline of the exploited species' abundance and a shift in size composition of their populations that results from the size-selective nature of limpet harvest [100]. This is a result of larger individuals being more visible, thus more prone to be caught, and due to their greater commercial value [22, 113, 114]. The loss of older and larger individuals results in cascading effects on the biology of these species and the affected populations, including changes in life-history parameters, demographics, reproductive success, and ecological interactions [98].

For instance, the decline of larger individuals in an exploited population of limpets might lead to the complete disappearance of the population's viable size as a consequence of a seriously diminished reproductive success, affecting different species in a differentiated manner, as observed by Martins et al. [115] in the Azores. Protandrous hermaphrodite species are particularly susceptible to changes in their population size composition that promote a decline of frequency of larger individuals, since it directly affects the sex ratio of the population resulting in a decrease in female specimens that in natural conditions occur with higher frequency in the larger size classes. Also, larger individuals represent a greater contribution to the reproductive effort in limpets [104], thus the harvest of larger individuals contributes to a decrease in the reproductive success of marine invertebrates such as reported for *P. ferruginea* [33] and may eventually result in the collapse of exploited populations [86, 116, 117].

Reduction of sizes and abundance of larger individuals in exploited populations of limpets have been reported for *Patella candei* d'Orbigny, 1840 [116] and *Patella candei crenata* [114] in the Canaries, *P. candei* e *P. aspera* in the Azores [115], *Helcion concolor* (Krauss, 1848) [44], and *P. ferruginea* in Algeria [118] and Spain [86], as well as for the species *C. oculus* in Southern Africa [88]. The overexploitation of limpets has prompted the implementation of management strategies in order to protect the exploited populations and mitigate human impacts in several parts of the world [26, 53]. The establishment of species-specific total allowable catch,

http://dx.doi.org/10.5772/67862

minimum size of capture, closed seasons, and closed areas has been the most common measures ensued with this objective. These strategies are thought to maintain sex ratios, preserve age structure, prevent sperm limitation, enhance yield, and restrict evolutionary changes in response to fishing, such as shifts to early maturation [119–122]. When considering limpets, due to the phenomenon of protandrous hermaphroditism, in addition to minimum size limits used to prevent recruitment overfishing, management policies should also consider minimum and maximum size limits [122].

For instance, in Madeira archipelago the harvest of *Patella candei* sensu lato and *P. aspera* is regulated since 2006, enforcing the maximum allowable commercial catch of 15 kg/person/day or 200 kg/boat/day and a minimum capture size of 40 mm. Additionally, the competent authorities became responsible for issuing harvest licenses, limiting the number of active fishermen involved in limpet harvest. A closed season was also implemented between November and February in order to prevent limpet harvest during the reproductive season. More recently, the closed season was modified in order to more effectively provide protection to these heavily exploited species, now lasting from December to March. In the Azores, the over-exploitation of limpets resulted in a drastic decline in population density and abundance of limpet populations, and in order to prevent a complete collapse of the stocks, regulation was implemented through the establishment of limpet protected zones that comprise stretches of coast of a few kilometers where the collection of limpets is strictly prohibited throughout the year, seasonal fishing closures, and minimum legal catch sizes [123].

Martins et al. [123] studied the effect of regulation on the recovery of the exploited populations of limpets in the Azores and concluded that the legislation and current levels of enforcement were insufficient to protect the exploited populations and greater levels of enforcement, such as the establishment of physical barriers and other protective strategies should be considered to protect limpet populations. The authors further elaborate that in the absence of adequate enforcement, a complementary approach that has had positive results is co-management [124], due to increasing awareness of the need to increase ownership of conservation areas and to involve all interested parties in the development of management schemes [125, 126].

## 4. Marine protected areas and their protective role in exploited limpet populations

Marine protected areas are frequently considered as a key tool in the conservation of marine biodiversity in coastal regions [127, 128] due to its ecosystem-level approach for exploited species. Reserves are supposed to restore and protect exploited marine organisms within their boundaries and have been shown to harbor denser populations, larger individuals, and higher biomass of exploited species [129].

MPAs potentially offer a way to conserve marine biodiversity by prohibiting harvest and at the same time sustaining fisheries by re-establishing natural conditions for reproduction [129–131]. Thus, protected populations would have higher densities and larger individuals leading to greater production of larvae that would eventually settle outside of the protected area [88, 132–134]. However, increase in recruitment outside reserves can be difficult to verify in the field [135, 136], and there is debate about whether marine reserves can benefit fisheries, as well as act as a conservation tool [137–139].

Human harvesting of limpets is usually size-selective with a strong preference for larger individuals [98] that may potentially alter the sex ratio and reduce the reproductive output of populations in successive hermaphrodite species [75, 140]. A reduction in the abundance of large limpet species, induced by high harvesting pressure, has been observed worldwide with several documented cases of drastic declines such as in the case of the endemic limpet *P. candei* in the Macaronesian Archipelagos [115, 141, 142], *P. ferruginea* considered one of the most endangered marine invertebrates on western Mediterranean rocky shores [118, 143] and *C. oculus* in South Africa [88]. In a more extreme case, the overexploitation as a food source and adornments [144], since pre-Columbian times [145] of *Scutellastra mexicana* (Broderip & G. B. Sowerby I, 1829), resulted in this species being thought extinct [146]. However, some populations of this species were reported to have survived and now the species is considered endangered [147, 148].

MPAs are zones where the harvest of marine organisms is interdicted and are considered a popular alternative to traditional marine resource management measures [149]. Exploited marine organisms in general achieve higher abundance, biomass, and size in MPAs [104, 150]. Halpern [129] reported that abundance and species diversity of marine invertebrates were significantly higher in MPAs regardless of their size.

Halpern and Warner [149] reported that establishing MPAs results in significant increases in the average level of density and biomass in a period of 3 years and that these values are persistent over time. Even though it is considerably difficult to predict the amount of time needed for a community to respond to MPA protection, evidence collected by some authors suggests that the response occurs within 2 years [151, 152]. The speed of response to MPA protection depends on the degree of exploitation to which the species is subjected. If exploitation levels are high, the species are more probable to respond rapidly to the MPA protection, when recruitment occurs at the required levels, as a consequence of the removal of the fishing activity that limits population size, demographics of the species [153–155], and the trophic level occupied by the species, since recruitment is associated to the species' life-history parameters.

In general, for marine invertebrates with a long lifespan and slow growth, it is assumed that the response to protection from MPAs occurs at a slower rate [149]. Some limpet species such as *P. candei* sensu lato and *P. aspera* are considered to have slow growth and relatively long lifespan, thus they are extremely vulnerable to size-selective harvest and would have a slower response to MPA protection [53] (P. Henriques, pers. comm.).

Another possible effect of MPAs is the enhancement of recruitment on adjacent exploited populations, since the higher densities and larger individuals in reserves are expected to lead to greater production of larvae than in nearby exploited areas [88]. Therefore, MPAs are expected to enhance adjacent fisheries through the export of larvae [132, 133]. However, it is still unclear how and to what extent reserves influence exploited populations regarding the renewal of recruitment on these populations, due to the export of larvae originated in MPAs

[137, 138, 156]. For instance, Hockey and Branch [157] found that limpet populations closer to protected areas benefit from an increase in juvenile individuals, suggesting a spillover of recruitment from MPAs. Nevertheless, the correlation between larvae production in MPAs and recruitment on exploited populations is difficult to predict, due to the difficulties in determining patterns of physical transport, especially at small scales [78, 79, 158].

According to Halpern [129], the average values of several biological variables are 20 to 30% higher in populations of MPAs when compared to exploited populations, independent of MPA size, indicating that small MPAs can also produce high values. Several studies have reported a pattern of better preserved populations of limpets in MPAs regarding abundance and biometric structure, for example *P. candei* in Fuerteventura [116], *P. candei crenata, P. aspera* and *P. rustica* in the Canaries archipelago [114], *P. ferruginea* in the Mediterranean [159], *C. oculus* in South Africa [88] as well as *H. concolor, Scutellastra longicosta* (Lamarck, 1819) and *Scutellastra granularis* (Linnaeus, 1758) in South Africa [26].

Núñez et al. [116] studied the abundance and size composition of eight populations of the heavily exploited *P. candei* in the island of Fuerteventura, two of which were included in two protected areas, and reported that these two populations were the best preserved in terms of abundance and size composition, while the areas closer to human settlement, thus more accessible, exhibited less abundance and smaller size individuals. Another study in the Canaries archipelago by Ramírez et al. [114] showed that the populations of limpets exposed to anthropogenic effects return lower levels of abundance and smaller size composition compared to more isolated populations; even when the populations are encompassed in an MPA, the non-enforcement of the imposed regulations and lack of surveillance may compromise their effectiveness. Coppa et al. [159] also reported that the impact of MPAs in the protection of the endangered limpet *P. ferruginea* in terms of population density, spatial distribution, and morphometric characteristics is inversely correlated to accessibility.

The effect of MPAs in population density, size structure, and biomass of the exploited limpet *C. oculus* in South Africa was assessed by Branch and Odendaal [88], resulting in important increases of the studied parameters in MPAs when compared to exploited populations. Also, survivability, sex ratio, and reproductive output were significantly higher in MPAs. Other examined parameters such as growth rate and age at maturity were apparently unaffected by the protection of MPAs. Conversely, recruitment was higher in exploited populations than in protected areas. These results clearly show the necessity for MPAs among the tools used for coastal management.

Nakin and McQuaid [26] reported the effect of MPAs in the populations of heavily exploited limpets *S. longicosta* and *H. concolor* and the less exploited *S. granularis*. The authors evidenced a subtle enhancement of population density and size structure, more evident in heavily exploited species. However, the effects of spatial and temporal variation allied to the existence of poaching activities appear to dilute the effect of marine reserves.

Even though these studies put in evidence the overall benefits of establishing MPAs in protecting the intertidal habitat and the species that inhabit it, they also raise important questions regarding their effectiveness. If on one side, MPAs allow exploited limpet populations to recover in regard to certain biological parameters, on the other hand their effectiveness is in some cases hindered by the lack of surveillance and poor enforcement of protection regulations. In fact, these two factors seem the most important in determining the effectiveness of MPAs. Nonetheless, the implementation of MPAs even when unable to fully stop illegal harvest of limpet species, results in direct improvements for the protected populations in terms of abundance, size structure, and population density and indirect effects regarding reproductive output of these broadcast spawners. For this reason, the implementation of MPAs has become one of the most widely advocated tools for the management and conservation of coastal marine ecosystems in the recent decades [160, 161].

Several factors affect the response of protected populations, thus comparison between different MPAs is somewhat difficult. In fact, the recovery indicators reported for protected populations may be a consequence not only of MPA protection but also of changes in environmental conditions, biological characteristics of the species and, level of exploitation to which they are subjected [162–164]. The degree of exposure to wave action, as well as the vertical distribution of the species is thought to play an important role in the recovery of limpet populations; limpets more exposed to wave action as well as species exposed for longer periods to desiccation have a less pronounced response to the protection given by MPAs as shown by Branch and Odendaal [88] for *C. oculus* in South Africa.

Unsatisfactory results generally occur in those MPAs that are affected by inappropriate planning, ineffective surveillance, poor acceptance by local communities, and the lack of political will to reinforce the importance of environmental protection [140, 165–167]. For instance, Coppa et al. [159] concluded that although the designation of MPAs as a tool to preserve the remaining populations of the heavily exploited *P. ferruginea* is of extreme importance, for these MPAs to fulfil their goal, additional measures must be considered. In 2015, Coppa et al. [163] suggested that without a joint effort toward the protection of intertidal habitats by enforcement bodies, regulators, researchers, and sea users, the MPAs will not be able to achieve their conservation objectives.

The effectiveness of MPAs' conservation of limpet populations could be enhanced through the implementation of several additional measures that encompass a broader view of these exploited populations and the biological and ecological factors that influence their capacity to recover. For instance, it is necessary to determine which actions are required to ensure the reproductive success of individuals, essential to maintain the genetic biodiversity of overexploited species, particularly in species with absent gene flow between populations, since inbreeding increases the extinction probability of wild populations [168]. Also, the reintroduction or reinforcement of recruitment of depleted populations with allochthonous specimens produced by artificial fertilization procedures could be considered as a strategy to further fulfil the MPAs' conservation objective [169].

The establishment of MPAs as a conservation tool of marine coastal habitats and species has returned valuable contributions over the years, particularly in terms of density, abundance, and size structure of exploited species. However, to overcome limitations a possible route to improve the success of conservation strategies could be the establishment of networks of MPAs based on solid scientific information that identifies the type of measures that need to be implemented. Planning should consider the number and size of MPAs, which should be large enough to ensure the recovery of protected populations but sufficiently spaced in order to allow the spillover of recruits and adults to the exploited populations. MPA planning should ultimately target the ecosystem and not a specific exploited species, since the success of a reserve depends not only on the recovery of a single species but on the recovery of the ecosystem to which the species belongs. Additionally, due to geographic specificities, the prevalent abiotic factors and how they influence the target ecosystem should be considered when planning MPAs. Also, continuous monitoring of the effects of MPAs on the exploited populations would allow for a more adequate management of MPAs, allowing for the adjustment of the protective measures as needed.

Besides adequate planning of MPAs, new conservation strategies are required to implement measures that raise public awareness and the political will of decision makers that would allow for innovative approaches involving not only decision makers but also the end users of these marine resources in the conservation effort of exploited species, particularly to avoid illegal poaching, which is one if not the greatest factors hindering MPA success.

## Acknowledgements

ARDITI (Agência Regional para o Desenvolvimento da Investigação, Tecnologia e Inovação) is acknowledged for providing financial support for this work in the framework of the grant ARDITI-OOM/2016/010 (M1420-01-0145-FEDER-000001-Observatório Oceânico da Madeira-OOM). The third author (RS) was supported by a grant from ARDITI-OOM/2016/010 (M1420-01-0145-FEDER-000001-Observatório Oceânico da Madeira-OOM).

## Author details

Paulo Henriques[1]*, João Delgado[2,3] and Ricardo Sousa[4]

*Address all correspondence to: paulomvh@gmail.com

1 University of Madeira, Campus da Penteada, Funchal, Madeira, Portugal

2 Service Directorate of Research (DSI)—Regional Directorate of Fisheries, Estrada da Pontinha, Funchal, Madeira, Portugal

3 Interdisciplinary Centre of Marine and Environmental Research (CIIMAR/CIMAR), Rua dos Bragas, Porto, Portugal

4 Oceanic Observatory of Madeira (OOM)—Edifício Madeira Tecnopolo, Funchal, Madeira, Portugal

## References

[1] Bowman RS, Lewis JR. Geographical variation in the breeding cycles and recruitment of *Patella* spp. Hydrobiologia. 1986;**142**:41-56. DOI: 10.1007/BF00026746

[2] Stearns SC. The evolution of life histories (1st ed.). Oxford: Oxford University Press; 1992. 249 p

[3] Begon M, Harper JL, Townsend CR. Ecology: individuals, populations and communities (2nd ed.). Oxford: Blackwell Scientific Publications; 1996. 945 p

[4] Ridgway SA, Reid DG, Taylor JD, Branch GM, Hodgson AN. A cladistic phylogeny of the family Patellidae (Mollusca: Gastropoda). Phil Trans R Soc B. 1998;**353**:1645-1671. DOI: 10.1098/rstb.1998.0316

[5] Koufopanou V, Reid DG, Ridgway SA, Thomas RH. A molecular phylogeny of the Patellid limpets (Gastropoda: Patellidae) and its implications for the origins of their antitropical distribution. Mol Phylogenet Evol. 1999;**11**(1):138-156. DOI: 10.1006/mpev.1998.0557

[6] WoRMS Editorial Board. World register of marine species [Internet]. 2017. Available from: http://www.marinespecies.org [Accessed: 24-01-2017]

[7] Branch GM. The ecology of *Patella* Linnaeus from the Cape Peninsula, South Africa. I. Zonation, movements, and feeding. Zool Afr. 1971;**6**:1-38. DOI: 10.1080/00445096.1971.11447402

[8] Christiaens J. Révision du genre *Patella* (Mollusca, Gastropoda). B Mus Natl Hist Nat. 1973;**182**(3):1305-1392

[9] Powell AWR. The patellid limpets of the world (Patellidae) (1st ed.). Delaware: Delaware Museum of Natural History; 1973. 132 p

[10] Espinosa F, Nakano T, Guerra-García JM, García-Gómez JC. Population genetic structure of the endangered limpet *Cymbula nigra* in a temperate Northern hemisphere region: influence of palaeoclimatic events? Mar Ecol. 2010;**32**:1-5. DOI: 10.1111/j.1439-0485.2010.00410.x

[11] Moore HB. The relation of shell growth to environment in *Patella vulgata*. Proc Malacol Soc Lond. 1934;**21**(3):217-222

[12] Davies PS. Effect of environment on metabolic activity and morphology of Mediterranean and British species of *Patella*. Pubbl Stn Zool. 1969;**37**:641-656

[13] Bannister JV. Shell parameters in relation to zonation in Mediterranean limpets. Mar Biol. 1975;**31**:63-67. DOI: 10.1007/BF00390648

[14] Boukhicha J, Kalthoum O, Hassine B, Tlig-Zouari S. Morphological evidence for adaptive diversification of sympatric Mediterranean *Patella* limpets. Rapp Comm Int Mer Medit. 2013;**40**:686

[15] Scuderi D, Eernisse DJ. A new alien limpet for the Mediterranean: *Lottia* sp. (Patellogastropoda: Lottiidae). Biodivers J. 2016;**7**(2):287-293

[16] Jenkins SR, Coleman RA, Burrows MT, Hartnoll RG, Hawkins SJ. Regional scale differences in determinism of limpet grazing effects. Mar Ecol Prog Ser. 2005;**287**:77-86. DOI: 10.3354/meps287077

[17] Coleman RA, Underwood AJ, Benedetti-Cecchi L, Aberg P, Arenas F, Arrontes J, Castro J, Hartnoll RG, Jenkins SR, Paula J, Della Santina P, Hawkins SJ. A continental scale evaluation of the role of limpet grazing on rocky shores. Oecol. 2006;**147**(3):556-564. DOI: 10.1007/s00442-005-0296-9

[18] Southward AJ. Limpet grazing and the control of vegetation on rocky shores. In: Crisp DJ, editor. Grazing in terrestrial and marine environments (1st ed.). Oxford: Blackwell Publications; 1964. pp. 265-273

[19] Hawkins SJ, Hartnoll RG. Grazing of intertidal algae by marine invertebrates. Oceanogr Mar Biol Annu Rev. 1983;**21**:195-282

[20] Boaventura D, Alexander M, Della Santina P, Smith ND, Ré P, da Fonseca LC, Hawkins SJ. The effects of grazing on the distribution and composition of low-shore algal communities on the central coast of Portugal and the southern coast of Britain. J Exp Mar Biol Ecol. 2002;**267**:185-206. DOI: 10.1016/S0022-0981(01)00372-0

[21] Menge B. Indirect effects in marine rocky intertidal interaction webs: patterns and importance. Ecol Monogr. 1995;**65**(1):21-74. DOI: 10.2307/2937158

[22] Kido JS, Murray SN. Variation in owl limpet *Lottia gigantea* population structures, growth rates and gonadal production on southern California rocky shores. Mar Ecol Prog Ser. 2003;**257**:111-124. DOI: 10.3354/meps257111

[23] Arrontes J, Arenas F, Fernández C, Rico JM, Oliveros J, Martínez B, Viejo RM, Alvarez D. Effect of grazing by limpets, on mid-shore species assemblages in northern Spain. Mar Ecol Prog Ser. 2004;**277**:117-133. DOI: 10.3354/meps277117

[24] Underwood AJ. The biology of gastropods. Adv Mar Biol. 1979;**16**:111-210

[25] Branch GM. Limpets: their role in littoral and sublittoral community dynamics. In: Moore PG, Seed R, editors. The ecology of rocky coasts (1st ed.). London: Hodder & Stoughton; 1985. pp. 97-116

[26] Nakin MDV, McQuaid CD. Marine reserve effects on population density and size structure of commonly and rarely exploited limpets in South Africa. Afr J Mar Sci. 2014;**3**:1-9. DOI: 10.2989/1814232X.2014.946091

[27] Farrell TM. Models and mechanisms of succession: an example from a rocky intertidal community. Ecol Monogr. 1991;**61**:95-113. DOI: 10.2307/1943001

[28] Anderson MJ, Underwood AJ. Effects of gastropod grazers on recruitment and succession of an estuarine assemblage: a multivariate and univariate approach. Oecol. 1997;**109**:442-453. DOI: 10.1007/s004420050104

[29] Menge BA, Freidenburg TL. Keystone species. In: Levin SA, editor. Encyclopedia of biodiversity (Vol. 4). New York: Academic Press; 2001. pp. 613-631

[30] Bree PJH. Homing-gedrag van *Patella vulgata* L. Verslagen van de gewone vergadering der Afdeeling natuurkunde K. Nederlandse Akademie van wetenschappen. 1959;**68**:106-108

[31] Cook A, Bamford OS, Freeman JDB, Teidman DJ. A study on the homing habit of the limpet. Anim Behav. 1969;**17**:330-339. DOI: 10.1016/0003-3472(69)90019-0

[32] Evans MR, Williams GA. Time partitioning of foraging in the limpet *Patella vulgata*. J Anim Ecol. 1991;**60**(2):563-575. DOI: 10.2307/5298

[33] Espinosa F, Guerra-García JM, Fa D, García-Gómez JC. Aspects of reproduction and their implications for the conservationof the endangered limpet, *Patella ferruginea*. Invertebr Reprod Dev. 2006;**49**:85-92. DOI: 10.1080/07924259.2006.9652197

[34] Sebastián CR, Steffani CN, Branch GM. Homing and movement patterns of a South African limpet *Scutellastra argenvillei* in an area invaded by an alien *mussel Mytilus galloprovincialis*. Mar Ecol Prog Ser. 2002;**243**:111-122. DOI: 10.3354/meps243111

[35] Branch GM, Cherry MI. Activity rhythms of the pulmonated limpet *Siphonaria capensis* Q. and G. as an adaptation to osmotic stress, predation and wave action. J Exp Mar Biol Ecol. 1985;**87**:153-168

[36] Iwasaki K. Intra- and interspecific variation in activity patterns of intertidal limpets. Venus Jpn J Malacol. 1994;**53**:85-104

[37] Mackay DA, Underwood AJ. Experimental studies on homing in the intertidal patellid limpet *Cellana tramoserica* (Sowerby). Oecol. 1977;**30**:215-237

[38] Branch GM. The response of South African patellid limpets to invertebrate predators. Zool Afr. 1978;**13**:221-232. DOI: 10.1080/00445096.1978.11447624

[39] Garrity SD, Levings SC. Homing to scars as a defence against predators in the pulmonate limpet *Siphonaria gigas* (Gastropoda). Mar Biol. 1983;**72**:319-324. DOI: 10.1007/BF00396838

[40] Iwasaki K. Analyses of limpet defence and predator offense in the field. Mar Biol. 1993;**116**:277-289. DOI: 10.1007/BF00350018

[41] Branch GM. Activity rhythms in *Siphonaria thersites*. In: Chelazzi G, Vannini M, editors. Behavioural adaptation to intertidal life NATO ASI Series (Vol. 151. 1st ed.). New York: Plenum Press; 1988. pp. 27-44

[42] Gray DR, Hodgson AN. Foraging and homing behaviour in the high-shore, crevice-dwelling limpet *Helcion pectunculus* (Prosobranchia: Patellidae). Mar Biol. 1998;**132**: 283-294. DOI: 10.1007/s002270050394

[43] Stimson J. Territorial behaviour of the owl limpet *Lottia gigantea*. Ecology. 1970;**51**:113-118. DOI: 10.2307/1933604

[44] Branch GM. Mechanisms reducing intraspecific competition in *Patella* spp.: migration, differentiation and territorial behaviour. J Anim Ecol. 1975;**44**:575-600

[45] Funke W. Heimfindevermögen und Ortstreue bei *Patella* L. (Gastropoda: Prosobranchia). Oecol. 1968;**2**:19-142. DOI: 10.1007/BF00394506

[46] Branch GM. The biology of limpets: physical factors, energy flow and ecological interactions. Oceanogr Mar Biol Annu Rev. 1981;**19**:235-380

[47] Kenny R. Growth studies of the tropical intertidal limpet *Acmaea antillarum*. Mar Biol. 1977;**39**:161-170. DOI: 10.1007/BF00387001

[48] Gray DR, Hodgson AN. Growth and reproduction in the high-shore South African limpet *Helcion pectunculus* (Mollusca: Patellogastropoda). Afr Zool. 2003;**38**(2):371-386

[49] Clarke A, Prothero-Thomas E, Beaumont JC, Chapman AL, Brey T. Growth in the limpet *Nacella concinna* from contrasting sites in Antarctica. Polar Biol. 2004;**28**:62-71. DOI: 10.1007/s00300-004-0647-8

[50] Espinosa F, Gonzáles AR, Maestre MJ, Fa D, Guerra-García JM, García-Gómez JC. Responses of the endangered limpet *Patella ferruginea* to reintroduction under different environmental conditions: survival, growth rates and life history. Ital J Zool. 2008;**75**: 371-384. DOI: 10.1080/11250000801887740

[51] Brêthes JC, Ferreyra G, de la Vega S. Distribution, growth and reproduction of the limpet *Nacella* (*Patinigera*) *concinna* (Strebel 1908) in relation to potential food availability, in Esperanza Bay (Antarctic Peninsula). Polar Biol. 1994;**14**:161-170. DOI: 10.1007/BF00240521

[52] Khow AS. Growth determination of tropical limpet *Cellana testudinaria* (Linnaeus, 1758) living on the rocky shore of Ohoiwait, Southeast Moluccas, Indonesia. J Coastal Dev. 2007;**10**(2):89-103

[53] Henriques P, Sousa R, Pinto AR, Delgado J, Faria G, Alves A, Khadem M. Life history traits of the exploited limpet *Patella candei* (Mollusca: Patellogastropoda) of the north-eastern Atlantic. J Mar Biol Assoc UK. 2012;**92**(1):1-9. DOI: 10.1017/S0025315411001068

[54] Vermeij GJ. Gastropod shell growth rate, allometry & adult size: environmental implications. In: Rhodes DC, Lutz RA, editors. Skeletal growth of aquatic organisms (1st ed.). New York: Plenum Press; 1980. pp. 379-394

[55] Janson K. Genetic and environmental effects on the growth rate of *Littorina saxatilus*. Mar Biol. 1982;**69**:73-78. DOI: 10.1007/BF00396963

[56] McQuaid CD. The establishment and maintenance of vertical size gradients in populations of *Littorina africana knysnaensis* (Philippi) on an exposed rocky shore. J Exp Mar Biol Ecol. 1981;**54**:77-89. DOI: 10.1016/0022-0981(81)90104-0

[57] Bosman AL, Hockey PAR. Life-history patterns of populations of the limpet *Patella granularis*: the dominant roles of food supply and mortality rate. Oecol. 1988;**75**:412-419. DOI: 10.1007/BF00376945

[58] Branch GM, Marsh AC. Tenacity and shell shape of six *Patella* species: adaptive features. J Exp Mar Biol Ecol. 1978;**87**:153-168. DOI: 10.1016/0022-0981(78)90035-7

[59] Thompson GB. Distribution and population dynamics of the limpet *Patella aspera* (Lamarck) in Bantry Bay. J Exp Mar Biol Ecol. 1979;**40**:115-135. DOI: 10.1016/0022-0981(79)90039-x

[60] Brown KM, Quinn JF. The effect of wave action on growth in three species of intertidal gastropods. Oecol. 1988;**75**(3):420-425

[61] Lewis JR, Bowman RS. Local habitat-induced variations in the population dynamics of *Patella vulgata* L. J Exp Mar Biol Ecol. 1975;**17**:165-203

[62] Branch GM. The ecology of *Patella Linnaeus* from the Cape peninsula, South Africa. 3. Growth rates. Trans R Soc S Afr. 1974;**41**:161-193

[63] Black R. Population regulation in the intertidal limpet *Patelloida alticostata* (Angas, 1865). Oecol. 1977;**30**:9-22. DOI: 10.1007/BF00344888

[64] Gray DR. Studies of the biology and ecology of the high shore South African limpet, *Helcion pectunculus* (Mollusca: Patellogastropoda) [thesis]. South Africa: Rhodes University; 1996. 304 p

[65] Etim L, Sankare Y. Growth and mortality, recruitment and yield of the fresh-water shrimp, *Macrobrachium vollenhovenii*, Herklots, 1851 (Crustacea, Palaemonidae) in the Fahe reservoir, Côte d'Ivoire, West Africa. Fish Res. 1988;**38**:211-223. DOI: 10.1016/S0165-7836(98)00161-1

[66] Pauly D, Munro JL. Once more on the comparison of growth in fish and invertebrates. ICLARM Fishbyte. 1984;**2**:21

[67] Sparre P, Ursin E, Venema SC. Introduction to tropical fish stock assessment, part 1 manual. Rome: FAO Fisheries Technical Paper; 1989. 337 p

[68] Orton JH, Southward AJ, Dodd JM. Studies on the biology of limpets II. The breeding of *Patella vulgata* L. in Britain. J Mar Biol Assoc UK. 1956;**35**:149-176. DOI: 10.1017/S0025315400009036

[69] Hyman LH. The invertebrates VI – mollusca I: aplacophora, polyplacophora, monoplacophora, gastropoda (1st ed.). New York: McGraw-Hill Book Company; 1967. 792 p

[70] Fretter V, Graham A. The prosobranch mollucs of Britain and Denmark. I. Pleurotomariacea, Fissurellacea and Patellacea. J Molluscan Stud Supp. 1976;**1**:1-37

[71] Orton JH. Observations on *Patella vulgata*. Part III. Habitat and habits. J Mar Biol Assoc UK. 1929;**16**:277-288. DOI: 10.1017/S0025315400029805

[72] Orton JH, Southward AJ. Studies on the biology of limpets IV. The breeding of *Patella depressa* pennant on the north Cornish coast. J Mar Biol Assoc UK. 1961;**41**(3):653-662. DOI: 10.1017/S0025315400016210

[73] Orton JH. Biology of *Patella* in Great Britain. Nature. 1946;**158**:173-174

[74] Henninger TO, Hodgson AN. The reproductive cycle of *Helcion pruinosus* (Patellogastropoda) on two South African boulder shores. J Molluscan Stud. 2001;**67**:385-394

[75] Hawkins SJ, Côrte-Real HBSM, Pannacciulli FG, Weber LC, Bishop JDD. Thoughts on the ecology and evolution of the intertidal biota of the Azores and other Atlantic Islands. Hydrobiologia. 2000;**440**:3-17. DOI: 10.1023/A:1004118220083

[76] Boaventura D, Fonseca LC, Hawkins SJ. Size matters: competition within populations of the limpet *Patella depressa*. J Anim Ecol. 2003;**72**:435-446. DOI: 10.1046/j.1365-2656.2003.00713.x

[77] Bowman RS, Lewis JR. Annual fluctuations in the recruitment of *Patella vulgata* L. J. Mar Biol Assoc UK. 1977;**57**:793-815. DOI: 10.1017/S0025315400025169

[78] Cowen RK, Paris CB, Srinivasan A. Scaling of connectivity in marine populations. Science. 2006;**311**:522-527. DOI: 10.1126/science.1122039

[79] Cowen RK, Sponaugle S. Larval dispersal and marine population connectivity. Ann Rev Mar Sci. 2009;**1**:443-466. DOI: 10.1146/annurev.marine.010908.163757

[80] Schmitt RJ, Holbrook SJ, Osenberg CW. Quantifying the effects of multiple processes on local abundance: a cohort approach for open populations. Ecol Lett. 1999;**2**:294-303. DOI: 10.1046/j.1461-0248.1999.00086.x

[81] Johnson MP. Is there confusion over what is meant by "open population?". Hydrobiologia. 2005;**544**:333-338. DOI: 10.1007/s10750-005-1698-8

[82] Orton JH. Observations on *Patella vulgata*. Part II. Rate of growth of shell. J Mar Biol Assoc UK. 1928;**15**:863-874. DOI: 10.1017/S0025315400009954

[83] Le Quesne WJF. The response of a protandrous species to exploitation, and the implications for management: a case study with Patellid limpets [thesis]. United Kingdom: University of Southampton; 2005. 203 p

[84] Robson G. Aspects of the biology of a new species of South African Patella. (Mollusca: Gastropoda: Patellidae) [thesis]. Pietermaritzburg: University of Natal; 1986

[85] Lindberg DR. Reproduction, ecology, and evolution of the Indo-pacific limpet *Scutellastra flexuosa*. Bull Mar Sci. 2007;**81**(2):219-234

[86] Espinosa F, Rivera-Ingraham G, García-Gómez JC. Gonochorism or protandrous hermaphroditism? Evidence of sex change in the endangered limpet *Patella ferruginea*. J Mar Biol Assoc Biodiv Rec. 2009;**2**:153. DOI: 10.1017/S1755267209990790

[87] Guallart J, Calvo M, Acevedo I, Templado J. Two-way sex change in the endangered limpet *Patella ferruginea* (Mollusca, Gastropoda). Invertebr Reprod Dev. 2013;**57**(3):247-253. DOI: 10.1080/07924259.2012.754794

[88] Branch GM, Odendaal F. The effects of marine protected areas on the population dynamics of a South African limpet, *Cymbula oculus*, relative to the influence of wave action. Biol Conserv. 2003;**114**:255-269. DOI: 10.1016/S0006-3207(03)00045-4

[89] Coe WR. Conditions influencing change of sex in mollusks of the genus *Crepidula*. J Exp Zool. 1938;**77**:401-424. DOI: 10.1002/jez.1400770305

[90] Hoagland KE. Protandry and the evolution of environmentally-mediated sex change: a study of the mollusca. Malacologia. 1978;**17**:365-391

[91] Wright WG. Intraspecific density mediates sex-change in the territorial patellacean limpet *Lottia gigantea*. Mar Biol. 1989;**100**:353-364. DOI: 10.1007/BF00391151

[92] Warner RR, Fitch DL, Standish JD. Social control of sex change in the shelf limpet, *Crepidula norrisiarum*: size-specific responses to local group composition. J Exp Mar Biol Ecol. 1996;**204**:155-167. DOI: 10.1016/0022-0981(96)02582-8

[93] Munday PL, Buston PM, Warner RR. Diversity and flexibility of sex-change strategies in animals. Trends Ecol Evol. 2006;**21**:89-95. DOI: 10.1016/j.tree.2005.10.020

[94] Marra S, de Lucia GA, Camedda A, Esinosa F, Coppa S. New records of the distribution and conservation status of the endangered limpet *Patella ferruginea* in Sardinia (Italy, W Mediterranean). Aquat Conserv Mar Freshw Ecosys. 2015;**26**(3):607-612. DOI: 10.1002/aqc.2615

[95] Durán LR, Castilla JC. Variation and persistence of the middle rocky intertidal community of central Chile, with and without human harvesting. Mar Biol. 1989;**103**:555-562. DOI: 10.1007/BF00399588

[96] Lasiak TA. Multivariate comparisons of rocky infratidal macrofaunal assemblages from replicate exploited and non-exploited localities on the Transkei coast of South Africa. Mar Ecol Prog Ser. 1998;**167**:15-23. DOI: 10.3354/meps167015

[97] Sagarin RD, Ambrose RF, Becker BJ, Engle JM, Kido J, Lee SF, Miner CM, Murray SN, Raimondi PT, Richards DV, Roe C. Ecological impacts on the limpet *Lottia gigantea* populations: human pressure over a broad scale on islands and mainland intertidal zones. Mar Biol. 2007;**150**:399-413. DOI: 10.1007/s00227-006-0341-1

[98] Fenberg PB, Roy K. Ecological and evolutionary consequences of size-selective harvesting: how much do we know? Mol Ecol. 2008;**17**:209-220. DOI: 10.1111/j.1365-294X.2007.03522.x

[99] Eekhout S, Raubenheimer CM, Branch GM, Bosman AL, Bergh MO. A holistic approach to the exploitation of intertidal stocks: limpets as a case study. Afr J Mar Sci. 1992;**12**:1017-1029

[100] Griffiths CL, Branch GM. The exploitation of coastal invertebrates and seaweeds in South Africa: historical trends, ecological impacts and implications for management. Trans R Soc S Afr. 1997;**52**:121-148. DOI: 10.1080/00359199709520619

[101] Pombo OA, Escofet A. Effect of exploitation on the limpet *Lottia gigantea*: a field study in Baja California (Mexico) and California (U.S.A.). Pac Sci. 1996;**50**:393-403. DOI: 10125/2914

[102] Turrero P, Munoz Colmenero AM, Prado A, García-Vázquez E. Long-term impacts of human harvesting on shellfish: North Iberian top shells and limpets from the upper palaeolithic to the present. J Mar Syst. 2014;**139**:51-57. DOI: 10.1016/j.jmarsys.2014.05.011

[103] McCoy M. Hawaiian limpet harvesting in historical perspective: a review of modern and archaeological data on *Cellana* spp. From the Kalaupapa Peninsula, Moloka'i Island. Pac Sci. 2008;**62**(1):28-38. DOI: 10.2984/1534-6188(2008)62[21:HLHIHP]2.0.CO;2

[104] Keough MJ, Quinn GP, King A. Correlations between human collecting and intertidal mollusc populations on rocky shores. Conserv Biol. 1993;**7**:378-390

[105] Lasiak TA. The susceptibility and/or resilience of rocky littoral molluscs to stock depletion by the indigenous coastal people of Transkei, southern Africa. Biol Cons. 1991;**56**:245-264. DOI: 10.1016/0006-3207(91)90060-M

[106] Moreno CA, Sutherland JP, Jara HF. Man as a predator in the intertidal zone of southern Chile. Oikos. 1984;**42**:155-160. DOI: 10.2307/3544787

[107] Santos SR, Hawkins SJ, Monteiro LR, Alves M, Isidro EJ. Marine research, resources and conservation in the Azores. Aquat Conservat Mar Freshwat Ecosyst. 1995;**5**:311-354. DOI: 10.1002/aqc.3270050406

[108] Moro L, Herrera R. Las lapas, un recurso en extincíon. Medio Ambiente Canarias. 2000;**16**:3

[109] Levitan DR. Influence of body size and population density on fertilization success and reproductive output in a free-spawning invertebrate. Biol Bull. 1991;**181**:261-268. DOI: 10.2307/1542097

[110] Tegner MJ, Basch LV, Dayton PK. Near extinction of an exploited marine invertebrate. Trends Ecol Evol. 1996;**11**:278-280. DOI: 10.1016/0169-5347(96)30029-3

[111] Dye AH. The effects of excluding limpets from the lower balanoid zone of rocky shores in Transkei, South Africa. S Afr J Mar Sci. 1995;**15**:9-15. DOI: 10.2989/025776195784156313

[112] Davies AJ, Johnson MP, Maggs CA. Limpet grazing and loss of *Ascophyllum nodosum* canopies on decadal time scales. Mar Ecol Prog Ser. 2007;**339**:131-141. DOI: 10.3354/meps339131

[113] Lindberg K, Estes JA, Warheit KI. Human influences on trophic cascades along rocky shores. Ecol Appl. 1998;**8**:880-890. DOI: 10.2307/2641274

[114] Ramírez R, Tuya F, Haroun R. Efectos potenciales del marisqueo sobre moluscos gasterópodos de interés comercial (*Osilinus* spp. y *Patella* spp.) en el Archipiélago Canario. Rev Biol Mar Oceanogr. 2009;**44**(3):703-714

[115] Martins GM, Thompson RC, Hawkins SJ, Neto AI, Jenkins SR. Rocky intertidal community structure in oceanic islands: scales of spatial variability. Mar Ecol Prog Ser. 2008;**356**:15-24. DOI: 10.3354/meps07247

[116] Núñez J, Brito MC, Riera R, Docoito JR, Monterroso Ó. Distribución actual de las poblaciones de *Patella candei* D'Orbigny, 1840 (Mollusca, Gastropoda) en las islas Canarias. Una especie en peligro de extinción. Bol Inst Esp Oceanog. 2003;**19**(1-4):371-377

[117] Guerra-García JM, Corzo J, Espinosa F, García-Gómez JC. Assessing habitat use of the endangered marine mollusk *Patella ferruginea* (Gastropoda, Patellidae) in the northern Africa: preliminary results and implications for conservation. Biol Cons. 2004;**16**: 319-326. DOI: 10.1016/S0006-3207(03)00201-5

[118] Espinosa F. Population status of the endangered mollusc *Patella ferruginea* Gmelin, 1791 (Gastropoda, Patellidae) on Algerian islands (SW Mediterranean). Anim Biodivers Conserv. 2009;**32**(1):19-28

[119] Alonzo SH, Mangel M. The effects of size selective fisheries on the stock dynamics of and sperm limitation in sex-changing fish. Fish Bull. 2004;**102**:1-13

[120] Baskett ML, Levin SA, Gaines SD, Dushoff J. Marine reserve design and the evolution of size at maturation in harvested fish. Ecol Appl. 2005;**15**:882-901. DOI: 10.1890/04-0723

[121] Heppell SS, Heppell SA, Coleman FC, Koenig CC. Models to compare management options for a protogynous fish. Ecol Appl. 2006;**16**:238-249. DOI: 10.1890/04-1113

[122] Hamilton SL, Caselle JE, Standish JD, Schroeder DM, Love MS, Rosales-Casian JA, Sosa-Nishizaki O. Size-selective harvesting alters life histories of a sex-changing fish. Ecol Appl. 2007;**17**:2268-2280. DOI: 10.1890/06-1930.1

[123] Martins GM, Jenkins SR, Hawkins SJ, Neto AI, Medeiros AR, Thompson RC. Illegal harvesting affects the success of fishing closure areas. J Mar Biol Assoc UK. 2011;**91**: 929-937. DOI: 10.1017/S0025315410001189

[124] Costello C, Gaines SD, Lynham J. Can catch shares prevent fisheries collapse? Science. 2008;**321**:1678-1681. DOI: 10.1126/science.1159478

[125] Baxter JM. Establishing management schemes on marine special areas of conservation in Scotland. Aquat Conserv Mar Freshw Ecosys. 2001;**11**:261-265. DOI: 10.1002/aqc.465

[126] Thompson RC, Crowe TP, Hawkins SJ. Rocky intertidal communities: past environmental changes, present status and predictions for the next 25 years. Environ Conserv. 2002;**29**:168-191. DOI: 10.1017/S0376892902000115

[127] Ballantine B. Marine reserves for New Zealand. Leigh Laboratory bulletin (Vol. 25). Auckland: University of Auckland; 1991. pp. 1-196

[128] Zann LP. Our sea, our future. Major findings of the state of the marine environment report for Australia (1st ed.). QLD: Great Barrier Reef Marine Park Authority; 1995

[129] Halpern BS. The impact of marine reserves: do reserves work and does reserve size matter? Ecol Appl. 2003;**13**:117-137. DOI: 10.1890/1051-0761(2003)013[0117:TIOMRD]2.0.CO;2

[130] Roberts CM, Hawkins JP. Extinction risk in the sea. Trends Ecol Evol. 1999;**14**:241-246

[131] Lubchenco J, Palumbi SR, Gaines SD, Andelman S. Plugging a hole in the ocean: the energy science of marine reserves. Ecol Appl. 2003;**13**:S3–S7. DOI: 10.1890/1051-0761(2003)013[0003:PAHITO]2.0.CO;2

[132] Gell FR, Roberts CM. Benefits beyond boundaries: the fishery effects of marine reserves. Trends Ecol Evol. 2003;**18**:448-455. DOI: 10.1016/S0169-5347(03)00189-7

[133] Halpern BS, Warner RR. Matching marine reserve design to reserve objectives. Proc R Soc Lond Ser B Biol Sci. 2003;**270**:1871-1878. DOI: 10.1098/rspb.2003.2405

[134] Pelc RA, Baskett ML, Tanci T, Gaines SD, Warner RR. Quantifying larval export from South African marine reserves. Mar Ecol Prog Ser. 2009;**394**:65-78. DOI: 10.3354/meps08326

[135] Pelc RA, Warmer RR, Gaines SD, Paris CB. Detecting larval export from marine reserves. Proc Natl Acad Sci USA. 2010;**107**:18266-18271. DOI: 10.1073/pnas.0907368107

[136] Cole VJ, McQuaid CD, Nakin MDV. Marine protected areas export larvae of infauna, but not of bioengineering mussels to adjacent areas. Biol Conserv. 2011;**144**:2088-2096. DOI: 10.1016/j.biocon.2011.04.030

[137] Stobutzki IC. Marine reserves and the complexity of larval dispersal. Rev Fish Biol Fish. 2001;**10**:515-518

[138] Gaylord B, Gaines SD, Siegel DA, Carr MH. Marine reserves exploit population structure and life history in potentially improving fisheries yields. Ecol Appl. 2005;**15**:2180-2191. DOI: 10.1890/04-1810

[139] Sale PF, Cowen RK, Danilowicz BS, Jones GP, Kritzer JP, Lindeman KC, Planes S, Polunin NV, Russ GR, Sadovy YJ, Steneck RS. Critical science gaps impede use of no-take fishery reserves. Trends Ecol Evol. 2005;**20**:74-80. DOI: 10.3410/f.1024416.288798

[140] Fenberg PB, Caselle JE, Claudet J, Clemence M, Gaines SD, García-Charton JA, Gonçalves EJ, Grorud-Colvert K, Guidetti P, Jenkins SR, Jones PJS, Lester SE, McAllen R, Moland E, Planes S, Sørensen TK. The science of European marine reserves: status, efficacy, and future needs. Mar Policy. 2012;**36**:1012-1021. DOI: 10.1016/j.marpol.2012.02.021

[141] Weber LI, Hawkins SJ. Evolution of the limpet *Patella candei* d'Orbigny (Mollusca: Patellidae) in the Atlantic archipelagos: human intervention and natural processes. Biol J Linn Soc. 2002;**77**:341-353. DOI: 10.1046/j.1095-8312.2002.00102.x

[142] Navarro PG, Ramírez R, Tuya F, Fernández-Gil C, Sánchez-Jerez P, Haroun RJ. Hierarchical analysis of spatial distribution patterns of patellid limpets in the Canary Islands. J Molluscan Stud. 2005;**71**:67-73. DOI: 10.1093/mollus/eyi009

[143] Ramos MA. Implementing the habitats directive for mollusc species in Spain. J Conchol Spec Publ. 1998;**2**:125-132

[144] Feinman GM, Nicholas LM. High-intensity household-scale production in ancient Mesoamerica – a perspective from Ejutla, Oaxaca. In: Feinman GM, Mazilla L, editors. Cultural evolution: contemporary viewpoints (1st ed.). New York: Kluwer Academic/Plenum Publisher; 2000. pp. 119-142

[145] Melgar-Tísoc E. Las ofrendas de concha de moluscos de la Pirámide de las Serpientes Emplumadas, Xochicalco, Morelos. Rev Mex Biodivers. 2007;**78**:83-92

[146] Simison WB. Evolution and phylogeography of new world gastropod faunas [thesis]. Berkeley: University of California; 1985. 214 p

[147] Ríos-Jara E, Pérez-Peña M, López-Uriarte E, Enciso-Padilla I, Juárez-Carillo E. Biodiversidad de moluscos marinos de la costa de Jalisco y Colima, con anotaciones sobre su aprovechamiento en la región. In: Jiménez-Quiroz MC, Espino-Barr E, editors. Los Recursos Pesqueros y Acuícolas de Jalisco, Colima y Michoacán. Sagarpa, México: Instituto Nacional de la Pesca, CRIP-Manzanillo; 2006. pp. 103-120

[148] Bastida-Zavala JR, García-Madrigal MS, Rosas-Alquicira EF, López-Pérez RA, Benítez-Villalobos F, Meraz-Hernando JF, Torres-Huerta M, Montoya-Márquez A, Barrientos-Luján NA. Marine and coastal biodiversity of Oaxaca. Mexico Check List. 2013;**9**:329-390. DOI: 10.15560/9.2.329

[149] Halpern BS, Warner RR. Marine reserves have rapid and lasting effects. Ecol Lett. 2002;**5**:361-366. DOI: 10.1046/j.1461-0248.2002.00326.x

[150] Hockey PAR, Bosman AL. Man as an intertidal predator in Transkei: disturbance, community convergence and management of a natural food resource. Oikos. 1986;**46**:3-14. DOI: 10.2307/3565373

[151] Roberts C. Marine fishery reserves for the Caribbean. Caribbean Parks Protect Area Bull. 1995;**5**(2):8-11

[152] Russ G, Alcala A. Natural fishing experiments in marine reserves 1983-1993: community and trophic responses. Coral Reefs. 1998;**17**:383-397. DOI: 10.1007/s003380050144

[153] Polacheck T. Year around closed areas as a management tool. Nat Resour Model. 1990;**4**:327-353

[154] Carr MH, Reed DC. Conceptual issues relevant to marine harvest refuges: examples from temperature reef fishes. Can J Fish Aquat Sci. 1993;**50**:2019-2028. DOI: 10.1139/f93-226

[155] Rowley RJ. Marine reserves in fisheries management. Aquat Conserv Mar Freshw Ecosys. 1994;**4**:233-254. DOI: 10.1002/aqc.3270040305

[156] Palumbi SR. Marine reserves and ocean neighborhoods: the spatial scale of marine populations and their management. Annu Rev Environ Resour. 2004;**29**:31-68. DOI: 10.1146/annurev.energy.29.062403.102254

[157] Hockey PAR, Branch GM. Conserving marine biodiversity on the African coast: implications of a terrestrial perspective. Aquat Conserv Mar Freshw Ecosys. 1994;**4**:345-362. DOI: 10.1002/aqc.3270040406

[158] Caley MJ, Carr MH, Hixon MA, Hughes TP, Jones GP, Menge BA. Recruitment and the local dynamics of open marine populations. Annu Rev Ecol Syst. 1996;**27**:477-500. DOI: 10.1146/annurev.ecolsys.27.1.477

[159] Coppa S, Lucia GA, Massaro G, Magni P. Density and distribution of *Patella ferruginea* in a marine protected area (western Sardinia, Italy): constraint analysis for population conservation. Mediterr Mar Sci. 2012;**13**(1):108-117. DOI: 10.12681/mms.27

[160] Carr MH. Marine protected areas: challenges and opportunities for understanding and conserving coastal marine ecosystems. Environ Conserv. 2000;**27**:106-109

[161] Claudet J, Guidetti P, Mouillot D, Shears NT, Micheli F. Ecological effects of marine protected areas: conservation, restoration and functioning. In: Claudel J, editor. Marine protected areas: a multidisciplinary approach (1st ed.). Cambridge, UK: Cambridge University Press; 2011. DOI: 10.1017/CBO9781139049382.005

[162] Boersma DP, Parrish JK. Limiting abuse: marine protected areas, a limiting solution. Ecol Econ. 1999;**31**:287-304. DOI: 10.1016/S0921-8009(99)00085-3

[163] Coppa S, de Lucia GA, Massaro G, Camedda A, Marra S, Magni P, Perilli A, Di Bitetto M, García-Gómez JC, Espinosa F. Is the establishment of MPAs enough to preserve endangered intertidal species? The case of *Patella ferruginea* in the Mal di Ventre Island (W Sardinia, Italy). Aquat Conserv Mar Freshw Ecosys. 2015. DOI: 10.1002/aqc.2579

[164] Sciberras M, Jenkins SR, Mant R, KaiserMJ, Hawkins SJ, Pullin AS. Evaluating the relative conservation value of fully and partially protected marine areas. Fish Fish. 2015;**16**:28-77. DOI: 10.1111/faf.12044

[165] Jameson SC, Tupper MH, Ridley JM. The three screen doors: can marine 'protected' areas be effective? Mar Pollut Bull. 2002;**44**:1177-1183. DOI: 10.1016/S0025-326X(02)00258-8

[166] Guidetti P, Milazzo M, Bussotti S, Molinari A, Murenu M, Pais A, Spanò N, Balzano R, Agardy T, Boero F, Carrada G, Cattaneo-Vietti R, Cau A, Chemello R, Greco S, Manganaro A, di Sciara GN, Russo GF, Tunesi L. Italian marine reserve effectiveness: does enforcement matter? Biol Conser. 2008;**141**:699-709. DOI: 10.1016/j.biocon.2007.12.013

[167] Camargo C, Maldonado JH, Alvarado E, Moreno-Sanchez R, Mendosa S, Manrique N, Mogollon A, Osorio JD, Grajales A, Sanchez JA. Community involvement in management for maintaining coral reef resilience and biodiversity in southern Caribbean marine protected areas. Biodivers Conserv. 2009;**18**:935-356. DOI: 10.1007/s10531-008-9555-5

[168] Clark SA, Richardson BJ. Spatial analysis of genetic variation as a rapid assessment tool in the conservation management of narrow-range endemics. Invertebr Syst. 2002;**16**:583-587. DOI: 10.1071/IT01041

[169] Guallart J, Peña JB, Pérez-Larruscain J. Primeras imágenes de una forma juvenil de la lapa ferruginosa. Quercus. 2013;**325**:52-53

Section 2

# Cellular and Molecular Malacology

Chapter 5

# Freshwater Mussels Exposed to Arsenic and Sulfate Show Contrasting Patterns of Gene Expression

Pawel Michalak, Lin Kang, Serena Ciparis,
William Henley, Jess Jones, Andrew Phipps and
Eric Hallerman

Additional information is available at the end of the chapter

http://dx.doi.org/10.5772/67674

**Abstract**

Freshwater mussels of the Clinch and Powell rivers of Virginia in the southeastern United States have been heavily impacted by runoff, leachates, or spills of materials related to coal extraction, processing, and use. Assays quantifying sublethal impacts of such wastes are needed. We assessed gene transcriptional markers in a laboratory study under controlled conditions, focusing upon arsenic (arsenate, As(V)) and sulphate, contaminants related to coal mining and processing. Pheasantshells *Actinonaias pectorosa* collected from the Clinch River were subjected to a 28-day chronic exposure to control or environmentally relevant concentrations of each compound. We compared gene expression in digestive gland among parasite-free, female pheasantshells among control and contaminant-exposed individuals using the Illumina HiSeq platform. Statistically significant differential expression of particular genes was observed among control mussels and those exposed to either arsenate or sulfate. Chemical stress was as likely to cause under-expression as it was to cause over-expression of particular genes. Arsenate and sulfate induced up- or down-expression of different suites of 50-100 genes. Our results provide proof-of-principle for using RNAseq technology to approach issues of toxicogenomics in freshwater mussels. The candidate markers could be validated for quantitative PCR assays for rapidly assessing single-gene responses to exposure to toxic compounds.

**Keywords:** arsenic, differential gene expression, sulfate, toxicogenomics, transcriptional markers

# 1. Introduction

## 1.1. Response of aquatic organisms to exposure to toxic substances

Exposure to toxic compounds has not only lethal but also important sublethal effects upon affected individuals. Successful identification of molecular mechanisms underlying response to toxic exposure depends upon development and use of a suitable set of assays, and several approaches are potentially available. While various biomarkers [1, 2] or gene expression [3, 4] assays have been demonstrated for marine mollusks in recent years, there has been no parallel work for freshwater mussels, many of which are of conservation concern. More fundamentally, adaptation of existing assays to freshwater mussels is not entirely attractive because a focus on candidate genes of known function certainly will miss genes or biochemical pathways that are not known. A preferable approach would be to screen global gene expression and thereby identify genes of interest not only in known but also in unknown pathways. Three approaches are available. Microarrays are available for several marine mollusks, and have in some cases been used for purposes of characterizing responses to environmental stressors including toxins or other bioactive compounds (e.g., marine mussels [5–8], Manila clam *Venerupis philippinarum* [9]). There are no existing microarrays for freshwater mussels, however, and construction of such screening platforms would require a large body of work developing and characterizing expressed sequence tags for hundreds to thousands of genes. Subtractive hybridization has been applied to other mollusks, such as the invasive zebra mussel *Dreissena polymorpha* [10], Pacific oyster *Crassostrea gigas* [11], and peppery furrow shell *Scrobicularia plana* [12], and has been shown a viable but technically challenging approach to identifying differentially expressed genes. Next-generation DNA-sequencing technology has made it possible to cost-effectively screen expression of all genes transcribed in tissues of interest, even in species for which prior knowledge of the genome is lacking. The approach is tantamount to sequencing all the RNAs produced in that tissue; hence, the approach is referred to as RNAseq. In our context, RNAseq makes quantitative comparison of gene expression in selected tissues among toxin-challenged and control individuals possible. Against this background, the goal of this study was to relate gene expression end points in a representative freshwater mussel, pheasantshell *A. pectorosa*, to exposure to arsenate and sulfate, two pollutants resulting from coal combustion and mining, respectively.

## 1.2. Freshwater mussels

Freshwater mussels (Class Bivalvia: Family Unionidae) have their center of biodiversity in the southeastern United States. However, many regionally important species face a variety of threats, including exposure to toxic compounds. In particular, mussels of the Clinch and Powell River systems of southwest Virginia have been heavily impacted by runoff, leachates, or spills of materials related to coal extraction, processing, and use for electric power generation. Given the continued operation of coal extraction and processing facilities, the shift from deep- to surface-mining practices, and the increase in coal-bed gas extraction wells in the Clinch and Powell River system, defensible biological assays for assessing the impacts on key components of this aquatic ecosystem must be developed. These assays could provide critical information for assessing the impacts of future toxic events and thereby allow characterization of potential long-term effects of resource extraction on freshwater mussel populations. This project was conducted to show the response of freshwater mussels to selected physiological stressors. Within this context, the objective of the study was to screen gene-transcriptional markers in a laboratory study under controlled conditions, focusing upon arsenate and sulfate, two contaminants related to coal mining and processing. Identification of such genetic markers could prove essential for use in future nonlethal determinations of contaminant impacts to federally listed mussels.

### 1.3. Contaminant selection

The coal industry is a potential source of numerous contaminants to aquatic environments. In several Virginia watersheds, habitat for imperiled freshwater mussel populations exists in close proximity to coal industry-associated activities. These activities include mining and coal-fired electric power plants with on-site storage of coal combustion residue (CCR). Two contaminants were selected for this study based on their association with surface mining (sulfate, $SO_4^{2-}$) or CCR storage (arsenate, As(V)). The contaminants As and $SO_4^{2-}$ were selected due to their environmental relevance and the potential for concentrations to become elevated due to activities related to coal mining and power generation.

Freshwater bivalves appear to be relatively tolerant to acute effects of As; the estimated 96-h $LC_{50}$ for Asian clam *Corbicula fluminea* is 20,740 μg/L As, with no mortality observed at concentrations up to 5000 μg/L for 21 days [13]. In surface waters, the majority of total As will be As(V) under oxygenated conditions. However, investigations of sublethal effects of As on bivalves have generally focused on As(III). Exposures of bivalves to As(III) have demonstrated histological effects including increased damage to digestive gland tissue [14] and biochemical effects including alteration of levels of adenosine triphosphate [15], inhibition of the detoxification enzyme catalase [14], and changes in the activity of glutathione-S-transferase (stimulation and inhibition) [14, 16]. Detoxification of As(III) and As(V) involves reduced glutathione and glutathione-dependent enzymes, and effects on this system in bivalves have been shown to be variable and dependent on both concentration and chemical speciation [14, 16]. The sublethal effects of environmentally relevant concentrations of As(V) on freshwater mussels are currently unknown.

Freshwater bivalves also appear to be relatively tolerant to acute effects of $SO_4^{2-}$. Soucek and Kennedy [17] determined a sulfate 96-h $LC_{50}$ for grooved fingernail clam *Sphaerium simile* of 2078 mg/L. However, exposure of *C. fluminea* to 1500 mg/L $SO_4^{2-}$ reduced feeding rates over a 4-week period [18].

For each contaminant, a relatively 'high' concentration was selected based on worst-case conditions previously measured in heavily impacted environments. The concentration of $SO_4^{2-}$ (1250 mg/L) was based on low-flow measurements up to 1200 mg/L in a notoriously polluted river in West Virginia, USA [19]. The high concentration of As (1000 μg/L) was based on measurements in pore-water downstream of a massive CCR spill (up to 1200 μg/L; [20]). For both chemicals, the selected concentrations were not expected to cause mortality based on results of acute and chronic toxicity studies conducted with other bivalves (*S. simile* and *C. fluminea;* [13, 17, 18, 21]). The majority of toxicity tests with As have been conducted with arsenite (As(III)), the more toxic form of As. However, due to the oxygenated environment in our exposure system, mussels were exposed to As as arsenate (As(V)). Hence, the purpose of this study was to determine whether As (as As(V)) and $SO_4^{2-}$ caused sublethal changes in freshwater mussels, the nature of these changes, and the relationships between biochemical, histological, and genetic markers.

## 2. Methods

### 2.1. Mussel collection and holding

Pheasantshell mussels (*A. pectorosa*) were collected from the Clinch River, Hancock County, Tennessee, on September 23, 2014. A total of 143 individuals ranging in size from 70 to 90 mm were collected by hand using snorkel and mask. Mussels were held for 14 days of acclimation at the Freshwater Mollusk Conservation Center (FMCC), Virginia Tech, Blacksburg, Virginia, in flow-through systems continuously replenished with unfiltered water from a man-made, on-site pond.

### 2.2. Chemicals

Sodium arsenate heptahydrate ($Na_2HAsO_4$*$7H_2O$, American Chemical Society (ACS)-certified reagent grade) was purchased from J.T. Baker Co. (Avantor Performance Materials, Center Valley, PA). A 100-mM (7.492 g/L) As (V) stock solution in ultrapure water was prepared at the beginning of the study. ACS-certified reagent-grade sodium sulfate ($Na_2SO_4$) was purchased from Fisher Scientific (Fair Lawn, NJ). Stock solutions were prepared weekly in ultrapure water.

### 2.3. Mussel exposure system

Mussels were exposed in 18-L downweller-bucket systems [22] modified to accommodate adult mussels [23]. Buckets were placed into a 757-L container filled with well water to serve as a temperature control bath. The target exposure temperature of 23°C was maintained in the water baths using aquarium heaters. Each bath held up to five buckets. In the bucket systems, unfiltered water from the FMCC pond was used for acclimation, control water, and make-up water for each treatment. Mussels were fed daily with a 1:1:1 algal cell ratio from three premixed commercial microalgae diets (Nanno 3600, Shellfish Diet 1800, and TP 1800; Reed Mariculture, Campbell, CA). A concentrated stock solution was prepared in deionized water and 1 mL of stock was added to each bucket to achieve a final concentration of 24,000 cells/mL.

### 2.4. A 7-day acute exposure

Thirty mussels were allocated to 10 buckets (three mussels each, arbitrarily chosen) on October 7, 2014 for a preliminary 7-day test of acute toxicity. This test was conducted to determine the potential for mussel survival in solutions of As(V) and $SO_4^{2-}$ over 28 days; treatments were not replicated. Five buckets ($n = 1$) were used to test acute toxicity of five $SO_4^{2-}$ concentrations: 1500,

1000, 500, 250, and 0 mg/L (control). Treatments were prepared via direct addition of appropriate volumes of a 607-mM (58.3-g/L) stock solution of $SO_4^{2-}$. Five buckets were used to test acute toxicity of five As(V) concentrations: 2000, 985, 515, 170, and 0 μg/L (control). Treatments were prepared via direct addition of the 100-mM stock solution of As(V). On Day 7, all 30 mussels were alive; there was no mortality in any of the As(V) or $SO_4^{2-}$ treatments or controls.

### 2.5. A 28-day chronic exposure

Sixty-four mussels were randomly selected on October 16, 2014. Four mussels were allocated to each of 16 buckets, and the length of each mussel was recorded using dial calipers. Mussels were allowed to acclimate in the buckets for 11 days prior to the start of the chronic exposure. During the acclimation period, dissolved oxygen and ammonia levels were monitored using the methods described below to ensure that the feeding rate was appropriate to maintain acceptable water quality.

On Day 0 of the exposure (October 27, 2014), each bucket received a 100% water exchange and was randomly assigned to one of four treatments/controls ($n$ = 4 for each treatment/control). Because of the availability of experimental units and mussels, two separate controls were used (control_1 and control_2); both consisted of 100% unfiltered pond water. The As(V) treatment (hereafter, HA) concentration of 1000 μg/L was achieved by adding appropriate volumes of 100 mM stock solution to unfiltered pond water. The $SO_4^{2-}$ treatment (hereafter HS) of 1250 mg/L was achieved by adding appropriate concentrations of a 569-mM (54.65-g/L) stock solution to unfiltered pond water. The final water volume in all treatment and control buckets was 17.5 L.

For each bucket, mussel mortality was checked daily and recorded, and a 100% water exchange was conducted weekly. Prior to water exchanges, temperature (°C), specific conductance (μS/cm), dissolved oxygen (DO; mg/L and % saturation), and pH were measured using a calibrated YSI 556 Multi-Probe Sensor (YSI Inc., Yellow Springs, OH), and water samples were collected. Total ammonia-nitrogen (TAN) was measured in unfiltered samples using a Hach DR/2400 meter following the manufacturer's methods. Concentrations of elemental As and S were measured in filtered (0.45-μm) water samples. Elemental quantification was performed by the Virginia Tech Soil Testing Laboratory using an inductively coupled plasma atomic emission spectrometer (ICP-AES, Spectro Analytical Instrumentation, Kleve, Germany) following the laboratory's standard operating procedures and quality assurance/quality control methods [24]. Sulfate concentrations were calculated from measured total S; all S were assumed to be present as $SO_4^{2-}$ due to the buckets' oxygenated environment. Although only total As was measured, all As are assumed to have been present as As (V) due to the oxygenated environment in the buckets. Alkalinity and hardness of the pond water were measured weekly as part of a concurrent study [24]. The 28-day exposure concluded on November 24, 2014.

### 2.6. Mussel dissection

At the end of the study on Day 28, all mussels were removed from the buckets. Each mussel was measured using dial calipers and opened using reverse pliers. Each mussel was evaluated for gravidity, that is, the presence of glochidia in the outer marsupial gills. Mussels were removed from their shells by severing the adductor muscles. The mantle tissue was removed and samples of gill, kidney, and gonad were excised and fixed in neutral buffered formalin for histological evaluation. The digestive gland was removed from the remainder of the tissue and divided into three portions, with one portion each preserved separately for histological evaluation and RNA extraction. The remaining portion was further divided into thirds for separate biochemical analyses. Samples for biochemical work were immediately frozen. Samples for screening of gene expression were preserved in RNALater (Qiagen). All dissection tools were sterilized in a concentrated bleach solution between dissections of each mussel.

### 2.7. Transcriptome analysis

Only parasite-free, female mussels (determined via histological analysis) were used for transcriptomic analysis. RNA isolated from digestive gland tissue samples was transcribed in vitro to double-stranded cDNA, yielding a library of transcripts for each individual mussel. Individual-specific adapters were ligated onto transcripts for each individual so that we could subsequently identify RNA sourced from the respective individuals. Barcoded cDNA samples—three from control and six from contaminant-exposed individuals (three per treatment)—were sequenced using the Illumina HiSeq platform. Adapters were removed from the raw sequencing reads. Duplicated and low-quality reads were discarded using FastqMcf [25] with default parameters. To exclude possible contamination, all reads were aligned to a bacterial database downloaded from NCBI (http://www.ncbi.nlm.nih.gov/), and only unmapped reads were used for assembly. Processed reads from all nine samples were merged together, assembled with Trinity [26] with parameter—trimmomatic, after duplications were removed. TransDecoder (Trinity package) was used to identify candidate-coding regions within assembled transcripts, and transcripts with open-reading frame (ORF) lengths less than 300 (100 amino acids) were removed from the assembly. The final transcript assembly was used as a reference for gene annotation and expression calculation. Transcripts were mapped to the nonredundant protein database [(NR database) from NCBI] using BLAST (v. 2.2.28). Alignments with threshold *e*-values greater than $1e^{-10}$ or identity less than 50% were discarded. The clean reads were mapped to the reference assembly using Bowtie v. 1.0.0 [27] with parameters set to '-l 25 -I 1 -X 1000 -a -m 200.' RSEM [28] was used to calculate the gene expression. Differentially expressed genes were calculated using the edgeR [29] package in R software (http://www.r-project.org/), and Benjamini-Hochberg adjusted *p*-values less than 0.05 were considered to be significant.

### 2.8. Histological analysis

Sections from gonads, digestive glands, and gills were stained with hematoxylin and eosin for routine histological evaluations, and sections from kidneys were stained with Long Ziehl-Neelsen stain for elaboration of lipofuscin [30]. Histological evaluations of tissues were conducted by light microscopy using point counting [31]. Evaluations determined fractions of reproductive acini containing mature and/or developing gametes, acini containing atretic and/or resorbing gametes, digestive gland diverticula cells with abnormal cytoplasm, gill filament termini without cilia, and kidney diverticula cells containing lipofuscin inclusions. Genders (female, male, hermaphrodite, and indeterminate) of mussels and incidences of parasitic infestation were recorded. Histological data were analyzed using generalized linear-mixed models for binomial data with SAS GLIMMIX.

### 2.9. Biochemical analysis

Digestive gland samples for biochemical analysis were placed in 1.5-mL tubes, immediately frozen, and held at -80°C. Analyses of biochemical parameters in mussel digestive glands were conducted in duplicate and expressed per mg of protein determined using a bicinchoninic acid method (Pierce BCA Protein Assay Kit, Thermo Scientific, Rockford, IL). Activities of catalase (CAT), glutathione S-transferase (GST), glutathione peroxidase (GPx), glutathione reductase (GR), malondialdehyde (MDA), total- and Na+, K+–ATPase, and lactate dehydrogenase (LDH), and the concentration of reduced glutathione (GSH) were determined.

### 2.10. Correlations with histological and biochemical markers

Correlations between biochemical measurements, histological variables, and gene expression were evaluated using rank-transformed data (Spearman's rho) to minimize effects of the different data types and scales for each class of variable. Due to the large number of variables examined, the criterion for statistical significance was reduced ($\alpha = 0.01$) and plots of all significant relationships were examined to ensure linearity. To select genes for assessment of correlations, genes with the lowest false-detection rate ($<9.67 \times 10^{-5}$) and $p$-values ($<9.5 \times 10^{-8}$) were identified among genes with differential expression between control, HS, and HA. All annotated genes with differential expression between control, HS, and HA females were also selected. For these 48 selected genes, gene expression was correlated with biochemical measurements and histological variables. Digestive glands from nine mussels were included in the analysis of gene expression, but only eight mussels were included in correlations with biochemical measurements of the glutathione antioxidant system; that is, one extract lost during the set of biochemical analyses was one of the mussels randomly selected for analyses of gene expression.

## 3. Results

### 3.1. Water chemistry

Mean-measured concentrations of $SO_4^{2-}$ and As were similar to nominal concentrations (**Table 1**). Background $SO_4^{2-}$ (~17 mg/L) was present in the pond water; thus, measured concentrations were slightly above nominal concentrations. There was no measureable background As in the pond water.

Water quality measurements were within acceptable ranges for toxicity tests with freshwater mussels. During the study, the pond had a mean measured hardness of 204 mg/L $CaCO_3$ and a mean measured alkalinity of 196 mg/L $CaCO_3$. Mean measured temperatures were similar between treatments, and temperature was generally within 1°C of the 23°C target temperature (**Table 2**). Measured pH was similar across all treatments and varied little over the course of the study. Specific conductivity was similar in the controls and As treatments, and was elevated in the $SO_4^{2-}$ treatments as expected. Dissolved oxygen concentrations remained relatively high for the duration of the study (lowest recorded value: 75% saturation, 6.20 mg/L, on Day 28). Dissolved oxygen concentrations were consistent between treatments but did show variation during the study, with a decrease in overall mean saturation from 86% on

| | $SO_4^{2-}$ (mg/L) | | As (μg/L) | |
|---|---|---|---|---|
| Treatment | Nominal | Mean | Nominal | Mean |
| Arsenic | - | 16.93 (1.44)[b] | 1000 | 1027 (111) |
| Sulfate | 1250 | 1276 (48) | 0 | <8.1[c] |
| Control[a] | - | 16.90 (1.56)[b] | 0 | <8.1[c] |

[a]Measurements were pooled for control_1 and control_2, since all replicates consisted of 100% unfiltered pond water.
[b]Background concentration of $SO_4^{2-}$ in the pond.
[c]Method detection limit for As.

**Table 1.** Nominal and mean (standard deviation) measured concentrations of $SO_4^{2-}$ and As in treatments ($n = 4$ each) and controls ($n = 8$) during the 28-day exposure.

| Treatment | Temp. (°C) | pH | Specific conductivity (mS/cm) |
|---|---|---|---|
| Arsenic | 22.8 (0.9) | 8.58 (0.03) | 0.436 (0.004) |
| Sulfate | 22.6 (0.8) | 8.64 (0.05) | 2.89 (0.10) |
| Control_1 | 22.6 (1.1) | 8.51 (0.27) | 0.431 (0.005) |
| Control_2 | 22.8 (1.2) | 8.61 (0.07) | 0.434 (0.004) |

**Table 2.** Mean (standard deviation) measured concentrations of water quality parameters in all treatments and controls ($n = 4$ each) during the 28-day exposure.

Day 7 (range 79.1–88.7%) to 78% on Day 28 (range 75–78.8%). TAN concentrations were also consistent between treatments, but varied over the course of the study. The overall mean TAN concentration was 0.001 mg/L on Day 7 (range 0–0.01 mg/L) and increased to 0.068 mg/L on Day 21 (range 0.05–0.08 mg/L). The maximum measured TAN concentration was 0.080 mg/L in several control and $SO_4^{2-}$ units, but this value is far below the 2013 acute and chronic water quality standards for ammonia (0.96 and 0.24 mg/L, respectively, at pH 8.6 and 23°C), derived based on the sensitivity of mussels [32].

### 3.2. Mussels

No mussels died during the course of the experiment. The mean length of all exposed mussels was 85.1 mm (range 79.0–99.3 mm). The mean lengths were not significantly different between treatments (GLIMMIX, $p = 0.22$). Mussel growth was not quantified in this study due to the use of adult mussels and the short duration of the exposure.

### 3.3. Reference assembly

The analytic program Trinity handled 61,774 transcripts, assembling them into the putative transcripts of 34,019 genes. The average length of these contiguous sequences, or "contigs," was 1012 base pairs.

### 3.4. Differential expression analysis

Statistically significant ($p < 0.05$) differential expression (DE) of particular genes was observed among control mussels (CT) and those exposed to either arsenate (HA) or sulfate (HS). Chemical stress was as likely to cause underexpression as it was to cause overexpression of genes relative to levels observed in control mussels. Contrasts also were observed among mussels exposed to the respective pollutants, indicating that pollutants induced up- or downexpression of different suites of genes. From 50 to 100 differentially expressed genes were found for each comparison. **Figures 1–3** compare levels of expression of particular genes among groups of mussels.

Comparing gene expression among control and arsenate-treated mussels (**Figure 1**), 18,572 transcripts were not affected by treatment, 28 were underexpressed in controls relative to treated mussels, and 30 were overexpressed.

Comparing gene expression among control and sulfate-treated mussels (**Figure 2**), 18,469 transcripts were not affected by treatment, 74 were underexpressed in controls relative to treated mussels, and 86 were overexpressed.

Comparing gene expression among pollutants, 18,444 transcripts were not differentially expressed, 128 were more highly expressed in sulfate-exposed mussels, and 57 more highly in arsenate-exposed mussels (**Figure 3**). Overall, the sulfate treatment was a ~3× greater stress factor in terms of the number of differentially expressed genes than was the arsenate (HA) treatment. We offer the explanation that in aquatic species, maintaining homeostasis in the face of ionic and osmotic stressors affects many cellular processes, while heavy metal toxicity

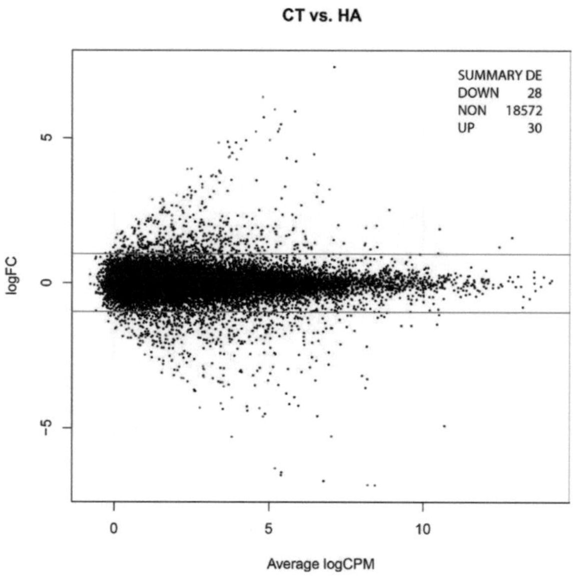


**Figure 1.** Summary of differential gene expression (DE) in control (CT) compared to arsenate-treated (HA) mussels. LogCPM is the logarithm of counts per million reads, and logFC is the log-transformed fold change in gene expression. Each dot in the figure represents the expression of one gene. Dots for genes whose expression is not significantly affected fall between the two horizontal lines. Dots for genes whose expression is upregulated relative to controls fall above the upper line. Dots for genes whose expression is downregulated relative to controls fall below the lower line.

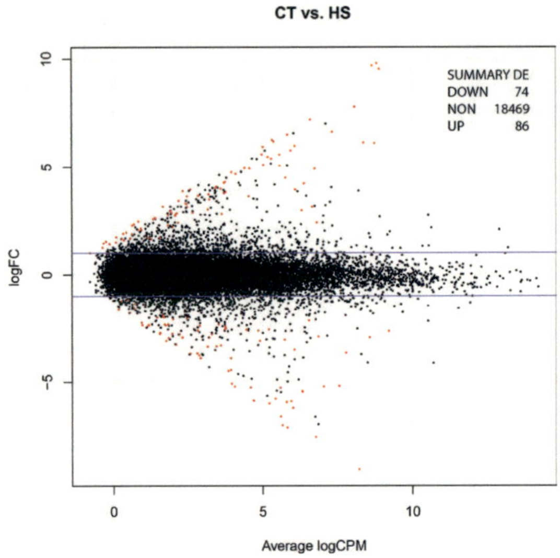


**Figure 2.** Summary of differential gene expression (DE) in control (CT) compared to sulfate-treated (HS) mussels. LogCPM is the logarithm of counts per million reads, and logFC is the log-transformed fold change in gene expression. Each dot in the figure represents the expression of one gene. Dots for genes whose expression is not significantly affected fall between the two horizontal lines. Dots for genes whose expression is upregulated relative to controls fall above the upper line. Dots for genes whose expression is downregulated relative to controls fall below the lower line.

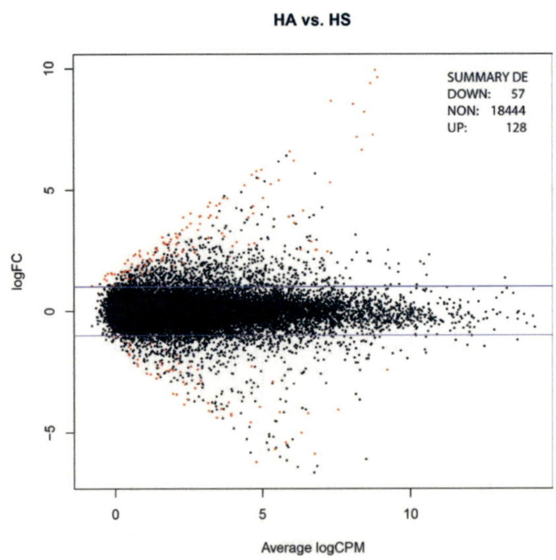


**Figure 3.** Summary of differential gene expression (DE) in high arsenate-(HA) compared to high sulfate-(HS) treated mussels. LogCPM is the logarithm of counts per million reads, and logFC is the log-transformed fold change in gene expression. Each dot in the figure represents the expression of one gene. Dots for genes whose expression is not significantly affected fall between the two horizontal lines. Dots for genes whose expression is upregulated relative to controls fall above the upper line. Dots for genes whose expression is downregulated relative to controls fall below the lower line.

affects relatively few. Comparing the effects of HS and HA directly, it was evident that HS causes >2 times more overexpression, again potentially pointing to a greater effect of HS. Interestingly, and as would be expected, the expression of different genes was affected. As discussed below, these genes are candidate markers for indicating exposure of mussels to the respective pollutants.

### 3.5. Annotation of expressed genes

The proportion of genes that we were able to annotate by reference to GenBank was relatively low (63%), as mollusks are nonmodel organisms and hugely underrepresented in genomic databases. We note that the proportions of genes that we annotated are very similar to those reported by Bai et al. [33, 34] for the freshwater pearl mussel *Hyriopsis cumingii.*

Among annotated underexpressed genes in the arsenate treatment relative to controls were guanylate binding protein 1 (*GBP1*) and poly [ADP-ribose] polymerase (BRAFLDRAFT_74879). In mammalian cells, *GBP1* has been shown to respond to exposure to multiple stress agents including paclitaxel and doxorubicin [35]. The enzymes poly(ADP-ribose) (*PAR*) and polymerase-1 (*PARP-1*) are central for cellular stress responses and for directing cells to specific fates (e.g., DNA repair vs. cell death) [36]. Among annotated overexpressed genes in arsenate relative to control were poly [ADP-ribose] polymerase 15-like (LOC101731886), and sodium- and chloride-dependent glycine transporter 2-like (LOC100899820).

Among annotated underexpressed genes in the sulfate treatment relative to controls were serine protease inhibitor dipetalogastin precursor and zinc-binding Sp-Hypp_8991. Interestingly, the expression of serine protease inhibitor seems to offer stress tolerance via delayed senescence [37]. Among annotated overexpressed genes in sulfate relative to control were NADPH-dependent alpha-keto amide reductase (YDL124W), and poly [ADP-ribose] polymerase 15-like (LOC101731886). Multiple aldo-keto reductases fulfill functionally redundant roles in stress response in yeast [38].

### 3.6. Correlations with histological, biochemical, and gene expression markers

Significant correlations ($r$, $p < 0.05$) were observed among genetics and histological data from the nine sampled females. Fractions of reproductive acini containing mature or developing oocytes were significantly negatively correlated with gene identification code (gene code) c140332_g1 (gene symbol = LOC101731886) ($r = -0.78$, $p = 0.01$). Fractions of digestive diverticula cells with lesions were significantly correlated with gene code c145102_g1 (none) ($r = 0.74$, $p = 0.04$) and negatively correlated with gene codes c103784_g1 (CGI_10002926) ($r = -0.76$, $p = 0.03$), c149150_g3 (none) ($r = -0.78$, $p = 0.02$), and c152424_g6 (none) ($r = -0.73$, $p = 0.04$). Fractions of kidney cells containing lipofuscin inclusions were significantly correlated with c138105_g1 (none) ($r = 0.76$, $p = 0.02$).

There were strong correlations ($r > 0.85$, $p < 0.01$) between the expression of five genes and biochemical measurements. Numerous weaker correlations ($0.05 > p > 0.01$) are not presented or discussed. The expression of one annotated gene, c150857_g2 (*GBP1*), was positively correlated with the activity of glutathione-S-transferase, but there was little separation between treatments and control_2 for both gene expression and enzyme activity. The expression of one annotated gene (c152797_g1; BRAFLDRAFT_74879) and one unannotated gene (c126914_g1) was positively correlated with the concentration of reduced glutathione, and there was good separation between treatments and control_2. The expression of two unannotated genes (c155139_g1 and c155139_g2) was negatively correlated with the concentration of reduced glutathione and there was good separation between the treatments and control_2. Correlation between the expression of a sixth gene (unannotated; c154720_g4) did not meet the $\alpha = 0.01$ criterion for strong correlation, but the plot demonstrated a similar pattern as the other two genes negatively correlated with reduced glutathione concentration, with good separation between treatments and control_2. For the eight mussels included in these correlations, concentrations of reduced glutathione were higher in the control and lower in the two treatments, whether these genes are directly related to reduced glutathione production or utilization is unknown. The small sample of mussels utilized for genetic analysis limits the interpretation of the relationships but suggests that these genes warrant further investigation, particularly as related to As(V) exposure.

## 4. Discussion

Gene expression profiling via next-generation sequencing has emerged as a tool to assess the biological impact of environmental pollutants and natural chemical stressors. This approach supports better understanding of whole-genome expression responses to chemical stress, aids in the identification of ecological and toxicological modes of action, and provides hundreds (or in some cases thousands) of rigorously tested markers for stress responses. Gene expression profiles thus provide a more comprehensive, sensitive, and actionable insight into toxicity than typical toxicological parameters such as morphological changes, altered reproductive capacity, or mortality [39]. For example, it has been demonstrated that chemicals from the same class of compounds give rise to discernible gene expression profiles that bear more similarity to each other than to patterns corresponding to exposure to compounds from a different class [40]. It would be useful for a database for gene expression responses to environmental pollutants to be developed for recognizing compound-specific expression profiles and molecular signatures of stress responses.

Functional annotations of genes may provide insights into affected regulatory and proliferative and repair/adaptive pathways, with immediate management and conservation implications. To improve the rate of functional annotations, expanded sequencing with more complete sequence information will be required. Recent bioinformatics developments, such as improvements to Trinotate (https://trinotate.github.io/), also hold promise of improved functional annotations. However, we note that complete annotations of affected genes are not necessary. Unannotated transcripts are potentially useful markers not only for compound-specific expression profiling but also candidates for cost-effective and rapid individual biomarker assays based on real-time polymerase chain reaction (PCR), which could be

http://dx.doi.org/10.5772/67674

developed for use in the laboratory or even in field settings. Despite the limited sampling, we found dozens of highly significant gene expression markers (after the correction for false discovery rate) for the two stressors tested. As many as 66 genes were expressed only in the presence of the two compounds, with zero transcripts (i.e., no sequences found) in the control group.

We sought quantitative differences between the RNA pools of control and arsenate- or sulfate-exposed mussels. We sought evidence of altered expression of genes known to be linked to toxin response (e.g., heat-shock proteins, glutathione transferase, metallothioneins, cytochrome P450, etc.), and confirmed that some of the pathways, such as poly(ADP-ribose), were indeed induced. We also sought to identify genes whose expression is influenced by toxin exposure, but of which we have no existing knowledge. We found many such putative markers of toxin-induced responses. In this latter case, more complete sequence data will be required to enhance functional annotation, but even if we do not know the functions of all the associated genes, they are still candidate markers for pollutant response.

### 4.1. Comparisons among biomarker types

Exposure to toxic compounds would be expected to affect gene expression, molecular flux through biochemical pathways, and ultimately histological structures in freshwater mussels. Further, the respective markers would be expected to be causally related to one another. Hence, we sought statistical indicators of such correlations among our data sets.

Correlations among histological and genetics data provide provisional bases for hypotheses regarding links among histological lesions and genes identified during this study. While many of the significant correlations involved genes without known function, a few involved genes of known function. Gene LOC101731886 was significantly negatively correlated with fractions of reproductive acini containing mature or developing gametes. The gene has been linked to the activity of poly [ADP-ribose] polymerase (15-like) (*PARP*) [41]. The *PARP* enzyme family is associated with physiological functions during cell division, transcriptional regulation, repair of DNA damage, and cytoplasmic cell response [42]. Females that provided the data for the correlations had completed oogenesis, and oocytes in acini were mostly atretic or resorbing; it seems logical that nucleus-related activities controlled by *PARP* in postgametogenic females would be negatively related to abundances of developing oocytes, since active gametogenesis had ceased. The positive relationship between abundances of lipofuscin in kidney cells and upregulation of gene LOC101731886 (expression of *PARP*) may be related to oxidative stress. Increased production of reactive oxygen species (ROS) has been related to DNA single-strand damage incurred during contaminant exposure and inadequate aquaculture holding conditions [43, 44]. Since abundances of intra-lysosomal lipofuscin granules are related to both contaminant exposure and oxidative stress [45, 46], it is logical that they are positively related to the expression of *PARP* activity to repair DNA.

Abundances of lesions in digestive cells were negatively correlated with genes BRAFL-DRAFT_118372 (histamine N-methyltransferase activity) and CGI_10002926 (uncharacterized protein with unknown function) [47–49]. A negative correlation between gene regulation of histamine methylation and lesions in digestive cells seems reasonable, because with increasing abundance of lesions, destruction of cytoplasmic organelles probably occurs. A positive correlation between the expression of Sp-Hypp_8991 and digestive lesions may be related to activities to repair nuclear damage incurred during treatment and captive holding.

### 4.2. Implications for future research

Although there has been some limited application in marine mollusks, and one study has identified candidate stress-response genes in the freshwater mussel *Elliptio complanata* [50], RNAseq has not been evaluated for its utility for monitoring of the responses of freshwater mussels to coal-related environmental contaminants. Successful execution of this project provides proof of principle for using RNAseq technology to approach issues of toxicogenomics in freshwater mussels. Among key findings, our results collectively support the view that substantial changes of gene expression occur before dramatic changes in biochemical and histological effects. Our understanding of how changes in gene expression result in physiological and histological effects on the organism is hampered by the current situation in which many freshwater mollusk genes remain unannotated.

RNAseq-based assessment of global gene expression identified candidate markers for quantitative polymerase chain reaction (qPCR) assays, which could be developed and validated for rapidly assessing single-gene responses to exposure to toxic compounds. As noted above, Hamadeh et al. [40] demonstrated that chemicals from the same class of compounds give rise to discernible gene expression profiles that bear more similarity to each other than to patterns corresponding to exposure to compounds from a different class [40]. It would be useful for a comprehensive database for gene expression responses to environmental pollutants to be developed toward recognizing compound-specific expression profiles and molecular signatures of stress responses. Additional, well-controlled laboratory studies would be appropriate for the purpose of defining defensible biomarkers for toxicant-induced harm in freshwater mussels. Ultimately, practical work would apply (modifications of) the respective approaches to a range of sites in the Clinch, Powell, and other aquatic ecosystems.

## Acknowledgements

This work was supported by a grant from the US Fish and Wildlife Service. The participation of EMH in this project was supported in part by the Virginia Agricultural Experiment Station under the US Department of Agriculture Hatch Program. Any use of trade, product, or firm names is for descriptive purposes only and does not imply endorsement by the US Government. The findings and conclusions in this article are those of the authors and do not necessarily represent the views of the US Fish and Wildlife Service.

## Author details

Pawel Michalak[1], Lin Kang[1], Serena Ciparis[2,3], William Henley[2], Jess Jones[2,3], Andrew Phipps[2] and Eric Hallerman[2*]

*Address all correspondence to: ehallerm@vt.edu

1 Virginia Biocomplexity Institute, Blacksburg, Virginia, USA

2 Department of Fish and Wildlife Conservation, Virginia Polytechnic Institute and State University, Blacksburg, Virginia, USA

3 U.S. Fish and Wildlife Service, Blacksburg, Virginia, USA

## References

[1] Damiens, G., His, E., Gnassia-Barelli, M., Quiniou, F., & Romeo, M. Evaluation of biomarkers in oyster larvae in natural and polluted conditions. Comparative Biochemistry and Physiology C. 2004; **138**:121-128, http://dx.doi.org/10.1016/j.cca.2004.05.010.

[2] Milan, M., Pauletto, M., Patarnello, T., Bargelloni, L., Marin, M.G., and Matozzo, V. Gene transcription and biomarker responses in the clam *Rudatipes philippinarum* after exposure to ibuprofen. Aquatic Toxicology. 2013; **126**:17-29, http://dx.doi.org/10.1016/j.aquatox.2012.10.007.

[3] Boutet, I., Tanguy, A., & Moraga, D. Molecular identification and expression of two non-P450 enzymes, monoamine oxidaseA and flavin-containing monooxygenase 2, involved in phase I of xenobiotic biotransformation in the Pacific oyster *Crassostrea gigas*. Biochimica et Biophysica Acta. 2004; **1679**:29-36, http://dx.doi.org/10.1016/j.bbaexp.2004.04.001.

[4] Cubero-Leon, E., Ciocan, C.M., Hill, E.M., Osada, M., Kishida, M., Itoh, N., Kondo, R., Minier, C., & Rotchell, J.M. Estrogens disrupt serotonin receptor and cyclooxygenase mRNA expression in the gonads of mussels (*Mytilus edulis*). Aquatic Toxicology. 2010; **98**:178-187, http://dx.doi.org/10.1016/j.aquatox.2010.02.007.

[5] Venier, P., De Pitta, C., Pallavicini, A., Marsano, F., Varotto, L., Romualdi, C., Dondero, F., Viarengo, A., & Lanffranchi, G. Development of mussel mRNA profiling: Can gene expression trends reveal coastal water pollution? Mutation Research. 2006; **602**:121-134, http://dx.doi.org/10.1016/j.mrfmmm.2006.08.007.

[6] Dondero, F., Dagnino, A., Jonsson, H., Capri, F., Gastaldi, L., & Viarengo, A. Assessing the occurrence of a stress syndrome in mussels (*Mytilis edulis*) using a combined biomarker/gene expression approach. Aquatic Toxicology. 2006; **78S**:S13–S24, http://dx.doi.org/10.1016/j.aquatox.2006.02.025.

[7] Manfrin, C., Dreos, R., Battistella, S., Beran, A., Gerdol, M., Varotto, L., Lanfranchi, G., Vernier, P., & Pallavicini, A. Mediterranean mussel gene expression profile induced by okadaic acid exposure. Environmental Science and Technology. 2010; **44**:8276-8283, DOI: 10.1021/es102213f.

[8] Fosnes, H. Gene transcription profiling in mussels (Mytilis sp.) exposed to the model compounds copper chloride and benzo[a]pyrene. M.S. thesis, Norwegian University of Life Sciences, As. 2010; http://hdl.handle.net/11250/189288.

[9] Milan, M., Coppe, A., Reinhardt, R., Cancela, L.M., Leite, R.B., Saavedra, C., Ciofi, C., Chelazzi, G., Patarnello, T., Bortoluzzi, S., & Bargelloni, L. Transcriptome sequencing and microarray development for the Manila clam, *Rudatipes philippinarum*: Genomic tools for environmental monitoring. BMC Genomics. 2011; **12**:2345, DOI: 10.1186/1471-2164-12-234.

[10] Bultelle, F., Panchout, M., Leboulenger, F., & Danger, J.M. Identification of differentially expressed genes in *Dreissena polymorpha* exposed to contaminants. Marine Environmental Research. 2002; **54**:385-389, http://dx.doi.org/10.1016/S0141-1136(02)00159-9.

[11] Boutet, I., Tanguy, A., & Moraga, D. Response of the Pacific oyster *Crassostrea gigas* to hydrocarbon contamination under experimental conditions. Gene. 2004; **329**:147-157, http://dx.doi.org/10.1016/j.gene.2003.12.027.

[12] Ciocan, C.M., Cubero-Leon, E., Peck, M.R. Langston, W.J., Pope, N., Minier, C., & Rotchell, J.M. Intersex in *Scrobicularia plana*: Transcriptomic analysis reveals novel genes involved in endocrine disruption. Environmental Science and Technology. 2012; **46**:12936-12942, DOI: 10.1021/es302948g.

[13] Liao, C.-M., Jau, S.-F., Chen, W.-Y., Lin, C.-M., Jou, L.-J., Liu, C.-W., Liao, V.-H.-C., & Chang, F.-J. Acute toxicity and bioaccumulation of arsenic in freshwater clam *Corbicula fluminea*. Environmental Toxicology. 2008; **23**:702-711, http://ntur.lib.ntu.edu.tw/handle/246246/245021.

[14] Chakraborty, S., Ray, M., & Ray, S. Cell to organ: Physiological, immunotoxic and oxidative stress responses of *Lamellidens marginalis* to inorganic arsenite. Ecotoxicology and Environmental Safety. 2013; **94**:153-163, http://dx.doi.org/10.1016/j.ecoenv.2013.04.012.

[15] Wu, H., Zhang, X., Wang, Q., Li, L., Jib, C., Liu, X., Zhao, J., & Yin, X. A metabolomic investigation on arsenic-induced toxicological effects in the clam *Ruditapes philippinarum* under different salinities. Ecotoxicology and Environmental Safety. 2013; **90**:1-6, http://dx.doi.org/10.1016/j.ecoenv.2012.02.022.

[16] Zhang, W., Guo, Z., Zhou, Y., Liu, H., & Zhang, L. Biotransformation and detoxification of inorganic arsenic in Bombay oyster *Saccostrea cucullata*. Aquatic Toxicology. 2015; **158**:33-40, http://dx.doi.org/10.1016/j.aquatox.2014.10.021.

[17] Soucek, D.J., & Kennedy, A.J. Effects of hardness, chloride, and acclimation on the acute toxicity of sulfate to freshwater invertebrates. Environmental Toxicology and Chemistry. 2005; **24**:1204-1210, DOI: 10.1897/04-142.1

[18] Soucek, D.J. Sodium sulfate impacts feeding, specific dynamic action, and growth rate in the freshwater bivalve *Corbicula fluminea*. Aquatic Toxicology. 2007; **83**:315-322, http://dx.doi.org/10.1016/j.aquatox.2007.05.006.

[19] Lindberg, T.T., Bernhardt, E.S., Bier, R., Helton, A.M., Merola, B., Vengosh, A., & Guilio, R.T. Cumulative impacts of mountaintop mining on an Appalachian watershed. Proceedings of the National Academy of Sciences USA. 2011; **108**:20929-20934, DOI: 10.1073/pnas.1112381108.

[20] Ruhl, L., Vengosh, A., Dwyer, G.S., Hsu-Kim, H., & Deonarine, A. Environmental impacts of the coal ash spill in Kingston, Tennessee: an 18-month survey. Environmental Science and Technology. 2010; **44**:9272-9278, DOI: 10.1021/es1026739.

[21] Costa, P.M., Santos, H.M., Peres, I., Costa, M.H., Alves, S., Capedo-Martinez, J.L., & Diniz, M.S. Toxicokinetics of waterborne trivalent arsenic to the freshwater bivalve *Corbicula fluminea*. Archives of Environmental Contamination and Toxicology. 2009; **57**:338-347, DOI: 10.1007/s00244-008-9267-6.

[22] Barnhart, M.C. Buckets of muckets: a compact system for rearing juvenile freshwater mussels. Aquaculture. 2006; **254**:227-233, http://dx.doi.org/10.1016/j.aquaculture.2005.08.028.

[23] Rogers, J.J. Assessment of mussel declines in the Clinch and North Fork Holston Rivers using histological evaluations of vital organs. M.S. thesis, Virginia Tech University, Blacksburg, Virginia. 2015, https://vtechworks.lib.vt.edu/handle/10919/63925.

[24] Ciparis, S., Phipps, A., Soucek, D.J., Zipper, C.E., & Jones, J.W. Effects of environmentally relevant mixtures of major ions on a freshwater mussel. Environmental Pollution. 2015; **207**:280-287, http://dx.doi.org/10.1016/j.envpol.2015.09.023.

[25] Aronesty, E. Comparison of sequencing utility programs. The Open Bioinformatics Journal. 2013; **7**:1-8, DOI: 10.2174/1875036201307010001.

[26] Grabherr, M.G., Haas, B.J., Yassour, M., Levin, J.Z., Thompson, D.A., Amit, I., Adiconis, X., Fan, L., Raychowdhury, R., Zeng, Q., Chen, Z., Mauceli, E., Hacohen, N., Gnirke, A., Rhind, N. di Palma, F., Birren, B.W., Nusbaum, C., Lindblad-Toh, K., Friedman, N., & Regev, A. Full-length transcriptome assembly from RNA-seq data without a reference genome. Nature Biotechnology. 2011; **29**:644-652, DOI:10.1038/nbt.1883.

[27] Langmead, B., Trapnell, C., Pop, M., & Salzberg, S.L. Ultrafast and memory-efficient alignment of short DNA sequences to the human genome. Genome Biology. 2009; **10**(3):R25, DOI: 10.1186/gb-2009-10-3-r25.

[28] Li, B., & Dewey, C.N. RSEM: accurate transcript quantification from RNA-Seq data with or without a reference genome. BMC Bioinformatics. 2011; **12**:323. DOI: 10.1186/1471-2105-12-323.

[29] Robinson, M.D., McCarthy, D.J, & Smyth, G.K. edgeR: a Bioconductor package for differential expression analysis of digital gene expression data. Bioinformatics. 2010; **26**(1):139-140. http://bioconductor.org/packages/release/bioc/html/edgeR.html.

[30] Bancroft, J.D., & Gamble, M. Theory and Practice of Histological Techniques. 2002; Churchill Livingston, London, 796 pp, ISBN 978-0-443-10279-0.

[31] Chalkey, H.W. Method for the quantitative morphology analysis of tissues. Journal of the National Cancer Institute. 1943; **4**:47-53.

[32] USEPA (U.S. Environmental Protection Agency). Aquatic life ambient water quality criteria for ammonia – freshwater. EPA-822-R-13-001, Office of Water, Office of Science and Technology, U.S. Environmental Protection Agency, Washington, DC. 2013.http://water.epa.gov/scitech/swguidance/standards/criteria/aqlife/ammonia/upload/AQUATIC-LIFE-AMBIENT-WATER-QUALITY-CRITERIA-FOR-AMMONIA-FRESHWATER-2013.pdf.

[33] Bai, Z., Yin, Y., Hu, S., Wang, G., Zhang, X., & Li, J. Identification of genes involved in immune response, microsatellite, and SNP markers from expressed sequence tags generated from hemocytes of freshwater pearl mussel (*Hyriopsis cumingii*). Marine Biotechnology. 2009; **11**(4):520-530, DOI: 10.1007/s10126-008-9163-0.

[34] Bai, Z., Zheng, H., Lin, J., Wang, G., & Li, J. Comparative analysis of the transcriptome in tissues secreting purple and white nacre in the pearl mussel *Hyriopsis cumingii*. PLoS One. 2013; 8(1):e53617, DOI:10.1371/journal.pone.0053617.

[35] Duan, Z., Foster, R., Brakora, K.A., Yusuf, R.Z., & Seiden, MV. GBP1 overexpression is associated with a paclitaxel resistance phenotype. Cancer and Chemotherapy. 2006; **57**:25-33, DOI: 10.1007/s00280-005-0026-3.

[36] Luo, X., & Kraus, K.L. On PAR with PARP: cellular stress signaling through poly(ADP-ribose) and PARP-1. Genes and Development. 2012; **26**:417-432, DOI: 10.1101/gad.183509.111.

[37] Joshi, R.S., Tanpure, R.S., Singh, R.K., Gupta, V.S., & Giri, A.P. Resistance through inhibition: ectopic expression of serine protease inhibitor offers stress tolerance via delayed senescence in yeast cell. Biochemistry and Biophysics Research Communications. 2014; **452**:361-368, http://dx.doi.org/10.1016/j.bbrc.2014.08.075.

[38] Chang, Q., Griest, T.A., Harter, T.M., & Petrash, J.M. Functional studies of aldo-keto reductases in *Saccharomyces cerevisiae*. Biochimica et Biophysica Acta. 2008; **1773**:321-329, http://dx.doi.org/10.1016/j.bbamcr.2006.10.009.

[39] Hamadeh, H.K., Bushel, P., Paules, R., & Afshari, C.A. Discovery in toxicology: mediation by gene expression array technology. Journal of Biochemical and Molecular Toxicology. 2001; **15**(5):231-242, DOI: 10.1002/jbt.10006.

[40] Hamadeh, H.K., Bushel, P.R., Jayadev, S., Martin, K., DiSorbo, O., Sieber, S., Bennett, L., Tennant, R., Stoll, R., Barrett, J.C., & Blanchard, K. Gene expression analysis reveals chemical-specific profiles. Toxicological Sciences. 2002; **67**(2):219-231, DOI: 10.1093/toxsci/67.2.219.

[41] National Center for Biotechnology Information. Gene. 2015; Accessed September 9, 2015. http://www.ncbi.nlm.nih.gov/gene/?item+loc101731886.

[42] Vyas, S., Matic, I., Uchima, L., Rood, J., Zaja, R., Hay, R.T., Ahel, I., & Chang, P. Family-wide analysis of poly(ADP-ribose) polymerase activity. Nature Communications. 2014; **5**(4426):1-13, DOI: 10.1038/ncomms5426.

[43] Livingstone, D.R. Oxidative stress in aquatic organisms in relation to pollution and aquaculture. Revue de Médecine Vétérinaire. 2003; **153**:427-430, http://www.revmedvet.com/2003/RMV154_427_430.pdf.

[44] Barzilai, A., & Yamamoto, K. DNA damage responses to oxidative stress. DNA Repair. 2004; **3**:1109-1115, http://dx.doi.org/10.1016/j.dnarep.2004.03.002.

[45] Terman, A., & Brunk, U.T. Lipofuscin. The International Journal of Biochemistry and Cell Biology. 2004; **36**:1400-1404, http://dx.doi.org/10.1016/j.biocel.2003.08.009.

[46] Franzellitti, S., Suratti, S., Donnini, F., & Fabbri, E. Exposure of mussels to a polluted environment: insights into the stress syndrome development. Comparative Biochemistry and Physiology, Part C. 2010; **152**:24-33, http://dx.doi.org/10.1016/j.cbpc.2010.02.010.

[47] Uniprot. BRAFLDRAFT_118372 – Putative uncharacterized protein from Florida lancelet. Accessed September 9, 2015. http://www.uniprot.org/uniprot/C3YYL8.

[48] Uniprot. CGI_10002926 – Uncharacterized protein from Pacific oyster. 2015. Accessed September 9, 2015. http://www.uniprot.org/uniprot/K1QSJ7.

[49] Uniprot. Sp-Hypp_8991 – Uncharacterized protein from Pacofoc oyster. 2015. Accessed September 9, 2015. http://www.uniprot.org/uniprot/W4Y7B4.

[50] Cornman, R.S., Robertson, L.S., Galbraith, H., & Blakeslee, C. Transcriptomic analysis of the mussel *Elliptio complanata* identifies candidate stress-response genes and an abundance of novel or non-coding transcripts. PLoS One. 2014; **9**(22):e112420, http://dx.doi.org/10.1371/journal.pone.0112420.

Chapter 6

# Spatiotemporal Neural Activities Involved in the Olfactory Processing of the Land Slug using Fluorescent-Imaging Technique

Minoru Saito

Additional information is available at the end of the chapter

http://dx.doi.org/10.5772/intechopen.69459

**Abstract**

The brain or central nervous system forms a network composed of so many neurons and their function is based on complex interactions among electric neural activities, intracellular calcium signals, intercellular communications by neurotransmitter, and so on. For multi-point measurement of neural activities, fluorescent-imaging technique using voltage-sensitive dyes or calcium-sensitive dyes can be a powerful technique. This technique has been applied to measure spatiotemporal neural activities involved in the olfactory processing of the land slug *Limax*. In *Limax*, the procerebral (PC) lobe, which is the olfactory center located in the lateral part of each cerebral ganglion, spontaneously produces a periodic oscillation of local field potential (LFP) of about 1 Hz, and the phase of the LFP oscillation is advanced at the distal region, resulting in periodic propagating waves of neural activity from the distal to proximal regions. The previous studies showed that odor stimuli change the LFP frequency and the wave propagation speed. In this article, we review the previous studies, as well as our recent studies, on the spatiotemporal neural activities of the land slug *Limax* using fluorescent-imaging technique.

**Keywords:** land slug, olfactory processing, spatiotemporal neural activity, fluorescent-imaging technique, voltage-sensitive dye, calcium-sensitive dye

## 1. Introduction

The conventional electrophysiological methods have been well used for the measurement of neural activities of vertebrate and invertebrate. The direct measurement from single neurons and neural areas is possible by the conventional methods using microelectrodes. However, the brain or central nervous system forms a network composed of so many neurons and their function is based on complex interactions among electric neural activities, intracellular calcium signals, intercellular communications by neurotransmitter, etc. The electrophysiological methods are not suitable for the simultaneous measurement from many neurons or neural areas. For multi-point measurement of neural activities, fluorescent-imaging technique using voltage-sensitive dyes or calcium-sensitive dyes can be a powerful technique. In this technique, the dyes, which change their fluorescent intensities due to the voltage or calcium ion concentration change [1], are loaded into the cells, and their fluorescence changes are acquired into a computer as a series of images.

The multi-point measurement using fluorescent-imaging technique has been reported in vertebrates [2–5] and invertebrates [6–16]. In invertebrates, for example, this technique has been applied to measure spatiotemporal neural activities involved in the olfactory processing of the land slug *Limax* [6, 7, 9–11, 13–16]. In *Limax*, the procerebral (PC) lobe, which is the olfactory center located in the lateral part of each cerebral ganglion, spontaneously produces a periodic oscillation of local field potential (LFP) of about 1 Hz. The LFP oscillation is well synchronized over the entire PC lobe, but the phase of oscillation is advanced at the distal region, resulting in periodic propagating waves of neural activity from the distal to proximal regions. The previous studies showed that odor stimuli change the LFP frequency and the wave propagation speed.

In this article, we review the previous studies, as well as our recent studies, on the spatiotemporal neural activities of the land slug *Limax* using fluorescent-imaging technique.

## 2. Fluorescent-imaging technique

In fluorescent-imaging technique, the dyes, which change their fluorescence intensities due to the voltage or calcium ion concentration change [1], are loaded into cells, and their fluorescence changes are acquired into a computer as a series of images. However, the voltage-sensitive dyes generally exhibit a relatively small change in the fluorescence intensities, resulting in a low signal-to-noise (S/N) ratio. On the other hand, the calcium-sensitive dyes exhibit a larger change in the fluorescence intensities than that of the voltage-sensitive dyes. Therefore, their fluorescence changes can be detected easily, which enables us to indirectly measure neural activities because the intracellular calcium concentration often increases with neural activities.

As the calcium-sensitive dyes, those bonded with the acetoxymethyl (AM) group are commonly used because they are permeable into cells. After the AM-bonded dyes permeate into cells, the AM group is dissociated by intracellular esterase, and then the dyes can be non-permeable and loaded into cells. Vertebrate neurons are well known to be easily loaded with the AM-bonded calcium-sensitive dyes. In invertebrates, however, the AM-bonded dyes cannot be easily loaded into neurons because the AM group is difficult to be dissociated due to their weak activity of intracellular esterase.

For these reasons, it has not been determined yet which type of dyes is better for fluorescent imaging of neural activities in invertebrates. In Section 4, we introduce some studies on the

http://dx.doi.org/10.5772/intechopen.69459

spatiotemporal neural activities of the land slug *Limax* using fluorescent-imaging technique with both types of dyes.

## 3. Neural systems for olfactory processing of the land slug

The land slug *Limax* has two pairs of tentacles, the superior and inferior tentacles. **Figure 1** shows a photograph of the central nervous system of *Limax* and the schematic illustration of the cerebral ganglion and superior tentacle. Each tentacle has an olfactory epithelium, which receives odors, at its tip. Olfactory receptor neurons are located at the tip regions of the digitate extensions of the tentacular ganglion (TG) (referring as "digitate ganglion" in **Figure 1B**) beneath the epithelium and terminate to synapse projection neurons in the TG that is the primary center of olfactory information processing (**Figure 1B**). The TG contains interneurons, as well as projection neurons that project to the cerebral ganglion [17]. The TG shows a spontaneous oscillation of LFP at about 2 Hz, which is usually measured by extracellular recording [18]. The oscillation is modulated by odor and air stimuli [19]. Several neurotransmitters such as serotonin, acetylcholine, glutamate, and gamma-aminobutyric acid (GABA), exist in the TG and modulate the spontaneous oscillation [20–22]. In addition, the olfactory epithelium also shows an oscillatory electro-olfactogram (EOG) in response to odor and air stimuli. The EOG oscillation interacts with the LFP oscillation in the TG [19].

TG neurons project to the cerebral ganglion via the tentacle or olfactory nerves. The major target of projection is the procerebral lobe, which is the lateral part of the cerebral ganglion (**Figure 1**). The PC lobe is a division of the cerebral ganglion unique to terrestrial slugs and snails that is specialized for the processing of olfactory information [23–25]. The PC lobe consists of three layers, the cell mass (CM), terminal mass (TM), and internal mass (IM) (**Figure 2**). The CM contains a large number of cell bodies of neurons, which are generally

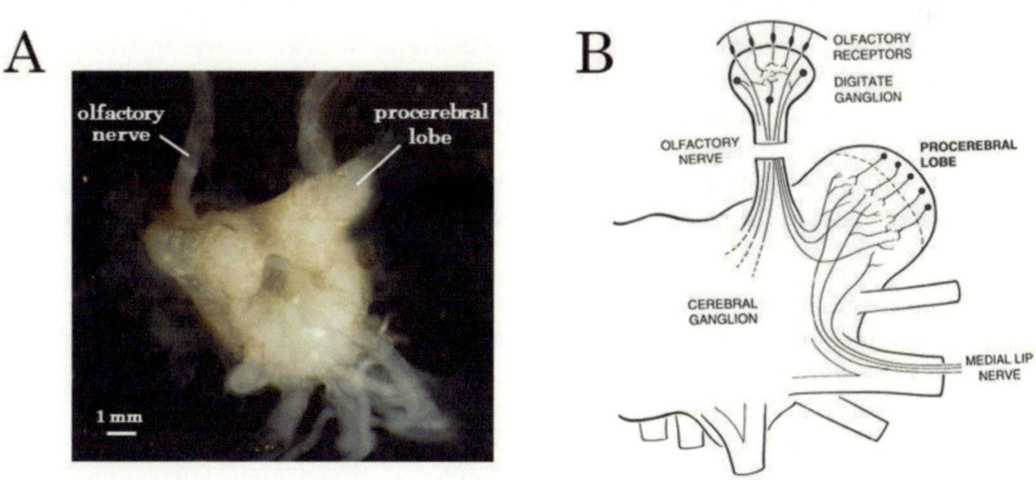


**Figure 1.** (A) Photograph of the central nervous system of the land slug *Limax* Color online and (B) schematic illustration of the cerebral ganglion and superior tentacle. This figure (B) is reproduced from Ref. [7] with permission.

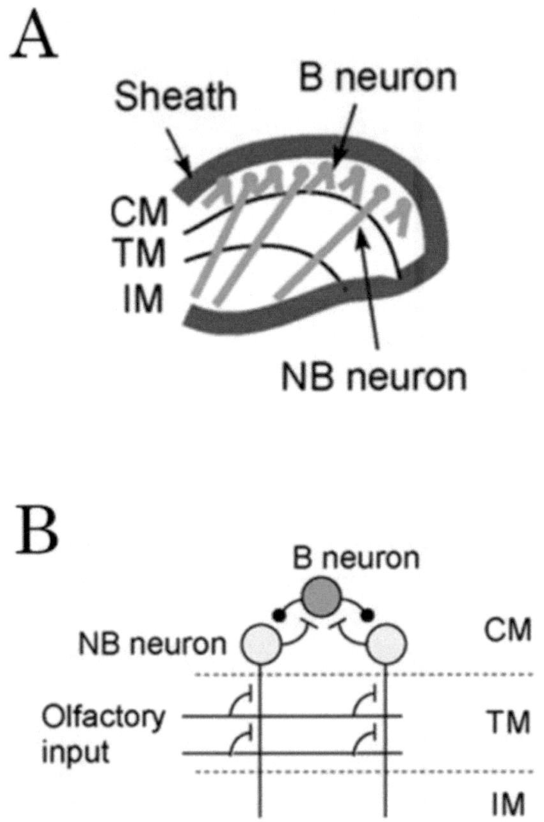


**Figure 2.** (A) Schematic illustration of the PC and (B) structure of the neural network of the PC. B neuron: bursting neuron, NB neuron: nonbursting neuron; CM: cell mass, TM: terminal mass, IM: internal mass. These figures are reproduced from Ref. [14] with permission.

small (5–8 μm) [23]. The TM and IM are neuropile layers. The TM receives input from the tentacle nerves [9]. The PC lobe spontaneously produces a periodic slow oscillation of LFP at about 1 Hz, which is usually measured by extracellular recording [26]. The LFP oscillation is well synchronized over the entire PC, but the phase of the oscillation is advanced at the distal region. This results in periodic propagating waves of neural activity from the distal to proximal regions, which has been clearly shown by fluorescent-imaging technique [6, 7, 9–11, 13–16].

Patch-clamp recording from single neurons showed that the neurons in the PC lobe are categorized as either bursting (B) or nonbursting (NB) neurons [6, 7] (**Figure 2**). B neurons are characterized by periodic bursting activity, and NB neurons are silent or fire at a low frequency. NB neurons are more numerous (about 90%) and have projection of neurites to the TM and IM [27]. NB neurons receive olfactory input from the tentacle nerves in the TM [28]. B neurons have extensive projection of neurites within the CM and inhibit NB neurons, while NB neurons excite B neurons. The spontaneous bursting activities of B neurons and inhibitory synaptic potentials in NB neurons are synchronized with the LFP oscillation.

The propagating waves of neural activity have also been found in mammalian cortex. However, the function and computational role are still unclear even in the land slug *Limax*. To elucidate it, fluorescent-imaging technique is expected to be a powerful technique as described in the subsequent section.

## 4. Fluorescent imaging of spatiotemporal neural activities involved in the olfactory processing of the land slug

As mentioned earlier, periodic waves of neural activity propagate from the distal to proximal regions in the PC lobe of the land slug *Limax*. Pioneer works of a group clearly showed this phenomenon for the first time in the PC lobe of *L. maximus* by fluorescent voltage imaging using a voltage-sensitive dye [6, 7]. They used di-4-ANEPPS, a substituted aminonaphthylethenylpyridinium [29], as the fluorescent voltage-sensitive dye. The fluorescence was detected with either an array of photodiodes or a cooled charge-coupled device (CCD) camera. **Figure 3** shows the experimental apparatus that they used. **Figure 4** shows the propagating waves of fluorescence change measured by the CCD camera. A band of depolarization, followed by a band of hyperpolarization, began at the distal tip and moved along the length of the PC lobe, resulting in a phase difference between the distal and proximal regions (**Figure 4A** and **C** ). The detailed shape of the optical signal as a function of time also varied between different regions (**Figure 4B**). Simultaneous recording of the optical signal and the LFP by extracellular recording showed that the oscillatory changes in the fluorescent intensity were synchronized to the LFP oscillation (**Figure 5**). Additionally, from simultaneous recording of the optical signal and the intracellular potential by whole-cell patch technique (**Figure 6**), the authors concluded that the major part of the optical signal can be interpreted as a superposition of the intracellular signals arising from the B and NB neurons.

They next examined the responses of spatiotemporal neural activity in the PC lobe to odor stimuli (**Figure 7**). The stimuli were applied to the epithelium of the superior tentacle. The phase difference of neural activity between the distal and proximal regions was not changed in response to the moist air (**Figure 7A**). However, in response to a component of natural potato odor (2-ethyl-3-methoxypyrazine; EMOP), a known appetitive stimulus to the naïve slugs, the phase difference was dramatically changed (**Figure 7B–F**). The stimulus caused a collapse of the phase difference, that is, an initial difference of $\Delta t > 250$ ms was reduced to $\Delta t < 50$ ms. Repeated stimulus also caused a collapse of the phase difference, although it recovered faster after cessation of the stimulus. The transient switch from the states with propagating waves to one with spatially uniform oscillations is hypothesized to allow nearest-neighbor interactions suppressed by the propagating wave activity [30]. This study suggested that olfactory information is encoded in the spatiotemporal neural activities in the PC lobe.

Thereafter, other groups have also studied the spatiotemporal neural activities of the PC lobe using fluorescent voltage imaging. Kimura et al. examined the responses of the PC lobe of *L. marginatus* to aversively conditioned odors [10]. In this study, di-4-ANEPPS was also used as the voltage-sensitive dye. First, they performed aversive conditioning procedures to the slugs using two odors (carrot and cucumber, which strongly attract the naïve slugs)

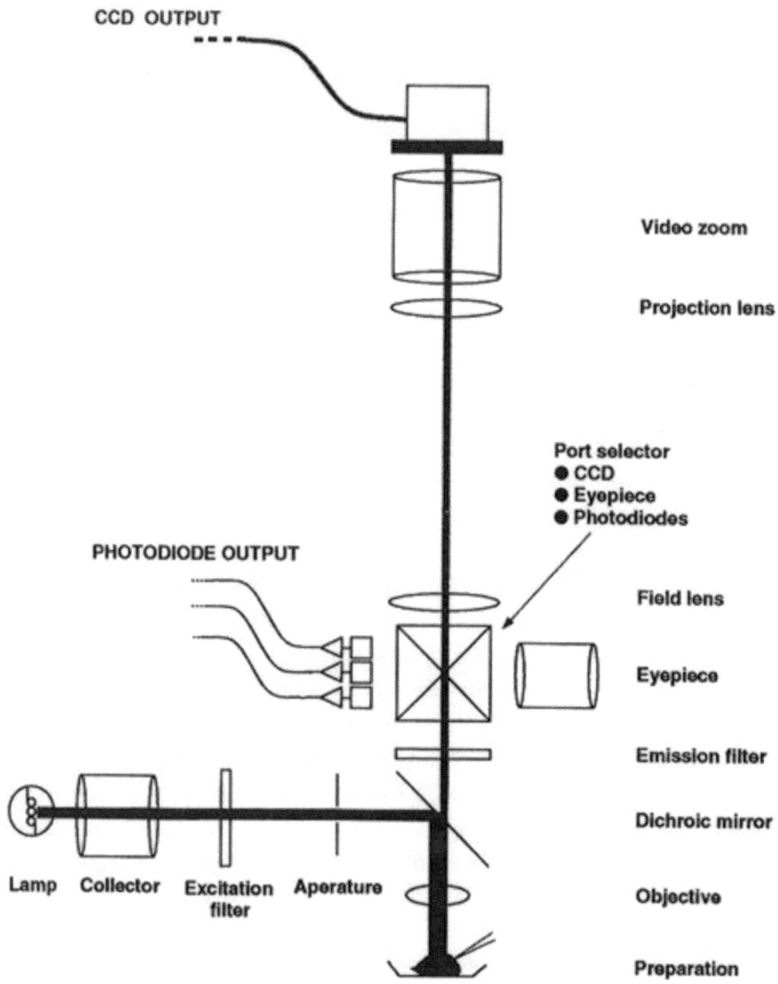


**Figure 3.** Schematic illustration of the experimental apparatus used for the optical measurement. This figure is reproduced from Ref. [7] with permission.

as conditioned stimuli (CS) and quinidine sulfate solution, which induces a strong aversive effect, as an unconditioned stimulus (UCS). The odor stimuli were applied to the epithelium of the inferior tentacle. When the conditioning is completed, the slugs avoid the CS odor and exhibit eating behavior at a lower frequency [31]. In the PC of the naïve slugs, they observed the propagating waves of fluorescence change similar to those reported in the previous study [6, 7] (**Figure 8**). In the PC lobe of the slugs aversively conditioned by carrot odor, the instantaneous frequency of the oscillatory fluorescence change was decreased from about 4 s after the onset of conditioned carrot odor presentation (**Figure 9a**). The optical signals recorded in an area of the PC lobe (area 3 in the upper image of **Figure 9**) showed depolarization in the basal potential (bottom level of each cycle of the oscillation) immediately after the onset of the odor presentation (**Figure 9d**). In another area of the PC lobe (area 4 in the upper image of **Figure 9**), hyperpolarization in the basal potential occurred from about 5 s after the carrot

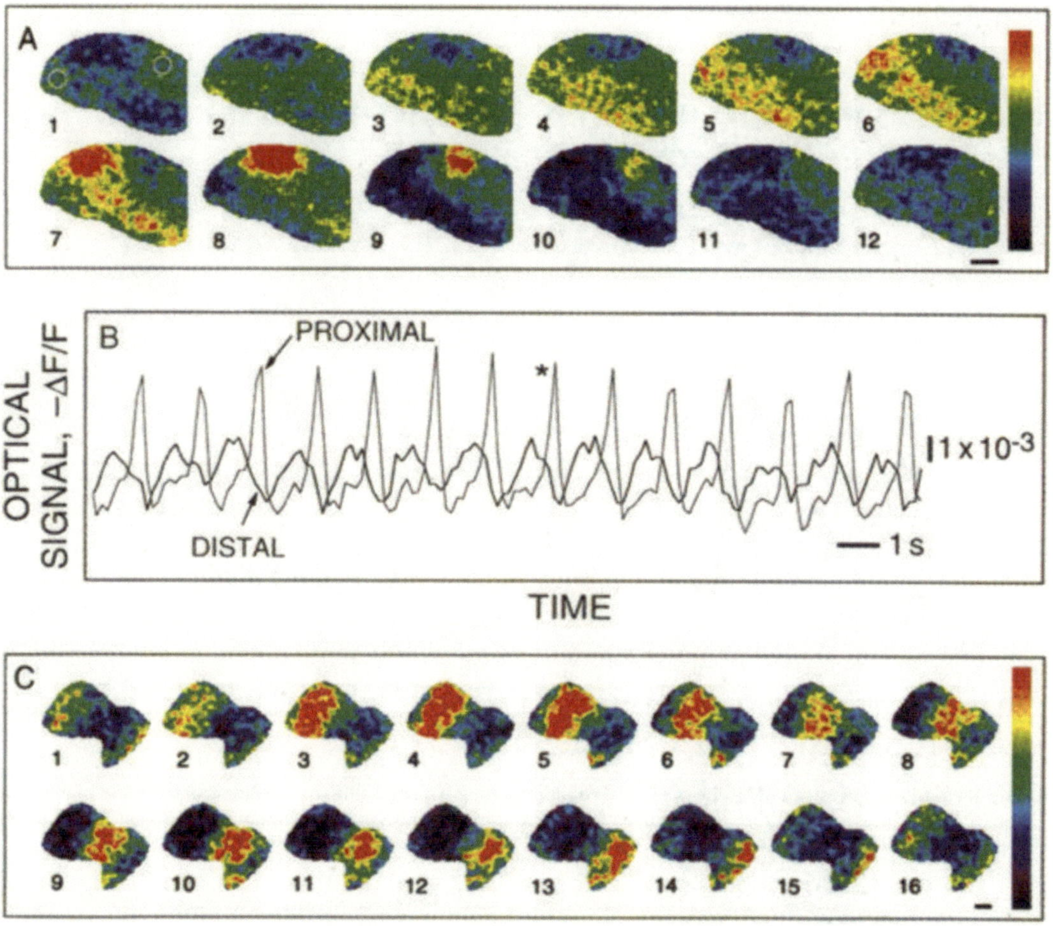


**Figure 4.** Propagation waves of fluorescence change in the PC lobe Color online. (A) Lateral view of fluorescence changes ($-\Delta F/F$) plotted as successive frames. The fluorescence images were acquired as 100 × 100 pixel images at a flame rate of 120 ms. The scale bar shows 100 μm. (B) Fluorescence changes versus time for the distal (thick line) and proximal (thin line) regions in the lobe. Each signal is the average signal from each circled region in (A). The asterisk (*) corresponds to flames 8 in (A). (C) Posterior view of fluorescence changes. The fluorescence images were acquired as 100 × 100 pixel images at a flame rate of 112 ms. The scale bar shows 100 μm. In (A) and (C), yellow/red indicate depolarization and blue/violet indicate hyperpolarization relative to the average voltage. These figures are reproduced from Ref. [6] with permission.

odor presentation (**Figure 9e**). On the other hand, the control cucumber odor did not significantly induce those changes (**Figures 9f–j**). **Figure 10** shows temporally average images of optical signals calculated in 10 different periods of two recordings shown in **Figure 9** (periods a–j in the upper graphs of **Figure 10**). This result shows that depolarization occurred especially in the neurons located in a belt-shaped region of the middle of the PC lobe (red-colored region in image b in **Figure 10**) immediately after the onset of conditioned carrot odor presentation. The belt-shaped region corresponded to area 3 in the upper image of **Figure 9**, in which depolarization in the basal potential occurred in the time profile, while the early phase depolarization was not observed for the control odor cucumber. After the depolarization was diminished, hyperpolarization occurred in the neurons in a relatively wide area of the PC lobe

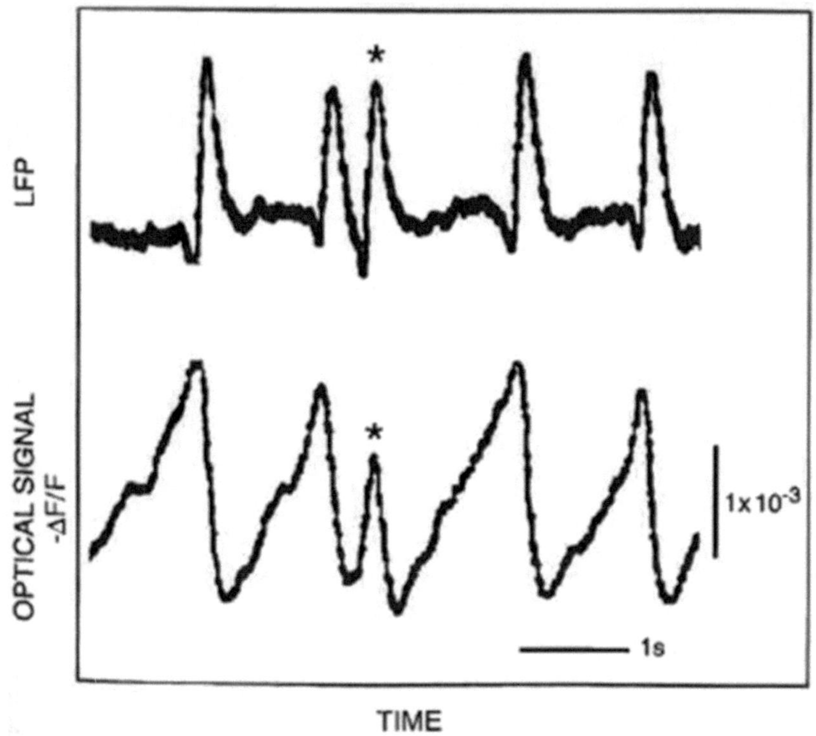


**Figure 5.** Simultaneous recording of the optical signal and the LFP by extracellular recording from a mid-distal region in the PC lobe. The asterisk (*) shows a "double event" that simultaneously presents in both recordings. This figure is reproduced from Ref. [7].

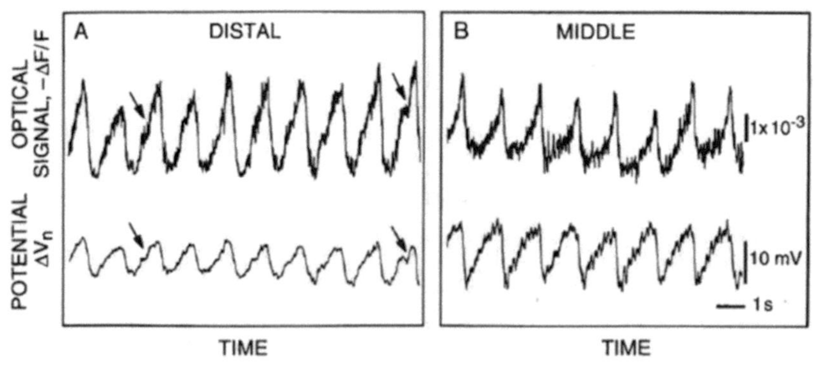


**Figure 6.** Simultaneous recording of the optical signal and the intracellular potential by whole-cell patch technique. (A) Recording from the distal region in the PC lobe. Two pairs of arrows show "fast events" that are simultaneously present in both recordings. (B) Recordings from the middle region in the PC lobe. These figures are reproduced from Ref. [7].

(blue-colored region in images c and d in **Figure 10**). Similar results were obtained in the PC lobe of the slugs aversively conditioned by cucumber odor. From these results, the authors concluded that in the PC lobe, learned odors are represented as spatiotemporal activity patterns, which is supported by activity-dependent dye uptake study [32].

http://dx.doi.org/10.5772/intechopen.69459

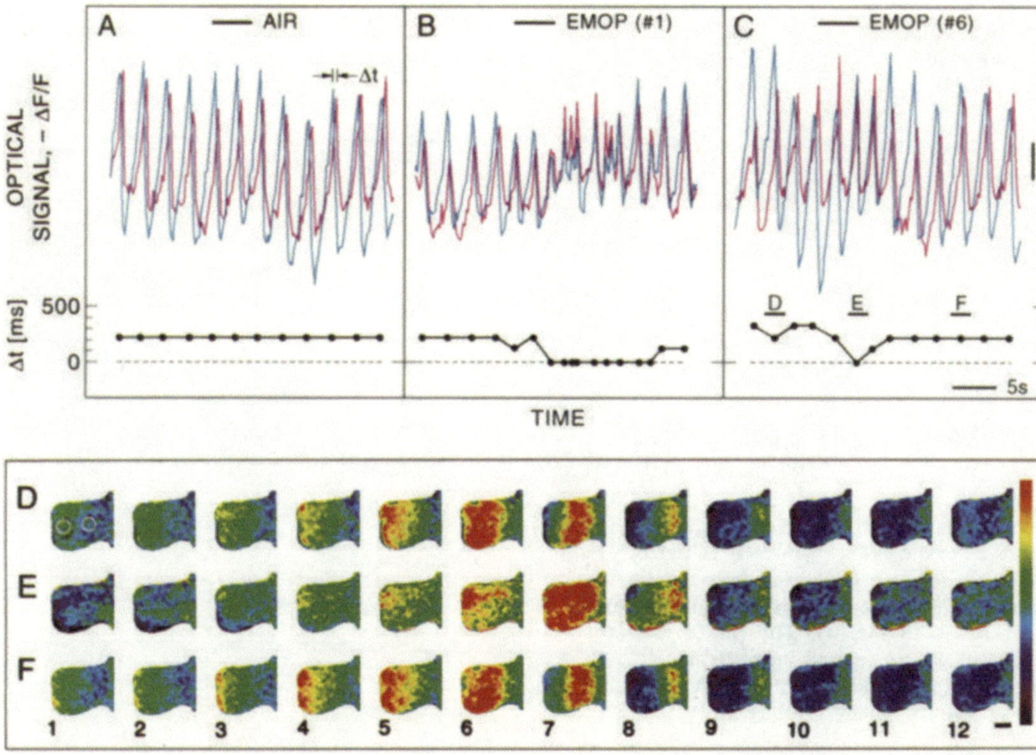


**Figure 7.** Fluorescent optical recordings of neural activity in the PC lobe in response to natural odor stimuli Color online. (A) Fluorescence changes ($-\Delta F/F$) versus time during the exposure to moist air (bar mark). The changes are shown for the distal (blue line) and proximal (red line) regions indicated by the circles in (D). (B) Fluorescence changes ($-\Delta F/F$) versus time during the first exposure to EMOP (bar mark). (C) Fluorescence changes ($-\Delta F/F$) versus time during the sixth exposure to EMOP (bar mark). Each lower plot in (A)–(C) shows the time difference (Δt) between successive peaks in the optical signal for each period. (D–F) Images of the fluorescence changes corresponding to the data in (C). They were taken (D) prior, (E) during, and (F) after the odor stimulation, which correspond to the letters D–F in (C). The fluorescence images were acquired as 100 × 100 pixel images at a flame rate of 112 ms. The scale bar shows 100 μm. These figures are reproduced from Ref. [6] with permission.

Watanabe et al. measured neural activities of other areas in the cerebral ganglion, metacerebrum/mesocerebrum (MC) (**Figure 9**), together with the PC lobe by fluorescent voltage imaging, and they analyzed the relationship between them using the correlation analysis [13]. In this study, di-4-ANEPPS was also used as the voltage-sensitive dye. **Figure 11** shows the fluorescent optical recording of spontaneous activity of the cerebral ganglion of *L. valentianus*. The image field covered the proximal part of the PC and the entire ipsilateral MC region. In all preparations, they observed oscillatory neural activity in the PC. Additionally, they found clear oscillatory activity in the MC with the same frequency as the PC oscillation in some preparations, although the amplitude of the MC oscillation was smaller than that of the PC oscillation. Their phase relationship was variable. In some preparations, the two oscillations were antiphase, while in the other preparations the oscillations were inphase. This study indicated that the oscillatory activity in the PC lobe, in which olfactory information is encoded as mentioned earlier, propagates to the MC that plays a role in motor command.

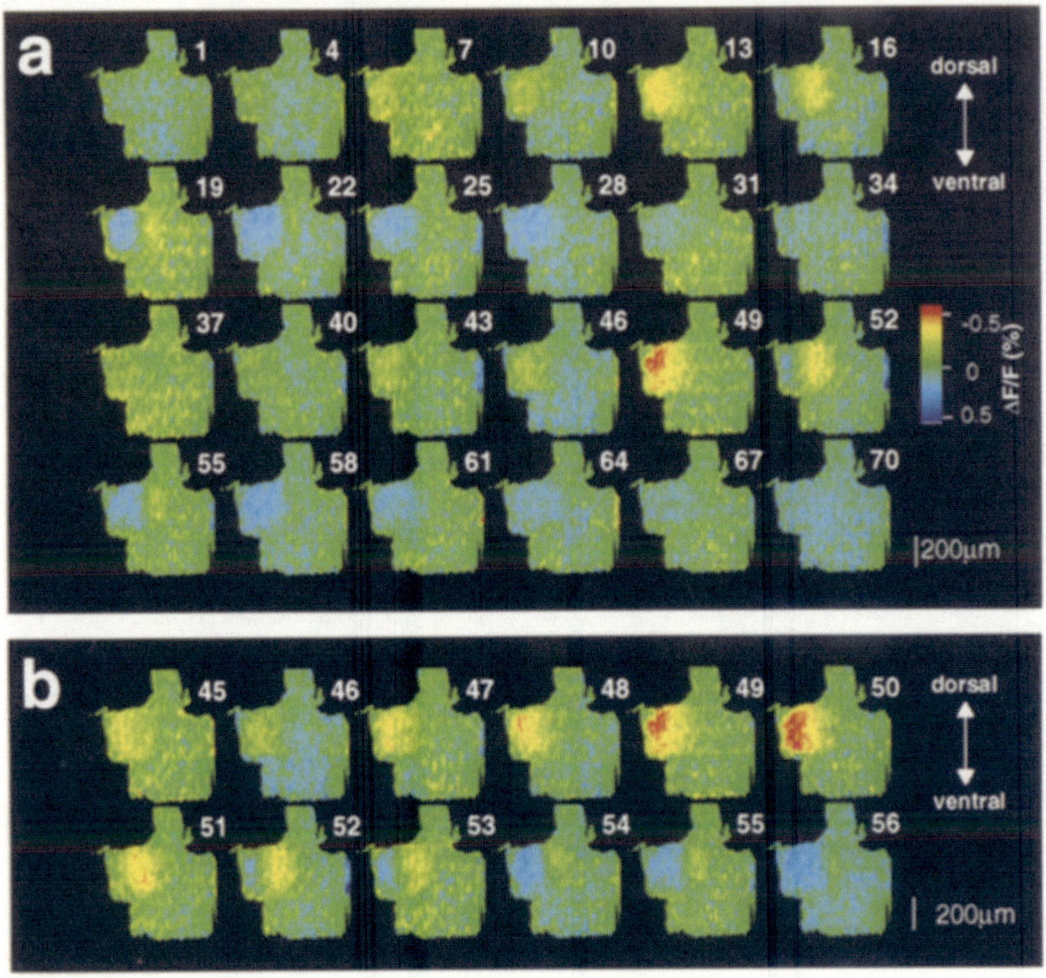


**Figure 8.** Propagation waves of fluorescence change in the PC lobe (posterior view) Color online. The fluorescence images were acquired by a photodiode array as 128 × 128 pixel images at a flame rate of 38.4 ms. Fluorescence changes ($\Delta F/F$) were plotted as successive frames. The $\Delta F/F$ value in each pixel is represented according to the color table. (a) Flames 1–70 with an interval of 115.4 ms. (b) Flames 45–56 with an interval of 38.4 ms. These figures are reproduced from Ref. [10] with permission.

The spatiotemporal neural activities in the PC lobe were also measured by fluorescent calcium imaging using calcium-sensitive dyes. Inoue et al. used Oregon Green-1, rhod-2, or Ca Orange bonded with AM group as the calcium-sensitive dye, and compared the fluorescent calcium imaging using each of them with the florescent voltage imaging using di-4-ANEPPS [11]. In fluorescent calcium imaging, propagation waves of fluorescence change were also observed in the PC lobe of *L marginatus* (**Figure 12**). The PC lobe was well stained by the calcium-sensitive dyes, even though AM group-bonded dyes were used. The calcium signals originated predominantly from the cell body layer, CM, while the voltage signals originated mainly from the neurophil layer, TM. The calcium-sensitive dyes gave 1–3% fluorescence change, which increased rapidly and decreased slowly, with a good signal-to-noise ratio (**Figure 13**). Among the calcium-sensitive dyes, rhod-2 gave the largest fluorescence oscillation. On the

http://dx.doi.org/10.5772/intechopen.69459

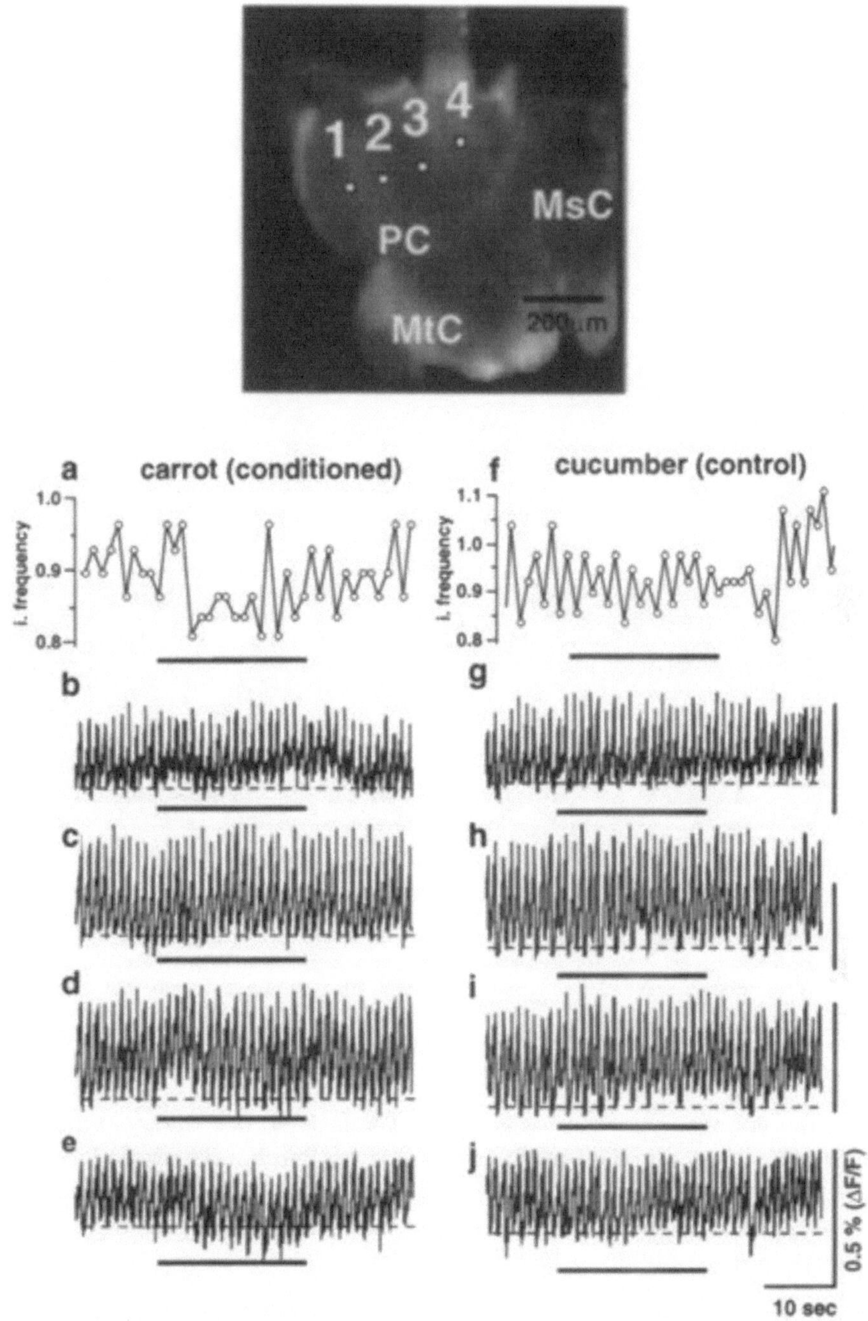


**Figure 9.** Olfactory responses in the fluorescence oscillation in the PC lobe. The preparation was isolated from a slug aversively conditioned by carrot odor. The optical signals were recorded from four regions (open squares 1–4 in the upper image) of the posterior surface. (a) Change in the oscillation frequency induced by conditioned carrot odor. (b–e) Fluorescence changes ($\Delta F/F$) versus time during the exposure to carrot odor (bar mark). (f) Change in the oscillation frequency induced by control cucumber odor. (g–j) Fluorescence changes ($\Delta F/F$) versus time during the exposure to cucumber odor (bar mark). (b) and (g), (c) and (h), (d) and (i), and (e) and (j) correspond to regions 1–4 in the upper image, respectively. MtC: metacerebrum, MsC: mesocerebrum. These figures are reproduced from Ref. [10] with permission.

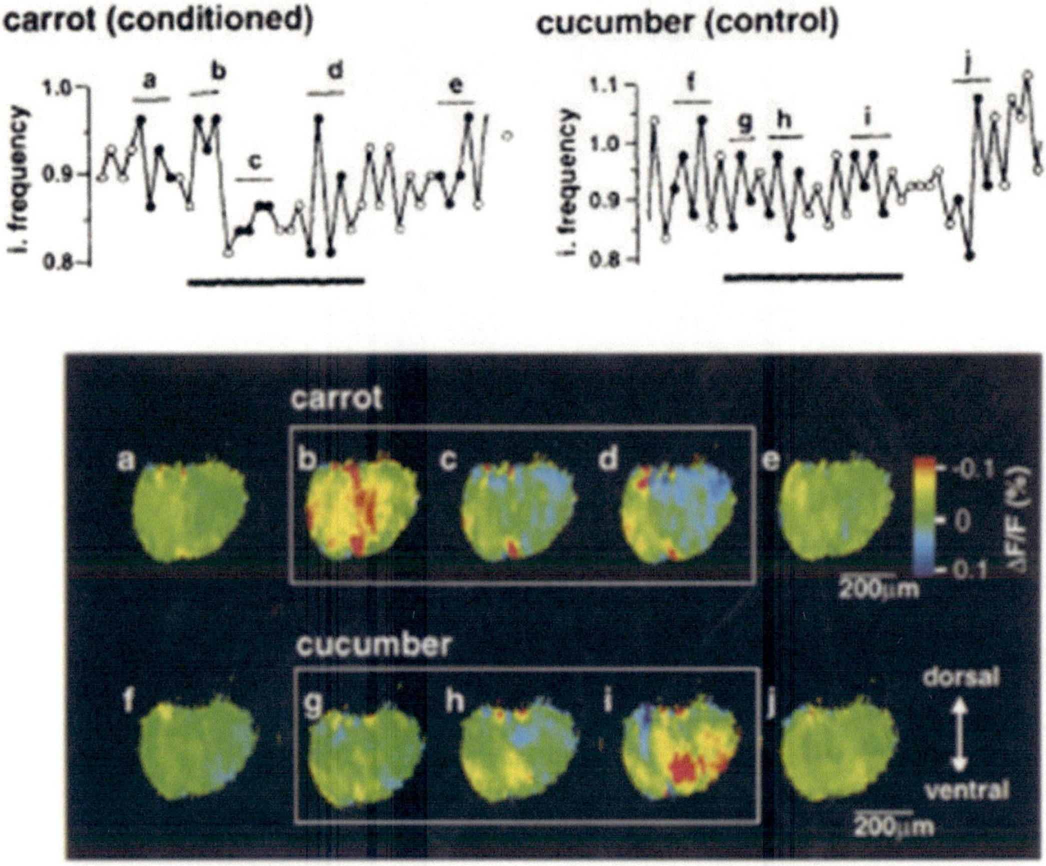


**Figure 10.** Spatiotemporal properties of olfactory responses in the PC lobe Color online. The upper two graphs show responses of the oscillation frequency to aversively conditioned carrot odor (left) and control cucumber odor (right). The lower images show temporally averaged $\Delta F/F$ images calculated from the optical recording shown in Figure 9. Successive images were averaged for 10 different periods (a–j in the upper figures, each of which includes three or four oscillatory cycles (•)). The images a–j corresponds to the periods a–j. These figures are reproduced from Ref. [10] with permission.

other hand, the voltage-sensitive dye gave 0.1–1% fluorescence change, which increased slowly and decreased rapidly (**Figure 13**). In this study, the authors concluded that the oscillatory signals of the calcium-sensitive dyes arise from B neurons of the CM and the neural activity of the CM can be monitored with them because the oscillatory signals were synchronized with the LFP (**Figure 13**).

Fluorescent calcium imaging has been limited with respect to the selectivity of staining of the dyes. Bath application of the dyes stains both B neurons and NB neurons. Even when we intend to observe the activity of NB neurons, which is thought to be involved in odor coding [33], it is obscured by stronger calcium signals from B neurons that exhibit periodic bursting activity. To observe the activity of NB neurons, Watanabe et al. attempted selective fluorescent calcium imaging for the PC lobe of *L. valentianus* by retrograde staining of NB neurons with rhod-2 [14]. When the sheath was removed only from the surface of the IM and the PC lobe was immersed in the dye solution, the NB neurons were expected to be stained

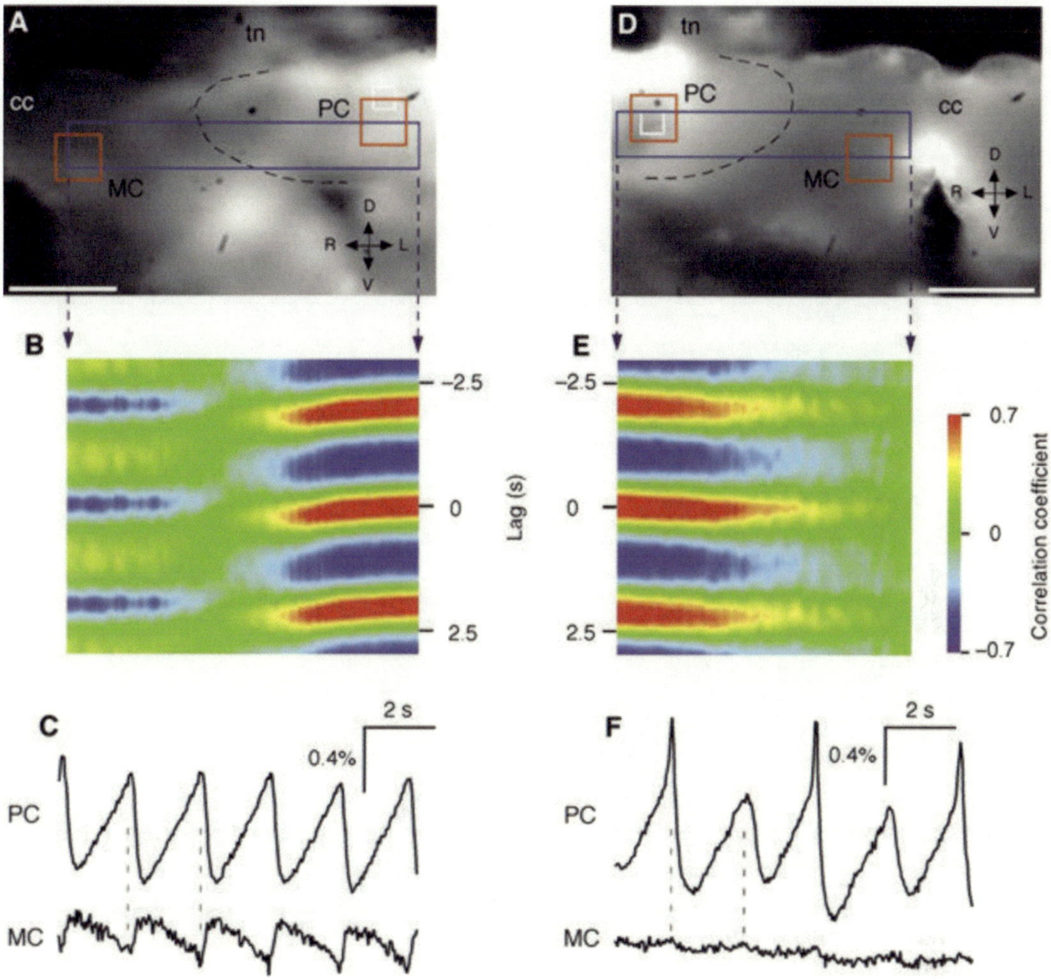


**Figure 11.** Fluorescent optical recording of spontaneous activity in the cerebral ganglion Color online. (A–C) A preparation that showed an antiphase MC oscillation. (A) Fluorescent image of the cerebral ganglion (left hemisphere) observed from the anterior surface. The scale bar shows 200 μm. tn: superior tentacle nerve, cc: cerebral commissure; D: dorsal, V: ventral, L: left, R: right. The broken line around the PC indicates the border between the PC and the MC. (B) Correlation profile along the area indicated by the blue rectangle in (A), with reference to the PC region (white square). The horizontal axis corresponds to the position along the area, and the vertical axis is the time lag. Positive values of the lag indicate delays relative to the PC. Pseudocolor presentation (the table on the right of (E)) refers to the amplitude of correlation against the PC oscillation. (C) Fluorescence changes versus time in the PC and MC regions indicated by the red squares in (A). Decrease in fluorescence intensity (depolarization of the membrane potential) is plotted upward. As the dotted lines indicate, the two traces show an antiphase relationship. (D–F) A preparation that showed an inphase MC oscillation. (D) Fluorescent image of the cerebral ganglion (right hemisphere) observed from the anterior surface. The scale bar shows 200 μm. Abbreviations are same as those in (A). The broken line around the PC indicates the border between the PC and the MC. (E) Correlation profile along the area indicated by the blue rectangle in (D), with reference to the PC region (white square). (F) Fluorescence changes versus time in the PC and MC regions indicated by the red squares in (D). As the dotted lines indicate, the two traces show an inphase relationship although the amplitude of the MC oscillation is small. These figures are reproduced from Ref. [13] with permission.

retrogradely through neurites projected to the IM (**Figure 14**). After nonselective staining, in which the sheath was removed from the entire surface of the PC lobe, a strong spontaneous oscillation was observed, which presumably arose from calcium oscillations of B neurons

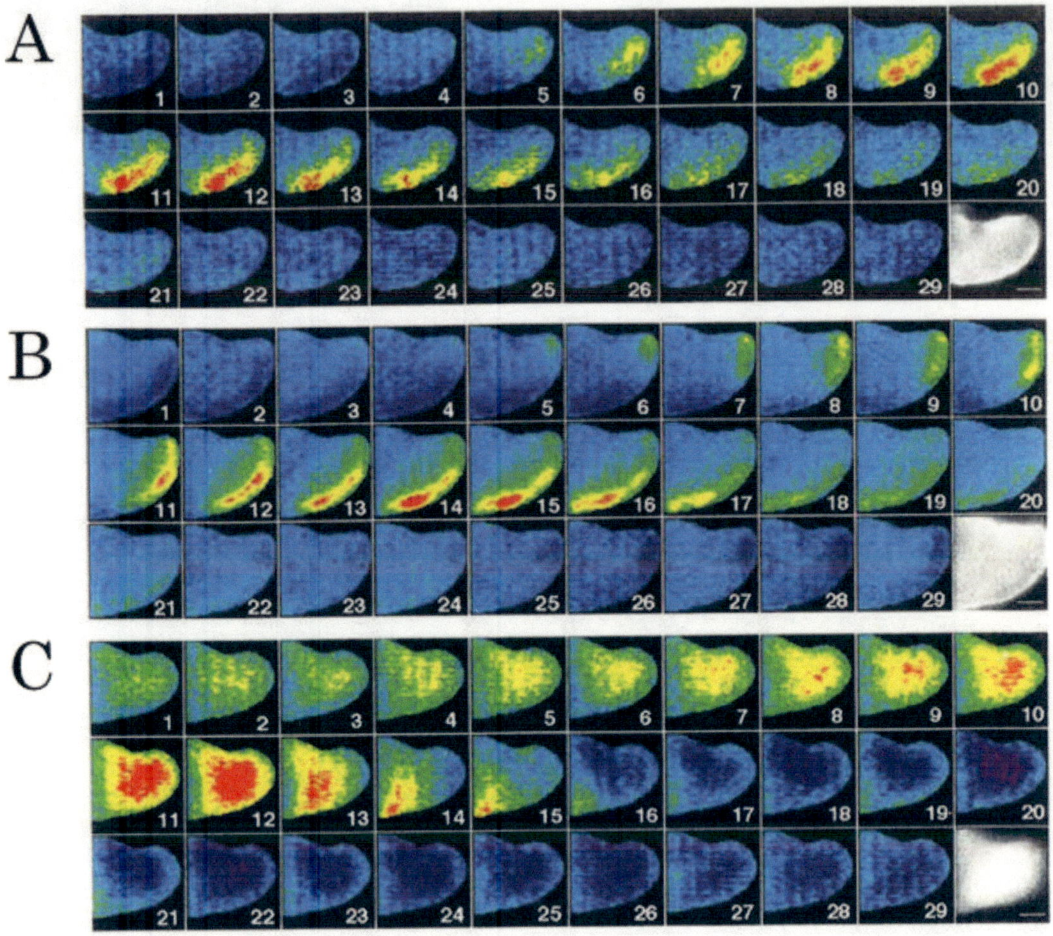


**Figure 12.** Propagation waves of fluorescence change in the PC lobe (dorsal view) stained with (A) Oregon Green-1, (B) rhod-2, or (C) di-4-ANEPPS Color online. The fluorescence images were acquired by a photodiode array as 128 × 128 pixel images at a flame rate of 38.4 ms. In (A) and (B), yellow/red indicates high calcium concentration and blue/violet indicates low calcium concentration. In (C), yellow/red indicates depolarization, and blue/violet indicates hyperpolarization. The scale bar shows 200 μm. These figures are reproduced from Ref. [11] with permission.

(**Figure 15**). On the other hand, the PC lobe showed a much smaller spontaneous oscillation after the selective staining, although a large number of cells were clearly stained (**Figure 15**). This result suggested that NB neurons are mostly silent in contrast to B neurons. However, NB neurons responded to electric stimulation of the superior tentacle nerve as shown in the inset of **Figure 15D**.

As described above, fluorescent-imaging technique is applicable for the measurement of the spatiotemporal neural activities involved in the olfactory processing of the land slug. Through the studies using it, especially, the function and computational roles of the propagation waves of neural activity in the PC lobe on odor discrimination and learning have been elucidated. Meanwhile, the experimental apparatus has also been improved. In the earlier reports of the fluorescent voltage imaging, consecutive images were acquired at a flame rate of about 100 ms

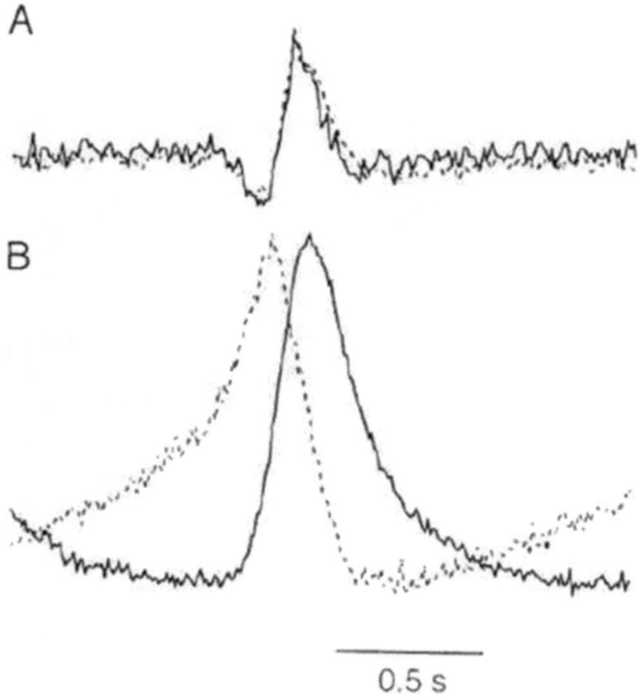


**Figure 13.** Phase relationship between the fluorescence oscillations of the calcium-sensitive dye (rhod-2) and the voltage-sensitive dye (di-4-ANEPPS) in the PC lobe. (A) LFP measured simultaneously with the optical recordings of the PC lobe stained with rhod-2 (solid line) or di-4-ANEPPS (dotted line). (B) Fluorescence changes of rhod-2 (solid line) and di-4-ANEPPS (dotted line) versus time. The fluorescence intensities were obtained by averaging over 10 × 10 pixels at the medial region of the CM. These figures are reproduced from Ref. [11] with permission.

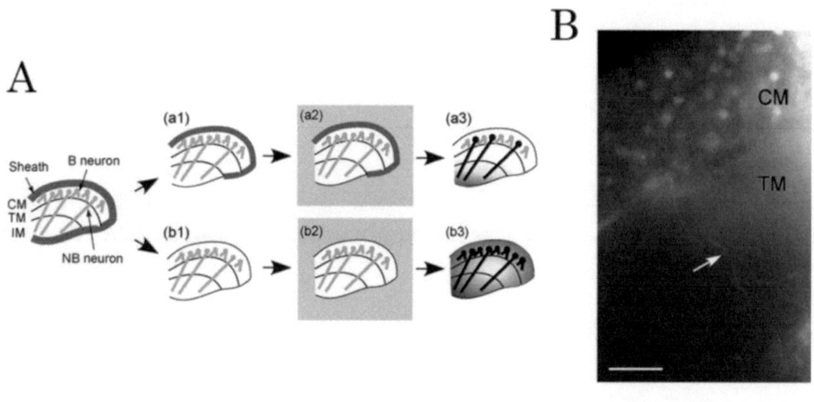


**Figure 14.** (A) Methods of staining. (a1–3) Selective staining of NB neurons. (a1) The sheath is removed only from the surface of the IM. (a2) The preparation is incubated in the rhod-2 solution. (a3) After staining, NB neurons but not B neurons are stained. (b1–3) Nonselective staining. (b1) The sheath is removed from the entire surface of the PC. (b2) The preparation is incubated in the rhod-2 solution. (b3) After staining, both B and NB neurons are stained. (B) Fluorescence image of a section of the PC lobe after staining. Cell bodies (bright spots) are stained in the CM. A neurite is also stained in the TM (arrow), through which the dye presumably diffused to stain the cell body. The scale bar shows 40 μm. These figures are reproduced from Ref. [14] with permission.

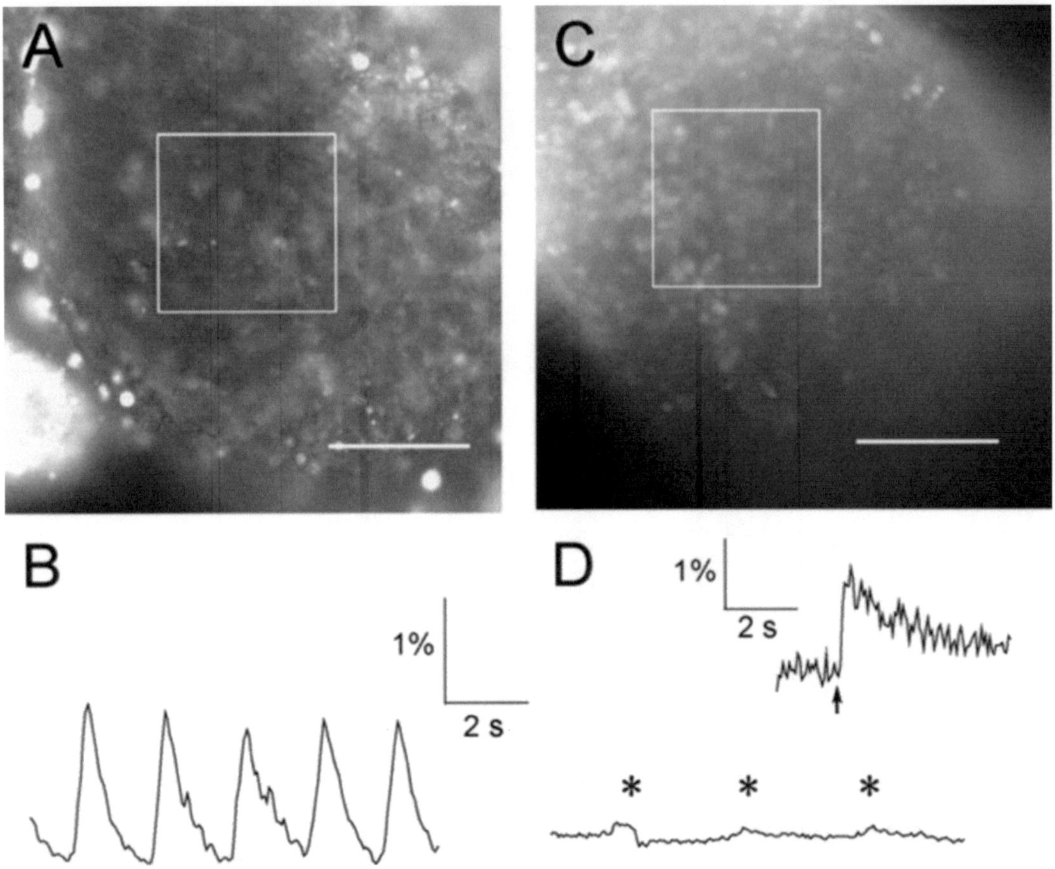


**Figure 15.** Comparison of selective and nonselective staining. Spontaneous activity was recorded from the surface of the CM in the PC lobe. (A and B) Nonselective staining. (A) Fluorescence image of the surface of the CM. (B) Time course of the averaged fluorescence from the square region (approximately 100 × 100 μm) in the center of (A). (C and D) Selective staining. (C) Fluorescence image of the surface of the CM. (D) Time course of the averaged fluorescence from the square region (approximately 100 × 100 μm) in the center of (C). The inset shows the time course of fluorescence in a cell in response to electric stimulus of the superior tentacle nerve (arrow). The scale bars in (A) and (C) show 100 μm. These figures are reproduced from Ref. [14] with permission.

for about 100 × 100 pixels in size [6, 7], while they can be acquired at a flame rate of about 10 ms for about 1000 × 1000 pixels in size in our recent apparatus as described in Section 5.

## 5. Our recent fluorescent-imaging apparatus and its application

We have also measured the spatiotemporal neural activities in the PC lobe of the land slug *Limax* by fluorescent voltage imaging [15, 16]. **Figure 16** shows our recent apparatus. We also use di-4-ANEPPS as the voltage-sensitive dye. The recording chamber, in which the stained preparation is placed, is mounted on the stage of a microscope (E-FN1, Nikon, Japan). The dye is excited by LED of 530 nm with a half width of 25 nm (LEX2-G, Brain Vision, Japan) through an excitation filter (EX510-560). The emitted fluorescence of 705 nm is detected through a dichroic

http://dx.doi.org/10.5772/intechopen.69459

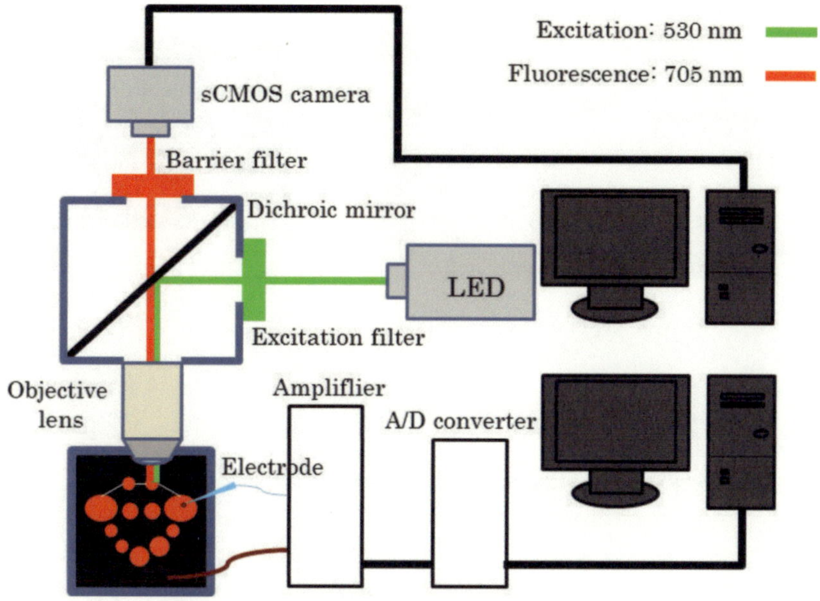


**Figure 16.** Schematic illustration of our recent fluorescent voltage-imaging apparatus Color online. The LFP can be simultaneously measured by extracellular recording. This figure is reproduced from Ref. [16] with permission.

mirror (DM575) and a barrier filter (BA590). The fluorescence images (1024 × 1024 pixels) are acquired at a flame rate of 10 ms by sCMOS camera (Zyla, Andor, Ireland). The acquired image sequences were stored into a personal computer (Dell Precision T5600, Dell, USA).

**Figure 17** shows the propagating waves of fluorescence change in the PC lobe of *L. valentianus* measured by our apparatus. The supplemental movie can be seen in the website (the URL is in the figure legend of **Figure 17**). Here, a 10× objective was used and the space resolution was about 0.6 μm/pixel. An aversive odor, isoamyl acetate, was applied to the epithelium of

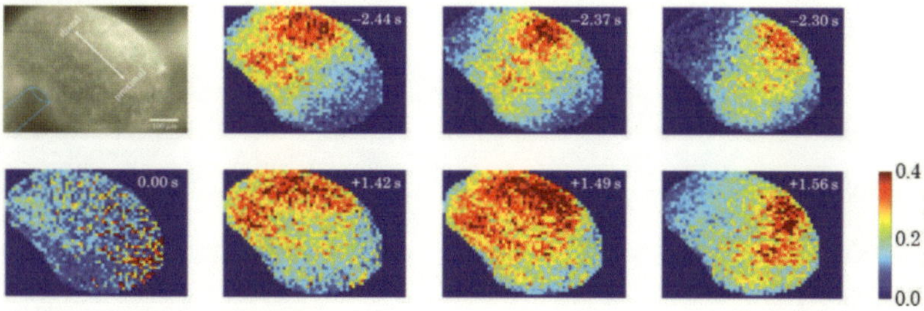


**Figure 17.** Propagation waves of fluorescence change in the PC lobe (upper left fluorescence image) measured by our recent apparatus Color online. The fluorescence images were acquired by a sCMOS camera as 1024 × 1024 pixel images at a flame rate of 10 ms. The space resolution was about 0.6 μm/pixel. Fluorescence changes ($\Delta F/F$) were plotted as successive frames. The $\Delta F/F$ values averaged over 16 × 16 pixels (yellow mesh in the upper left image) were represented according to the color table. The LFP was simultaneously measured by the glass electrode (blue figure in the upper left image). The odor (isoamyl acetate) was applied at 0.00 s, and the time before or after the odor stimulus is shown in each flame. See also the supplemental movie (http://saitolab-chaos.com/Papers/Malacology/Supplemental-movie.html; 1/4 speed reproduction).

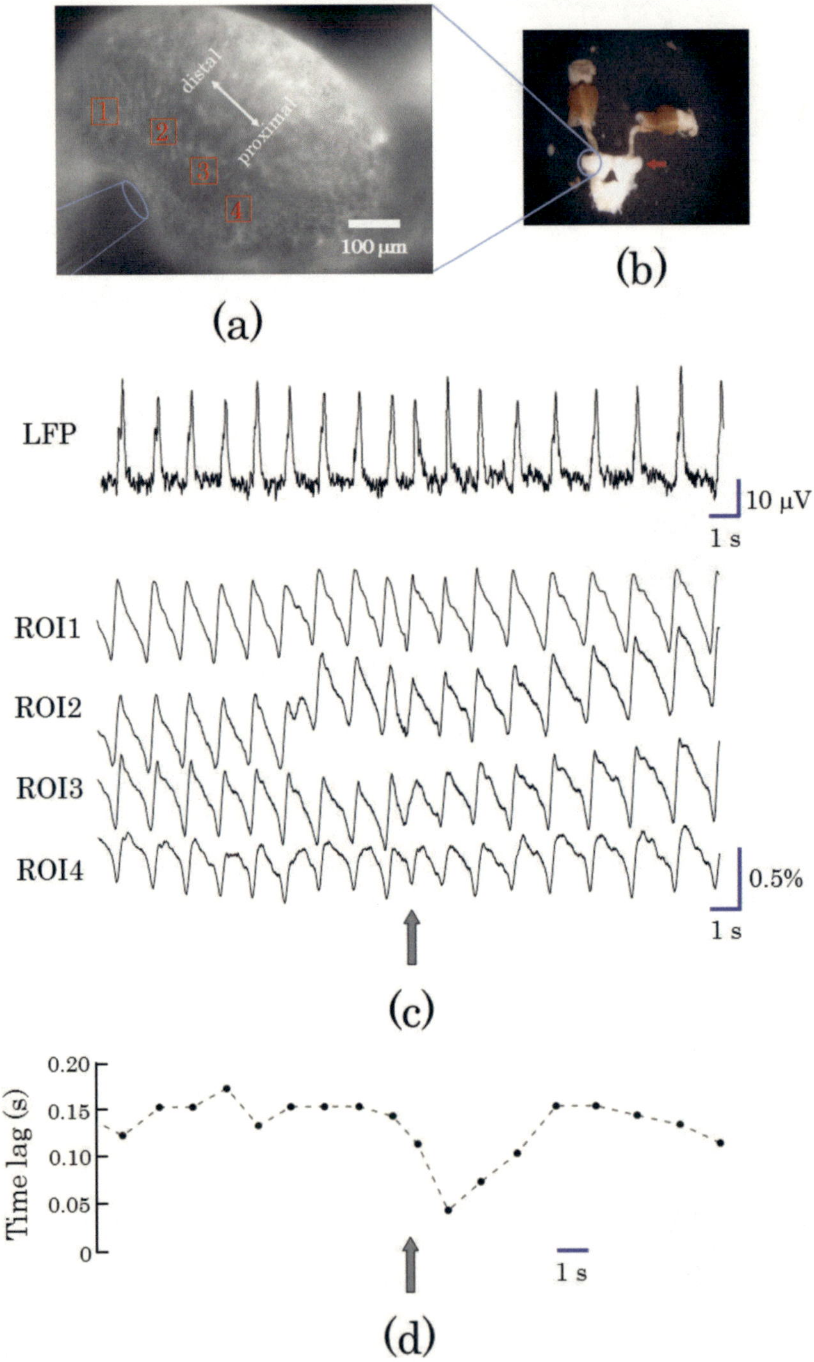


**Figure 18.** (a) Fluorescence image of the PC lobe Color online. The red squares (1–4) show the ROIs for fluorescence change analysis and the blue figure shows the glass electrode for LFP measurement. (b) Preparation of the central nervous system with tentacles. The blue circle and the red arrow show the left and right PC lobes, respectively. (c) Time courses of the LFP (upper figure) and the fluorescence change in the ROIs (lower figure). (d) Time lag estimated from the peaks of fluorescence oscillations in the ROI1 and ROI4. The odor (isoamyl acetate) was applied at the time shown by gray arrows in (c) and (d). These figures are reproduced from Ref. [16] with permission.

the superior tentacle. **Figure 18** shows the fluorescence changes in four ROIs. The time lag between two peaks at the distal and proximal regions significantly decreased, that is, the propagation speed significantly increased when the odor was applied. In our experiments, this phenomenon occurred only for the aversive odors, not for the appetitive odors.

We also examined the involvement of nitric oxide (NO) in the odor-induced changes in wave propagation by our apparatus. NO, which is a gaseous neurotransmitter, has been reported to function specifically in the PC lobe. Gelperin et al. showed that the excitability of B neurons is modulated by NO [34] and suppression of NO synthesis blocks the LFP oscillation and wave propagation in the PC lobe [35]. In our study [36], the stimulation of another aversive odor, hexanol, to the epithelium of the superior tentacle increased NO concentration measured by an NO electrode applied to the PC lobe of *L. valentianus*. This was blocked by NO synthase inhibitor L-NAME (3.7 mM), although it did not block the LFP oscillation and wave propagation in the PC lobe at this concentration. Additionally, the LFP frequency and wave propagation speed were increased by the odor (hexanol) stimulation or electrical stimulation of the superior tentacle nerve, and these phenomena also disappeared by L-NAME, but not by its inactive enantiomer D-NAME (**Figure 19**). On the other hand, L-NAME did not block synaptic transmission from the superior tentacle nerve to NB neurons in the PC lobe. These results suggested that NO is released from NB neurons diffuses into the CM to depolarize B neurons, which modifies the network activity, presumably the main effect of suppressing NB neurons (**Figure 20**). Based on this study as well as previous studies, the authors speculated that the effects of NO released during olfactory perception underlie the mechanism of precise odor discrimination and learning in the PC lobe of the land slug.

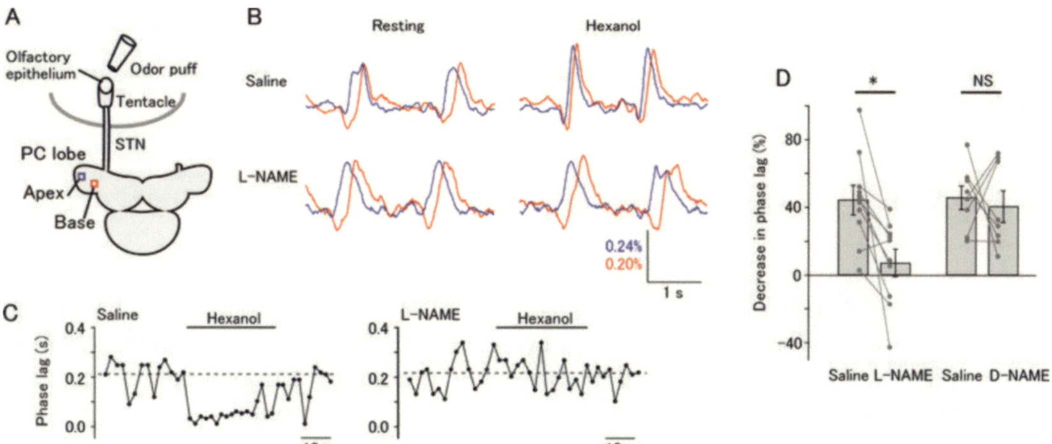


**Figure 19.** Effects of NO on the spatiotemporal neural activities in the PC lobe Color online. (A) Schematic illustration of the experiment. (B) Normalized fluorescence changes in the ROIs in the distal (apical) and proximal (basal) regions. Before the odor (hexanol) stimulation, the peaks of fluorescence changes had a time lag between the distal and proximal regions (upper left). The lag decreased during the odor stimulation (upper right). In the saline containing L-NAME, the odor stimulation did not decrease the lag (lower). (C) Time courses of the lag in saline (left) and L-NAME (right). The dotted lines indicate the average of lag before the stimulus. (D) Summary of the responses of lag to the odor stimulation. Average and individual data are shown. L-NAME significantly reduced the decrease in the lag (*$p < 0.05$, $N = 10$), whereas D-NAME did not significantly change the response (NS, not significant; $N = 8$). These figures are reproduced from Ref. [36] with permission.

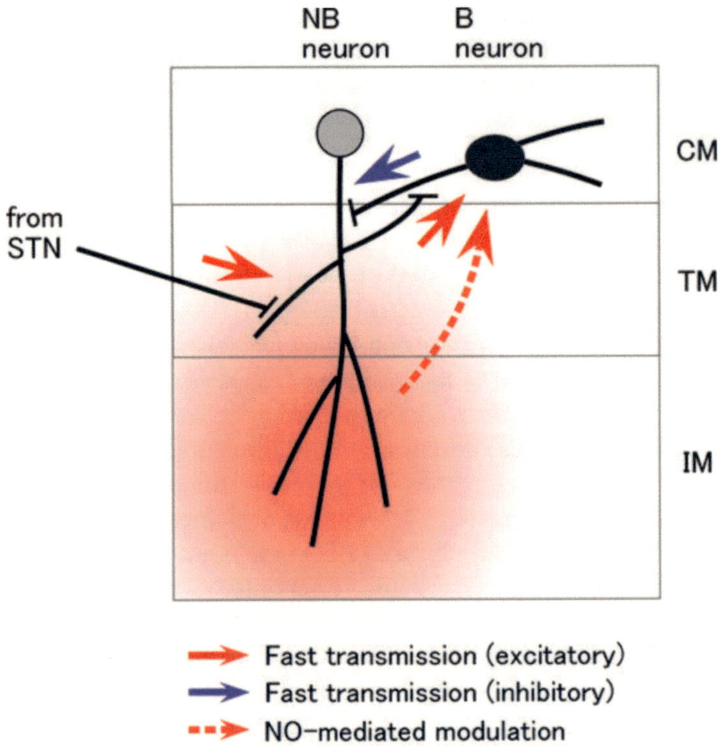


**Figure 20.** Schematic illustration of the pathways that transmit olfactory information to the PC lobe Color online. B and NB neurons have the cell bodies in the CM. The superior tentacle nerve projects in the TM and makes synapses on NB neurons. NB neurons produce spikes that propagate afferently and activate synapses on B neurons. At the same time, spikes also propagate efferently into the IM, where NO will be released. NO diffuses into the CM and depolarizes B neurons, which modifies the network activity, presumably by suppressing NB neurons. This figure is reproduced from Ref. [16] with permission.

## 6. Conclusions

Fluorescent voltage imaging is applicable for the measurement of the spatiotemporal neural activities involved in the olfactory processing of the land slug *Limax*. Fluorescent calcium imaging is also applicable for it, even though generally the AM group-bonded calcium-sensitive dyes cannot be easily loaded into invertebrate neurons. Through the studies using the fluorescent-imaging technique, especially, the function and computational roles of the propagation waves of neural activity in the PC lobe on odor discrimination and learning have been elucidated. Meanwhile, the experimental apparatus has also been improved. By our recent apparatus, consecutive images can be acquired at a flame rate of about 10 ms for about 1000 × 1000 pixels in size, and some new findings on the olfactory processing were obtained. Such an improvement of the experimental apparatus will enable us to obtain further information not only on the olfactory processing of the land slug but also on various biological functions.

## Author details

Minoru Saito

Address all correspondence to: msaito@chs.nihon-u.ac.jp

Department of Biosciences, College of Humanities and Sciences, Nihon University, Tokyo, Japan

## References

[1] Molecular Probes TM Handbook; A Guide to Fluorescent Probes and Labeling Technologies. 11th ed. Invitrogen USA 2010

[2] Orbach HS, Cohen LB, Grinvald A. Optical mapping electrical activity in rat somatosensory and visual cortex. Journal of Neuroscience. 1985;**5**(7):1886-1895

[3] Kudo Y, Nakamura T, Ito E. A 'macro' image analysis of Fura-2 fluorescence to visualize the distribution of functional glutamate receptor subtypes in hippocampal slices. Neuroscience Research. 1991;**12**(3):412-420

[4] Sato S, Osanai H, Monma T, Harada T, Hirano A, Saito M, Kawato S. Acute effect of corticosterone on *N*-methyl-D-aspartate receptor-mediated $Ca^{2+}$ elevation in mouse hippocampal slices. Biochemical and Biophysical Research Communications. 2004;**321**(2): 510-513

[5] Ikegaya Y, Bon-Jero ML, Yuste R. Large-scale imaging of cortical network activity with calcium indicators. Neuroscience Research. 2005;**52**:132-138

[6] Dalaney KR, Gelperin A, Fee MS, Flores JA, Gervais R, Tank DW, Kleinfeld D. Waves and stimulus-modulated dynamics in an oscillating olfactory network. Proceedings of the National Academy of Sciences of the United States of America. 1994;**91**(2):669-673

[7] Kleinfeld D, Delaney KR, Fee MS, Flores JA, Tank DW, Gelperin A. Dynamic of propagating waves in the olfactory network of a terrestrial mollusk: An electrical and optical study. Journal of Neurophysiology. 1994;**72**(3):1402-1419

[8] Okada K, Kanzaki R, Kawachi K. High-speed voltage-sensitive dye imaging of an *in vivo* insect brain. Neuroscience Letters. 1996;**209**(3):197-200

[9] Kawahara S, Toda S, Suzuki Y, Watanabe S, Kirino Y. Comparative study of neural oscillation in the procerebrum of the terrestrial slugs *Incilaria bilineata* and *Limax marginatus*. Journal of Experimental Biology. 1997;**200**(13):1851-1861

[10] Kimura T, Toda S, Sekiguchi T, Kawahara S, Kirino Y. Optical recording analysis of olfactory response of the procerebral lobe in the slug brain. Learning & Memory. 1998;**4**: 389-400

[11] Inoue T, Kawahara S, Toda S, Watanabe S, Kirino Y. Selective optical recording of the neural activity in the olfactory center of the land slug using a calcium indicator dye. Bioimages. 1998;**6**(2):59-67

[12] Yoshida R, Iwamoto A, Nagahama T. Calcium imaging for detection and estimation of spike activities in *Aplysia* neurons. Zoological Science. 2001;**18**(5):631-643

[13] Watanabe S, Shimozono S, Kirino Y. Optical recording of oscillatory neural activities in the molluscan brain. Neuroscience Letters. 2004;**359**(3):147-150

[14] Watanabe S, Kirino Y. Selective calcium imaging of olfactory interneurons in a land mollusk. Neuroscience Letters. 2007;**417**(3):246-249

[15] Hamasaki Y, Hosoi M, Nakada S, Shimokawa T, Saito M. Fluorescent voltage imaging technique for measurement of molluscan neural activities. Open Journal of Biophysics. 2013;**3**:54-58

[16] Hamasaki Y, Shimokawa T, Ishida K, Komatsuzaki Y, Watanabe S, Saito M. Coherency evaluation of spatiotemporal neural activities in the molluscan olfactory center applying extracellular recording with wavelet analysis. Open Journal of Biophysics. 2013;**3**:291-297

[17] Ito I, Nakamura H, Kimura T, Suzuki H, Kirino Y, Ito E. Neuronal components of the superior and inferior tentacles in the terrestrial slug, *Limax marginatus*. Neuroscience Research. 2000;**37**(3):191-200

[18] Ito I, Watanabe S, Kimura T, Kirino Y, Ito E. Distribution of γ-aminobutyric acid immunoreactive and acetylcholinesterase-containing cells in the primary olfactory system in the terrestrial slug *Limax marginatus*. Zoological Science. 2003;**20**(11):1337-1346

[19] Ito I, Watanabe S, Kirino Y. Air movement evokes electro-olfactogram oscillations in the olfactory epithelium and modulates olfactory processing in a slug. Journal of Neurophysiology. 2006;**96**(4):1939-1948

[20] Inokuma Y, Inoue T, Watanabe S, Kirino Y. Two types of network oscillations and their odor responses in the primary olfactory center of a terrestrial mollusc. Journal of Neurophysiology. 2002;**87**(6):3160-3164

[21] Ito I, Watanabe S, Kimura T, Kirino Y, Ito E. Negative relationship between odor-induced spike activity and spontaneous oscillations in the primary olfactory system of the terrestrial slug *Limax marginatus*. Zoological Science. 2003;**20**(11):1327-1335

[22] Ito I, Kimura T, Watanabe S, Kirino Y, Ito E. Modulation of two oscillatory networks in the peripheral olfactory system by γ-aminobutyric acid, glutamate, and acetylcholine in the terrestrial slug *Limax marginatus*. Journal of Neurobiology. 2004;**59**(3):304-318

[23] Ratté S, Chase R. Morphology of interneurons in the procerebrum of the snail *Helix aspersa*. Journal of Comparative Neurology. 1997;**384**(3):359-372

[24] Ratté S, Chase R. Synapse distribution of olfactory interneurons in the procerebrum of the snail *Helix aspersa*. Journal of Comparative Neurology. 2000;**417**(3):366-384

[25] Zaitseva OV. Projectional connections and a hypothetical scheme of structural organization of procerebrums of terrestrial molluscs. Journal of Evolutionary Biochemistry and Physiology. 2000;**36**(5):604-620

[26] Gelperin A, Tank DW. Odor-modulated collective network oscillations of olfactory interneurons in a terrestrial mollusc. Nature. 1990;**345**(6274):437-440

[27] Watanabe S, Kawahara S, Kirino Y. Morphological characterization of the bursting and nonbursting neurons in the olfactory centre of the terrestrial slug *Limax marginatus*. Journal of Experimental Biology. 1998;**201**(7):925-930

[28] Inoue T, Watanabe S, Kawahara S, Kirino Y. Phase-dependent filtering of sensory information in the oscillatory olfactory center of a terrestrial mollusk. Journal of Neurophysiology. 2000;**84**(2):1112-1115

[29] Loew LM, Cohen LB, Dix J, Fluhler EN, Montana V, Salama G, Jian-young W. A naphthyl analog of the aminostyryl pyridinium class of potentiometric membrane dyes shows consistent sensitivity in a variety of tissue, cell, and model membrane preparations. Journal of Membrane Biology. 1992;**130**(1):1-10

[30] Ermentrout GB, Wang JW, Flores J, Gelperin A. Model for olfactory discrimination and learning in *Limax* procerebrum incorporating oscillatory dynamics and wave propagation. Journal of Neurophysiology. 2001;**85**(4):1444-1452

[31] Kimura T, Toda S, Sekiguchi T, Kirino Y. Behavioral modulation induced by food odor aversive conditioning and its influence on the olfactory responses of an oscillatory brain network in the slug *Limax marginatus*. Learning & Memory. 1998;**4**:365-375

[32] Kimura T, Suzuki H, Kono E, Sekiguchi T. Mapping of interneurons that contribute to food aversive conditioning in the slug brain. Learning & Memory. 1998;**4**:376-388

[33] Murakami M, Watanabe S, Inoue T, Krino Y. Odor-evoked responses in the olfactory center neurons in the terrestrial slug. Journal of Neurobiology. 2004;**58**(3):369-378

[34] Gelperin A. Nitric oxide mediates network oscillations of olfactory interneurons in a terrestrial mollusc. Nature. 1994;**369**(6475):61-63

[35] Gelperin A, Flores J, Raccuia-Behling F, Cooke IRC. Nitric oxide and carbon monoxide modulate oscillations of olfactory interneurons in a terrestrial mollusc. Journal of Neurophysiology. 2000;**83**(1):116-127

[36] Watanabe S, Takahashi F, Ishida K, Kobayashi S, Kitamura Y, Hamasaki Y, Saito M. Nitric oxide-mediated modulation of central network dynamics during olfactory perception. PLoS One. 2015;**10**(9):e0136846

# Lipid Composition Modifications in the Blue Mussels (*Mytilus edulis* L.) from the White Sea

Natalia N. Fokina, Tatiana R. Ruokolainen and
Nina N. Nemova

Additional information is available at the end of the chapter

http://dx.doi.org/10.5772/67811

**Abstract**

Studying biochemical indicators in response to various environmental factors allows revealing the metabolic adaptive strategy of the organism's tolerance and survival under a variety of environmental impacts. This review analyses both the authors' own data and the available literature on the problem of biochemical adaptations of the lipid composition in marine bivalves, particularly blue mussels, *Mytilus edulis* L., to various environmental impacts. Modifications in the composition of lipids and their fatty acids in blue mussels caused by short-term (under laboratory conditions) and chronic (field monitoring) exposure to natural and human factors indicate that homeostasis is maintained in cell membranes and the organism's energy requirements and facilitate the adaptation and tolerance of the mussels to environmental disturbances. The lipid and fatty acid composition indices in White Sea intertidal mussels which reflect their chronic exposure to a wide variety of environmental factors are discussed and compared to data on changes in the lipid composition of blue mussels exposed to some environmental factors (salinity, anoxia, metals) in aquarium experiments. The lipid profile plays an important role in the adaptation of blue mussels to new conditions in the habitat, and it can be used as a biochemical marker for indicating the organism's physiological state.

**Keywords:** lipids, fatty acids, biochemical adaptation, environmental factors, *Mytilus edulis*

## 1. Introduction

Biochemical processes underlie the development of cell metabolic responses to environmental impacts and allow an organism to adapt and survive in a changing environment [1]. Metabolic modifications up to the level of physiological and morphological disorders are reflected in the changes of various biochemical indicators, which allow determining the adaptive strategy of an organism's tolerance and survival under both natural and human impacts. Lipid molecules, which are involved in all the essential physiological-biochemical processes [2], play a major role in the organism's adaptive responses to various factors in the environment [1]. The primary response to stress is modification of the physical state of cell membranes (mainly fluidity), which triggers a change of their lipid and fatty acid composition [3, 4]. The main lipid components of biological membranes are phospholipids and cholesterol. The ratio of phospholipids and cholesterol is considered as an indicator of membrane fluidity. Cholesterol is known to increase the order of the phospholipid fatty acid chains in membranes [5]. Membrane phospholipids in different molecular species and molecular shapes as well as their interaction with cholesterol and membrane proteins determine membrane fluidity and subsequently regulate the activity of membrane-bound enzymes and the functioning of ion channels, pumps and receptors [2, 6, 7]. Besides their effects on membrane fluidity, membrane phospholipids are also a source of bioactive compounds and messengers [8]. In particular, eicosapentaenoic acid (EPA, 20:5n-3) and arachidonic acid (AA, 20:4n-6) are released from phosphatidylcholine (PC), phosphatidylethanolamine (PE) and phosphatidylinositol (PI) by phospholipase A2 and serve as precursors of short-lived hormone-like substances called eicosanoids (prostaglandins, thromboxanes, leukotrienes, etc.). The bioactive molecules have a wide range of physiological actions, including immune response, inflammatory response, neural function, reproduction and enhancement of an organism's adaptation to environmental stress [2, 8]. PI is also a source of such messengers as diacylglycerols and inositol phosphates (namely, inositol trisphosphate and others). These messengers, as well as the phospholipid phosphatidylserine (PS), are involved in regulating the activity of protein kinase C, which controls many cell functions, such as differentiation, proliferation, metabolism and apoptosis [2, 8, 9]. Moreover, fatty acids are the most labile components of lipid molecules, quickly and accurately reflecting environmental impacts and activating an organism's adaptive abilities. For example, a well-known biochemical response of poikilothermic organisms to low temperature is increased fatty acid unsaturation of both membrane and storage lipids [7, 10, 11]. In addition to membrane lipids, an important role in the adaptive response of organisms to various environmental factors belongs to high-energy storage lipids, chiefly triacylglycerols and their fatty acids [12–15], which cover the energy costs needed for maintaining homeostasis under the new environmental conditions. Since long-chain polyunsaturated fatty acids, particularly such essential fatty acids as EPA and docosahexaenoic acid (DHA, 22:6n-3), cannot be de novo synthesized in marine mussels [16, 17], their incorporation and elimination in membrane and storage lipids are strongly regulated [18, 19]. Thus, lipid and fatty acid composition as a key component of various metabolic pathways that are linked to processes important for survival and tolerance reflects the adaptive response of an organism to environmental effects. It is assumed that lipid composition may be used as a biochemical marker for indicating the organism's physiological state in environmental assessments and biomonitoring.

The blue mussels, *Mytilus edulis* L., are used worldwide as marine sentinel organisms in biomonitoring programmes due to their longevity, sessile nature, global distribution and ability to bioaccumulate high concentrations of pollutants [20–22]. In the White Sea, *M. edulis* L. is the dominant species of coastal (intertidal) ecosystems. Numerous studies on White Sea mussels' response to various environmental effects have identified adaptive mechanisms on molecular, biochemical, cellular, physiological and behavioural levels of biological organization [23–38].

This paper summarizes the results of research on lipid composition effects in White Sea blue mussels, *M. edulis* L. (1758), in response to environmental factors such as temperature, salinity, short-term anoxia, change of nutrition source, metals and oil pollution. Phenotype-specific features of the lipid composition in White Sea blue mussels from different habitat conditions (intertidal zone and aquaculture), as well as compensatory modifications of the lipid composition in intertidal mussels under chronic stress in the natural habitat and under short-term exposure to stress in laboratory experiments, are discussed.

## 2. Environmental factors

### 2.1. Temperature

Being a major environmental factor, influences all aspects of life of an organism, especially poikilotherms [1]. The literature offers quite detailed descriptions of the contribution of the lipid composition to thermal adaptations in bivalves [11, 39–43]. Often, the response involving the lipid composition depends on the duration of exposure to ambient temperatures. Moreover, organ-specific distribution of lipids in Bivalvia causes differences in lipid composition response to temperature effects, depending on the studied organ. Thus, gills of bivalves, which are the location of primary contact with the environment, contain high concentrations of membrane lipids, chiefly cholesterol [42]. It is well-known cholesterol is necessary for membrane stabilizing and maintaining the permeability of membranes [2, 5]. It was demonstrated that when exposed to rapid (several hours) temperature fluctuations gills of bivalves experience modifications in cholesterol levels, whereas prolonged (several weeks) temperature impacts induce changes in the amount of phospholipids enriched in polyunsaturated fatty acids [41, 43]. Whereas the gill lipid composition response in White Sea mussels in the temperature experiment was the opposite: a significant rise of the cholesterol concentration in response to prolonged (14 days) impact of both low and high temperatures, while short-term temperature stress (1 day) influenced the content of phospholipids and their fatty acids [36]. Since the synthesis of polyunsaturated fatty acids in bivalves is limited as well as an involvement of essential fatty acids in/from membrane lipids is strictly regulated [18, 19], presumably, the mussels' adaptive strategy is to use a less energy-intensive mechanism for maintaining optimal membrane fluidity by redistributing polyunsaturated fatty acids among storage and membrane lipid fractions (in the absence of additional cholesterol synthesis) that bivalves employ to adapt to rapid temperature changes [35, 36]. The role of minor membrane phospholipids (namely, phosphatidylserine and sphingomyelin) in the acclimation of mussels to elevated ambient temperature was also demonstrated in gills. They are believed to facilitate adaptive modifications of the fluidity and permeability of cell membranes in response to elevation of the ambient temperature [35, 36]. Let us remark that a similar effect involving these phospholipids was observed in gills of mussels acclimating to variable seawater salinity [44]. Elevated ambient temperature is known to produce a destabilizing effect on cell membranes in poikilotherms [3]. Apparently, seawater salinity variations, primarily reduction of salinity, cause an analogous response of the membrane physicochemical properties which, in turn, initiates compensatory modifications of the composition of membrane lipids and their fatty acids similar to those observed in response to elevated temperature. Digestive glands in bivalves contain higher concentrations of triacylglycerols (TAGs) required for energy metabolism during their acclimation to new environmental conditions as well as for reproductive processes. In particular, it was shown that bivalves from thermally different habitats differed in TAG metabolism in the digestive glands during overwintering. Blue mussels, *M. edulis*, adapted to harsh winters, accumulate TAGs enriched in 20:5n-3 fatty acids (EPA) in the digestive glands, while the oyster, *Crassostrea virginica*, which generally occurs in warmer habitats, on the contrary, did not reserve TAG before overwintering [42].

### 2.2. Salinity

Is one of the key abiotic factors in the marine environment. Marine and freshwater molluscs were shown to be alike in the content of total and neutral lipids, but the levels of individual fractions of phospholipids such as PC, PI, PS and PE, as well as PE plasmalogen form, were found to be dependent on the ambient salinity [45]. In gills of bivalves, salinity stress induces an increase in negatively charged phospholipids, chiefly PI and cardiolipin. This may supposedly be one of the intracellular mechanisms to bind excessive cations lest they degrade the cell's enzymatic systems [46]. Exposure of White Sea mussels to reduced seawater salinity (15 psu) in aquarium experiments (25 psu as a control) resulted in an increase in the concentration of phospholipids, mainly PC and PS, in gills [44], alongside a reduction in the levels of cholesterol and storage lipids (triacylglycerols and cholesterol esters) in gills and digestive glands [34, 47]. The reduced levels of cholesterol and storage lipids indicate an adverse effect of low salinity on the physiological state of the mussels and their metabolism. We know that when salinity is drawn down from 25 psu (normal values of salinity in the White Sea) to 14 psu, the functional activity of *M. edulis* is suppressed [23, 48], and the mussels' tolerance of low salinity is ensured by cellular volume regulation using organic and inorganic osmolytes [32, 49, 50]. Thus, the acclimation of molluscs to reduced salinity apparently implies that storage lipids are utilized not only as sources of metabolic energy but also as substrates for the synthesis of organic osmolytes [34]. Moreover, it was shown that the action of increased salinity (from 25 psu to 35 and 45 psu) leads to various responses of the lipid composition of White Sea mussels, *M. edulis*, depending on the studied organ. Although mantle edge and gills are sites for primary contact with external environment, the effect of increased seawater salinity on both intertidal and cultured mussels caused organ-specific reactions in the cholesterol level: the level rose in the mantle edge [51] but declined in gills [44]. Apparently, the different cholesterol content in gills and mantle edge of the mussels reflects differences in membrane fluidity and ion permeability in response to increased salinity effect. Some authors have pointed out the lack of distinctions between marine and freshwater bivalves in the fatty acid composition [45, 52, 53], although some papers have reported elevated concentrations of C20 and C22 unsaturated fatty acids, predominantly 20:5n-3, 22:5n-3 and 22:6n-3, in marine molluscs [54] as well as high level of monounsaturated fatty acids and arachidonic acid (20:4n-6) in freshwater molluscs [55]. Yet, the high variability of fatty acid content observed in both freshwater and marine mollusc species is primarily due to the factors of nutrition and ambient temperature [14, 15, 41–43, 45, 56–58]. Nonetheless, a lower level of n-6 polyunsaturated fatty acids was found in gills of marine bivalves acclimated to high salinity as compared to the individuals exposed to low salinity [46]. White Sea mussels acclimated to different seawater salinities also manifested considerable modifications of the lipid fatty acid composition in gills and mantle edge. Thus, in gills of intertidal mussels, the level of n-6 polyunsaturated fatty acids (mainly owing to AA, 20:4n-6) increased in response to seawater salinity reduction (5 psu) and elevation (45 psu). At the same time, cultured mussels collected from aquaculture substrates responded with a decrease in n-3 polyunsaturated fatty acid (PUFA) content and an increase in saturated fatty acid level in gills both to a reduction (to 5 and 15 psu) and an elevation (to 35 and 45 psu) of seawater salinity. Remarkably, notwithstanding the considerably different lipid composition of gills in intertidal and aquaculture mussels, they both responded to critically low salinity (5 psu) with similar modifications of the lipid composition, indicative of non-specific defence reaction in bivalves—closure of shell valves, reduction of total metabolism and transition to anaerobic metabolic pathways [44]. Varied response of fatty acid composition to salinity effects was detected in the mantle edge of mussels from different tidal zones (intertidal and aquaculture) [51]. Thus, it was shown that the concentration of non-methylene-interrupted fatty acids (NMIFA) in mantle edge increased in the intertidal mussels exposed to 5, 35 and 45 psu (25 psu as a control), whereas in cultured mussels exposed to 5, 35 and 45 psu salinity, there was an increase in n-3 PUFA content. It is known that NMIFA can be synthesized in marine bivalve molluscs in the case of a lack of usual n-3 PUFA [53]. Probably, n-3 PUFA deficiency in intertidal mussels is the result of their utilization to generate energy required for mussel acclimation to different salinities, whereas additional synthesis of NMIFA is essential for maintaining the unsaturated state of membrane phospholipids as well as fluidity and permeability of membranes in mantle edge.

### 2.3. Short-term anoxia

Blue mussels, *M. edulis,* living in the marine coastal (intertidal) zone are facultative anaerobes tolerant of short-term anoxia during low tide [30, 59, 60]. The main sources of energy for bivalves in the anaerobic metabolism conditions are glycogen and proteins [1, 61], whereas lipids are utilized to provide for gametogenesis [61]. The role of storage lipids (chiefly triacylglycerols) in the adaptive reactions of mussels under anoxic conditions was demonstrated in our studies [62, 63]. In White Sea blue mussels, we observed a rise in the levels of cholesterol and PC within total lipids of soft tissues, which are known to have a stabilizing effect on membranes and thus reduce their permeability. It is known that anoxia may reduce the permeability of cell membranes, thus causing modifications in their lipid composition [64]. On the other hand, elevated concentrations of polyunsaturated fatty acids (in particular, arachidonic acid) and non-methylene-interrupted fatty acids within total lipids of soft tissues balance the stabilizing effect of membrane lipids and probably facilitate the functioning of membrane-bound proteins (enzymes, ion channels and receptors) [63]. Additional research is needed to determine organ-specific reaction of the lipid and fatty acid composition in blue mussels under short-term anoxia effect.

### 2.4. The nutrition factor

It is a known fact that lipid composition, especially the fatty acid profile of filter-feeding mussels, like in any consumers, is a trophic marker of the composition of their food and includes the biochemical markers of all seston components, namely, phytoplankton, zooplankton and bacteria (detritus) [14, 15, 56, 58, 65–68]. The study of modifications in the composition of lipids and their fatty acids in gills and digestive glands of White Sea mussels, *M. edulis,* induced by their acclimation to laboratory conditions where they were fed with artificial feed ("Coraliquid", Sera) revealed organ-specific patterns in the assimilation and modification of lipids, primarily concerning their fatty acid profile [69]. It was shown that change of the food source caused alterations in the lipid and fatty acid composition, mainly in the digestive gland. Elevated content of high-energy lipids (triacylglycerols) enriched in saturated fatty acids (namely, myristic 14:0 acid), as well as vaccenic 18:1n-7 acid in the feed, promoted the storage of these lipids in mussel gills and digestive gland. At the same time, the fact that phospholipids in the feed lacked essential fatty acids, EPA and DHA, which are known to be derived from phytoplankton, told considerably on the fatty acid profile of triacylglycerols in the mussels [69].

**Intertidal habitats** are the most variable in terms of such abiotic environmental factors as temperature, salinity, aerial exposure and concentrations of suspended nutritive material [70]. Life under such heavily variable environmental conditions reflects both on the mussels' physiological (including growth rate) and metabolic processes and on the size and age structure of mussel beds (local populations), their abundance and biomass [27, 29, 30, 71]. Ecologo-biochemical monitoring during 2009–2014 years of two intertidal mussel beds located in different parts of the Gulf of Kandalaksha, White Sea, and differing in hydrological characteristics, including seawater salinity, revealed the features of the lipid composition in gills and digestive glands which reflect the chronic impact of salinity drops [38]. Frequent salinity drops in one of the investigated sites (Site 1) are due not only to discharge from streams but also to human activities (namely, unregulated freshwater discharges from hydropower plant). The effect of reduced seawater salinity (approximately 9.7–14.0 psu) on mussels from Site 1 appears not only in the level of some phospholipids (in particular, phosphatidylserine, phosphatidylethanolamine and phosphatidylcholine), and the ratio of n-3/n-6 polyunsaturated fatty acids in the molluscs' gills and digestive gland, but also in some ecological characteristics of the mussel beds, i.e. its size-age structure, abundance and biomass [38]. The elevated content of the named phospholipids and the prevalence of n-3 polyunsaturated fatty acids over n-6 polyenes apparently serve to keep membranes permeable to ions and maintain the functioning of membrane-bound enzymes involved in cellular volume regulation in response to low seawater salinity. These data agree with the results of previous aquarium experiments on the effect of low seawater salinity on the lipid composition of intertidal and cultured mussels [44], which suggest that the mussels' lipid and fatty acid composition is adapted to secure the survival of the molluscs under low seawater salinity. The elevated ratio of n-3/n-6 polyunsaturated fatty acids in the mussels chronically exposed to salinity drops may also be a result of high metabolic rate of n-6 polyunsaturated fatty acids (chiefly AA, 20:4n-6). Arachidonic acid is a precursor for the synthesis of physiologically active hormone-like molecules, eicosanoids (such as prostaglandins), which are known to build up bivalves' resistance to stress, including various seawater salinities [8, 72–75]. One must mention that intertidal mussels living in a habitat with relatively stable salinity conditions (away from freshwater discharges, Site 2 where seawater salinity is 20.1–22.5 psu) feature an elevated content of cholesterol and n-6 polyunsaturated acids within total lipids of both gills and digestive glands [38]. Our monitoring studies of the lipid and fatty acid composition in intertidal mussels from different habitats in the White Sea showed that the fatty acid composition of digestive glands, unlike their content in gills, reflects the adaptive features of the lipid metabolism in the mussels under chronic effect of a wide range of environmental factors [38].

There is a lot of research on the study of the differences in physiological (energetic, growth rate, clearance rate, ingestion rate, absorption rate, respiration rate) and biochemical indices between mussels collected in intertidal (rocky shore) and subtidal (aquaculture) environments [13–15, 76–78]. It is considered that these origin-related differences in physiological rates have to do with features of the energy distribution, namely, in intertidal mussels more energy is directed to the formation of a thicker shell, while in subtidal mussels the energy is spent on tissue growth [78]. Simultaneously, biochemical differences between these mussel groups are associated with various concentrations and quality of seston in the intertidal and subtidal zones [13, 14, 76]. In particular, frequent exposure to air (during low tide) has very high effect on the mussels' energy reserves including triacylglycerols, saturated fatty acids and some polyunsaturated fatty acids, similarly to the effect of starvation [13, 14]. We have also studied origin-related lipid composition differences in gills of littoral (intertidal) and cultured (sublittoral) mussels after 2 weeks of acclimation to laboratory conditions [44, 79]. It was demonstrated that gills of intertidal mussels differ from those kept under the fairly stable conditions of aquaculture (cultured mussels) in that the former have a higher level of lipids that stabilize membrane structure (cholesterol and saturated fatty acids), as well as n-6 polyunsaturated fatty acids (chiefly AA, 20:4n-6), which arguably contribute to the establishment of suitable membrane permeability and regulate the activity of membrane-bound enzymes, ion channels and receptors. High level of AA in the whole body as well as in gills of intertidal mussels appears to be due to selective retention of the fatty acid required for eicosanoid synthesis [14, 15]. In addition, unlike for mussels collected in August (where mussels were on reproductive stage IIIc or stage 0, resting), increased content of the fatty acid in the whole body of the mussels collected in June (where mussels were on reproductive stage IIIb, spawning) is needed for reproductive processes [79]. Although no differences were found in the sterol content in mussels originating from the two habitats (rocky shore and subtidal) [13], a significant excess in cholesterol level in the whole body and gills of intertidal mussels from the White Sea is probably due to the effect of severe fluctuations in temperatures (up to subzero temperatures). These features of the lipid composition are assumed to be one of the biochemical adaptation mechanisms providing for the phenotypic plasticity and survival of blue mussels in a frequently changing coastal environment. On the other hand, high level of triacylglycerols as well as elevated concentrations of n-3 polyunsaturated fatty acids, primarily of phytoplanktonic origin, EPA and DHA, in mussels collected from artificial substrates evidences high food availability (phytoplankton) and relatively stable environmental conditions in aquaculture [44, 79]. These origin-related differences of blue mussel lipid composition reflect the important role of lipids in adaptation to a changing environment.

## 3. Pollution effect

Natural habitats of marine aquatic organisms may also be negatively affected by human impact. Seawater is contaminated by organic and inorganic chemical substances (such as metals, petroleum hydrocarbons, pesticides) from municipal and industrial discharges. Since pollutants of various nature get either directly or indirectly involved in lipid peroxidation reactions [80–82], it is obvious that a characteristic sign of their impact on cell membranes is the disruption of lipid bilayer packing, which in its turn triggers modifications in the composition of membrane lipid components (cholesterol, phospholipids and their fatty acids) [83]. Thus, mussels from contaminated sites had an elevated level of triacylglycerols and an increased triacylglycerol/phospholipid ratio, which implies a reduced rate of mobilization of triacylglycerols into the phospholipid pool with serious consequences for the structure and function of cell membranes. There was also a substantial decrease in phospholipids, apparently in connection with membrane destruction [84–86]. Some papers have reported the modifications in lipid and fatty acid composition of hydrobionts, including marine mussels, in response to organic and inorganic pollutants' effect [34, 37, 87–93]. Since the lipid metabolism plays an important role in living organism, it is believed that the lipid and fatty acid profile may be used to indicate the organism's health under stress conditions of pollutant effect. Exposure of blue mussels from the White Sea to various concentrations of oil products in an aquarium experiment led to an increase in the level of phospholipids and a reduction of cholesterol concentration in gills and mantle, i.e. the gateway organs for external impacts [25, 34]. These modifications in membrane lipids are believed to make cell membranes more permeable to oil products and create the conditions for their accumulation in these organs for further detoxification. A significant decrease in the level of membrane lipids—phospholipids (mainly at the expense of PC and PE) and cholesterol—simultaneously with an increase in triacylglycerols was observed in gills and digestive glands of mussels exposed to various concentrations of cadmium [37]. These modifications of the lipid profile reflect the destructive effect of cadmium on cell membranes realized through the activation of lipid peroxidation processes. It is worth noting that a significant decrease of the cholesterol concentration under the impact of oil products, mainly their high concentrations, was observed in all the studied organs (gills, mantle, mantle edge and foot) of *M. edulis* [25, 34], as well as under the impact of cobalt on *Mytilus galloprovincialis* [94]. One of the presumed examples of the toxic effect of oil products, as well as some heavy metals on bivalves, is the inhibition of cholesterol synthesis, leading to high membrane permeability. At the same time, the effect from exposure to copper as an essential metal was the opposite (significant increase of cholesterol concentration), probably meant to stabilize the membranes under the metal's oxidative action and to reduce their permeability. It was noted also that when exposed to cadmium and copper [37], as well as to relatively low concentrations of oil products [95], mussels demonstrated an elevated level of arachidonic acid. Apparently, AA involvement in the synthesis of eicosanoids ensures high resistance of the mussels to these xenobiotic impacts. On the other hand, when the concentrations of oil products were high, the level of this acid in the mussels decreased, probably due to inhibition of its biosynthesis, given the observed elevated concentrations of linoleic acid, its metabolic precursor [95]. The results of studies on the lipid composition of gills and digestive glands of intertidal blue mussels, *M. edulis*, collected from different sites in the Gulf of Kandalaksha, White Sea, prove that the composition of lipids and their fatty acids depends not only on the hydrological conditions in the habitat but also on the degree of human impact on it [38]. To wit, the fatty acid profile of the intertidal mussels living in habitats with high human impact is noted for the prevalence of oleic (18:1n-9) acid among total lipids of gills and digestive glands. A similar effect in the fatty acid composition of total lipids was observed in the mussels exposed to various doses of copper in an aquarium experiment [37]. Elevated content of non-essential oleic acid in bivalves may be associated with its additional synthesis under the toxic effect of pollutants and have the goal of binding and detoxifying xenobiotic substances. Unsaturated fatty acids are known to be capable of forming complexes with metal ions and thus to contribute to the accumulation and detoxification of these xenobiotic substances in mussels [96].

## 4. Conclusions

The lipid profile of White Sea blue mussels, *M. edulis* L., is modified in response to various environmental factors in order to protect cell membranes, maintain or recover their homeostasis, replenish the cell's energy and metabolic resources and thus to secure the mussels' adaptation to the change in environmental conditions. Organ-specific distribution of lipids and fatty acids in White Sea blue mussels, as well as the dependence of the lipid and fatty acid composition response on the effect of various environmental factors on the studied organ, was detected. Modifications in the lipid composition predominantly in gills reflect the acute effect of environmental factors in aquarium experiment conditions, whereas changes in the lipid composition of digestive glands represent an adaptation of the lipid metabolism in response to chronic exposure to ambient factors (field monitoring). The composition of lipids and their fatty acids in intertidal mussels evidences their chronic exposure to abiotic environmental factors and human impact and is in agreement with data on the modifications of the lipid profile in White Sea blue mussels subjected to such environmental factors (namely, salinity, short-term anoxia, heavy metal and oil pollution) in aquarium experiments. The data discussed above prove that the lipid profile plays an important role in the adaptation of blue mussels, *M. edulis,* to new conditions in the habitat. Assessment of the lipid composition in intertidal and cultured mussels helps disclose the metabolic strategy to ensure resistance and adaptation of the organisms to environmental impacts of different nature and can be used as a biochemical marker for indicating the organism's physiological condition. This knowledge is necessary for environmental safety assessment under both natural and human impacts, as well as to predict an organism's and population's status in biomonitoring.

## Acknowledgements

The results on the lipid composition of White Sea blue mussels were obtained using the facilities of the Equipment Sharing Centre of the Institute of Biology KarRC RAS (Petrozavodsk, Russia).

The authors are grateful to the administration and staff of the "Kartesh" White Sea Biological Research Station of the Zoological Institute RAS, especially Drs. V.Ja. Berger, V.V. Khalaman and A.A. Sukhotin, for the permission and help with aquarium experiments at the station, as well as to researchers from the Institute of Biology KarRC RAS I.N. Bakhmet, PhD, and I.V. Sukhovskaya, PhD, for the assistance in setting up the experiments and sampling biological material. The authors also acknowledge the help of the Kandalaksha Strict Nature Reserve staff, as well as the professor at the Petrozavodsk State University Dr. G.A. Shklyarevich in collecting material from the nature reserve territory.

The study was funded by the federal budget under state-ordered project № 0221-2014-0033.

## Author details

Natalia N. Fokina*, Tatiana R. Ruokolainen and Nina N. Nemova

*Address all correspondence to: fokinann@gmail.com

Institute of Biology, Karelian Research Centre of Russian Academy of Sciences, Petrozavodsk, Russia

## References

[1] Hochachka PM, Somero GN. Biochemical Adaptation. Oxford: Princeton University Press; 2002. 480 p.

[2] Vance DE, Vance JE, editors. Biochemistry of Lipids, Lipoproteins and Membranes. 4th ed. Amsterdam: Elsevier; 2002. 610 p.

[3] Los DA, Murata N. Membrane fluidity and its role in the perception of environmental signals. Biochimica et Biophysica Acta. 2004;**1666**(1-2):142-157. DOI: http://dx.doi.org/10.1016/j.bbamem.2004.08.002

[4] Lopez CS, Alice AF, Heras H, Rivas EA, Sanches-Rivas C. Role of anionic phospholipids in the adaptation of *Bacillus subtilis* to high salinity. Microbiology. 2006;**152**:605-616. DOI: 10.1099/mic.0.28345-0

[5] Crockett EL. Cholesterol function in plasma membranes from ectotherms: membrane-specific roles in adaptation to temperature. American Zoologist. 1998;**38**(2):291-304. DOI: http://dx.doi.org/10.1093/icb/38.2.291

[6] Fodor E, Jones RH, Buda C, Kitajka K, Dey I, Farkas T. Molecular architecture and biophysical properties of phospholipids during thermal adaptation in fish: an experimental and model study. Lipids. 1995;**30**(12):1119-1126. DOI: 10.1007/BF02536612

[7] Logue JA, de Vries AL, Fodor E, Cossins AR. Lipid compositional correlates of temperature-adaptive interspecific differences in membrane physical structure. Journal of Experimental Biology. 2000;**203**:2105-2115.

[8] Parrish CC. Essential fatty acids in aquatic food webs. In: Arts MT, Brett MT, Kainz M, editors. Lipids in Aquatic Ecosystems. New York: Springer; 2009. pp. 309-326. DOI: 10.1007/978-0-387-89366-2_13

[9] Di Paolo G, De Camilli P. Phosphoinositides in cell regulation and membrane dynamics. Nature. 2006;**443**(7112):651-657. DOI: 10.1038/nature05185

[10] Cossins AR. Homeoviscous adaptation of biological membranes and its functional significance. In: Cossins AR, editor. Temperature Adaptation of Biological Membranes. London: Portland Press; 1994. pp. 63-76.

[11] Hazel JR, Williams EE. The role of alterations in membrane lipid-composition in enabling physiological adaptation of organisms to their physical environment. Progress in Lipid Research. 1990;**29**:167-227. DOI: 10.1016/0163-7827(90)90002-3

[12] Sangiao-Alvarellos S, Laiz-Carrion R, Guzman JM, Martin del Rio MP, Miguez JM, Mancera JM, Soengas JL. Acclimation of *S. aurata* to various salinities alters energy metabolism of osmoregulatory and nonosmoregulatory organs. American Journal of Physiology. Regulatory, Integrative and Comparative Physiology. 2003;**285**:R897–R907. DOI: 10.1152/ajpregu.00161.2003

[13] Freites L, Fernandez-Reiriz MJ, Labarta U. Lipid classes of mussel seeds *Mytilus galloprovincialis* of subtidal and rocky shore origin. Aquaculture. 2002;**207**:97-116.

[14] Freites L, Fernandez-Reiriz MJ, Labarta U. Fatty acid profiles of *Mytilus galloprovincialis* (Lmk) mussel of subtidal and rocky shore origin. Comparative Biochemistry and Physiology Part B: Biochemistry and Molecular Biology. 2002;**2**:453-461.

[15] Freites L, Labarta U, Fernández-Reiriz MJ. Evolution of fatty acid profiles of subtidal and rocky shore mussel seed (*Mytilus galloprovincialis*, Lmk.). Influence of environmental parameters. Journal of Experimental Marine Biology and Ecology. 2002;**268**(2):185-204. DOI: http://dx.doi.org/10.1016/S0022-0981(01)00377-X

[16] De Moreno JEA, Moreno VJ, Brenner RR. Lipid metabolism of the yellow clam, *Mesodesma mactroides*: 2-polyunsaturated fatty acid metabolism. Lipids. 1976;**11**:561-566. DOI: 10.1007/BF02532902

[17] De Moreno JEA, Moreno VJ, Brenner RR. Lipid metabolism of the yellow clam, *Mesodesma mactroide*: 3-saturated fatty acids and acetate metabolism. Lipids. 1977;**12**:804-808. DOI: 10.1007/BF02533268

[18] Delaunay F, Marty Y, Moal J, Samain JF. The effect of monospecific algal diets on growth and fatty acid composition of *Pecten maximus* (L) larvae. Journal of Experimental Marine Biology and Ecology. 1993;**173**:163-179. DOI: 10.1016/0022-0981(93)90051-O

[19] Soudant P, Marty Y, Moal J, Masski H, Samain JF. Fatty acid composition of polar lipid classes during larval development of scallop *Pecten maximus* (L.). Comparative Biochemistry and Physiology Part A: Molecular & Integrative Physiology. 1998;**121**(3):279-288. DOI: http://dx.doi.org/10.1016/S1095-6433(98)10130-7

[20] Goldberg ED. The mussel watch concept. Environmental Monitoring and Assessment. 1986;**7**(1):91-103. DOI: 10.1007/BF00398031

[21] Widdows J, Donkin P. Mussels and environmental contaminants: bioaccumulation and physiological aspects. In: Gosling E, editor. The Mussel Mytilus: Ecology, Physiology, Genetics and Aquaculture. Amsterdam: Elsevier; 1992. pp. 383-424.

[22] Gosling E. The Mussel Mytilus: Ecology, Physiology, Genetics and Culture. Amsterdam: Elsevier; 1992. 589 p.

[23] Berger VJa, Kharazova AD. Mechanisms of salinity adaptations in marine molluscs. Hydrobiologia. 1997;**355**:115-126. DOI: 10.1007/978-94-017-1907-0_12

[24] Bakhmet IN, Berger VJa, Khalaman VV. The effect of salinity change on the heart rate of *Mytilus edulis* specimens from different ecological zones. Journal of Experimental Marine Biology and Ecology. 2005;**318**(2):121-126. DOI: http://dx.doi.org/10.1016/j.jembe.2004.11.023

[25] Bakhmet IN, Fokina NN, Nefedova ZA, Nemova NN. Physiological-biochemical properties of blue mussel *Mytilus edulis* adaptation to oil contamination. Environmental Monitoring and Assessment. 2009;**155**:581-591. DOI: 10.1007/s10661-008-0457-5

[26] Bakhmet IN, Khalaman VV. Heart rate variation patterns in some representatives of Bivalvia. Biology Bulletin. 2006;**33**:276-280. DOI: 10.1134/S1062359006030101

[27] Sukhotin AA, Abele D, Portner H-O. Growth, metabolism and lipid peroxidation in *Mytilus edulis*: age and size effects. Marine Ecology Progress Series. 2002;**226**:223-234. DOI: 10.3354/meps226223

[28] Sukhotin AA, Lajus DL, Lesin PA. Influence of age and size on pumping activity and stress resistance in the marine bivalve *Mytilus edulis* L. Journal of Experimental Marine Biology and Ecology. 2003;**284**:129-144. DOI: http://dx.doi.org/10.1016/S0022-0981(02)00497-5

[29] Sukhotin AA, Portner H-O. Age-dependence of metabolism in mussels *Mytilus edulis* (L.) from the White Sea. Journal of Experimental Marine Biology and Ecology. 2001;**257**:53-72. DOI: http://dx.doi.org/10.1016/S0022-0981(00)00325-7

[30] Sukhotin AA, Pörtner HO. Habitat as a factor involved in the physiological response to environmental anaerobiosis of White Sea *Mytilus edulis*. Marine Ecology Progress Series. 1999;**184**:149-160. DOI: 10.3354/meps184149

[31] Lesin PA, Khalaman VV, Flyachinskaya LP. Heterogeneity of the byssal threads in *Mytilus edulis* L. on different stages of attachment. Proceedings of the Zoological Institute of the Russian Academy of Sciences. 2006;**310**:107-112.

[32] Lysenko LA, Kantserova NP, Käiväräinen EI, Krupnova MJu, Nemova NN. Osmotic balance in marine organisms: adaptation through protein degradation. Comparative Biochemistry and Physiology Part A: Molecular & Integrative Physiology. 2012;**163**: S29–S30.

[33] Khalaman VV, Lezin PA. Clumping behavior and byssus production as strategies for substrate competition in *Mytilus edulis*. Invertebrate Biology. 2015;**134**(1):38-47. DOI: 10.1111/ivb.12075

[34] Fokina NN, Bakhmet IN, Shklyarevich GA, Nemova NN. Effect of seawater desalination and oil pollution on the lipid composition of blue mussels *Mytilus edulis* L. from the White Sea. Ecotoxicology and Environmental Safety. 2014;**110**:103-109. DOI: http://dx.doi.org/10.1016/j.ecoenv.2014.08.010

[35] Fokina NN, Lysenko LA, Sukhovskaya IV, Vdovichenko EA, Borvinskaya EV, Kantserova NP, Krupnova MYu, Ruokolainen TR, Smirnov LP, Vysotskaya RU, Bakhmet IN, Nemova

NN. Biochemical response of blue mussels *Mytilus edulis* L. from the White Sea to rapid changes in ambient temperature. Journal of Evolutionary Biochemistry and Physiology. 2015;**51**(5):378-387. DOI: 10.1134/S0022093015050038

[36] Fokina NN, Ruokolainen TR, Bakhmet IN, Nemova NN. Lipid composition in response to temperature changes in blue mussels *Mytilus edulis* L. from the White Sea. Journal of the Marine Biological Association of the United Kingdom. 2015;**95**(08):1629-1634. DOI: http://dx.doi.org/10.1017/S0025315415000326

[37] Fokina NN, Ruokolainen TR, Nemova NN, Bakhmet IN. Changes of blue mussels *Mytilus edulis* L. lipid composition under cadmium and copper toxic effect. Biological Trace Element Research. 2013;**154**(2):217-225. DOI: 10.1007/s12011-013-9727-3

[38] Fokina NN, Shklyarevich GA, Ruokolainen TR, Nemova NN. Effect of low salinity on intertidal blue mussels, *Mytilus edulis* L., from the White Sea: lipids and their fatty acid composition as a biochemical marker. In: Snyder M, editor. Aquatic Ecosystems: Influences, Interactions and Impact on the Environment. New York: Nova Science Publishers Inc; 2016. pp. 87-124.

[39] Dey I, Farkas T. Temperature shifts induce adaptive changes in the physical state of carp (*Cyprinus carpio* L.) erythrocyte plasma membranes in vitro. Fish Physiology and Biochemistry. 1992;**10**(4):347-355. DOI: 10.1007/BF00004484

[40] Williams E, Somero G. Seasonal-, tidal-cycle- and microhabitat related variation in membrane order of phospholipid vesicles from gills of the intertidal mussel *Mytilus californianus*. Journal of Experimental Biology. 1996;**199**:1587-1596.

[41] Parent GJ, Pernet F, Tremblay R, Sevigny JM, Ouellette M. Remodeling of membrane lipids in gills of adult hard clam *Mercenaria mercenaria* during declining temperature. Aquatic Biology. 2008;**3**(2):101-109. DOI: http://doi.org/10.3354/ab00073

[42] Pernet F, Tremblay R, Comeau L, Guderley H. Temperature adaptation in two bivalve species from different thermal habitats: energetics and remodelling of membrane lipids. Journal of Experimental Biology. 2007;**210**(17):2999-3014. DOI: 10.1242/jeb.006007

[43] Pernet F, Tremblay R, Gionet C, Landry T. Lipid remodeling in wild and selectively bred hard clams at low temperatures in relation to genetic and physiological parameters. Journal of Experimental Biology. 2006;**209**(23):4663-4675. DOI: 10.1242/jeb.02581

[44] Nemova NN, Fokina NN, Nefedova ZA, Ruokolainen TR, Bakhmet IN. Modifications of gill lipid composition in littoral and cultured blue mussels *Mytilus edulis* L. under the influence of ambient salinity. Polar Record. 2013;**49**(03):272-277. DOI: http://dx.doi.org/10.1017/S0032247412000629

[45] Kashin AG. Influence of Ambient Salinity on Lipid Composition of Some Water Invertebrates [dissertation]. Samara: Samara State University; 1997. 14 p.

[46] Glemet HC, Ballantyne JS. Influences of environmental salinity on the structure and function of gill mitochondrial membranes of an osmoconforming invertebrate, *Crassostrea virginica*. Marine Biology. 1995;**121**:673-683. DOI: 10.1007/BF00349303

[47] Fokina NN, Bakhmet IN, Nemova NN. Cooperative effect of crude oil and low salinity on the digestive glands lipid composition of the White Sea blue mussels *Mytilus edulis*. Proceedings ZIN. 2016;**320**(3):357-366.

[48] Berger VJa. On the minimal terms of triggering the processes of phenotypic adaptation. Doklady Biological Sciences. 2005;**400**(1):57-60. DOI: 10.1007/s10630-005-0043-8

[49] Pagliarani A, Bandiera P, Ventrella V, Trombetti F, Pirini M, Borgatti AR. Response to alkyltins of two $Na^{+}$-dependent ATPase activities in *Tapes philippinarum* and *Mytilus galloprovincialis*. Toxicology in Vitro. 2006;**20**:1145-1153. DOI: http://dx.doi.org/10.1016/j.tiv.2006.02.006

[50] Pierce SK. Invertebrate cell volume control mechanisms: a coordinated use of intracellular amino acids and inorganic ions as osmotic solute. The Biological Bulletin. 1982;**163**(3):405-419.

[51] Fokina NN, Nefedova ZA, Ruokolainen TR, Bakhmet IN, Nemova NN. Effects of various salinity on the White Sea blue mussels *Mytilus edulis* lipid composition. Proceedings ZIN. Supplement 3. 2013;**3**:55-63.

[52] Pollero RJ, Irazú CE, Brenner RR. Effect of sexual stages on lipids and fatty acids of *Diplodon delodontus*. Comparative Biochemistry and Physiology Part B: Comparative Biochemistry. 1983;**76**(4):927-931. DOI: 10.1016/0305-0491(83)90414-5

[53] Zhukova NV. Non-methylene-interrupted fatty acids of marine bivalves: distribution in tissues and lipid classes. Journal of Evolutionary Biochemistry and Physiology. 1992;**28**(4):434-440.

[54] Gardner D, Riley JP. The component fatty acids of the lipids of some species of marine and freshwater molluscs. Journal of the Marine Biological Association of the United Kingdom. 1972;**52**(04):827-838. DOI: http://dx.doi.org/10.1017/S0025315400040571

[55] Fried B, Rao KS, Sherma J, Huffman JE. Fatty acid composition of *Goniobasis virginica, Physa* sp. and *Viviparus malleatus* (Mollusca: Gastropoda) from lake Musconetcong, New Jersey. Biochemical Systematics and Ecology. 1993;**21**(8):809-812.

[56] Sushchik NN. Role of essential fatty acids in trophometabolic interactions in the freshwater ecosystems (a review). Zhurnal obshchei biologii. 2007;**69**(4):299-316.

[57] Khardin AS, Aizdaicher NA, Latyshev NA. Changes in the fatty acid composition of hepatopáncreas of the mollusk *Mytilus trossulus* fed on microalgae. Russian Journal of Marine Biology. 2003;**29**(6):378-382. DOI: 10.1023/B:RUMB.0000011706.89867.ec

[58] Alkanani T, Parrish CC, Thompson RJ, McKenzie CH. Role of fatty acids in cultured mussels, *Mytilus edulis*, grown in Notre Dame Bay, Newfoundland. Journal of Experimental Marine Biology and Ecology. 2007;**348**:33-45. DOI: http://dx.doi.org/10.1016/j.jembe.2007.02.017

[59] Alyakrinskaya IO. Resistance to drying in aquatic molluscs. Biology Bulletin. 2004;**31**:299-309. DOI: 10.1023/b:bibu.0000030153.33353.77

[60] Gromosova SA, Shapiro AZ. The Main Biochemical Patterns of Energy Metabolism in Mussels. Moscow: Legkaya I pishchevaya promyshlennost; 1984. 120 p.

[61] Gabbott PA. Developmental and seasonal metabolic activities in marine molluscs. In: Hochachka P, editor. Mollusca Vol. 2 Environmental Biochemistry and Physiology. New York: Academic press Inc; 1983. pp. 165-217.

[62] Fokina NN, Nefedova ZA, Nemova NN. Lipid Composition of *Mytilus edulis* L. Mussels from the White Sea. Effect of Some Environmental Factors. Petrozavodsk: Karelian Research Centre of RAS; 2010. 243 p. DOI: 10.13140/2.1.2154.8322

[63] Fokina N, Nemova N, Nefedova Z. Fatty acid composition of mussels *Mytilus edulis* under short-term anoxia. Chemistry and Physics of Lipids. 2007;**149**:S60–S60.

[64] Hochachka PW, Bick LT, Doll CJ, Land SC. Unifying theory of hypoxia tolerance: molecular/metabolic defense and rescue mechanisms for surviving oxygen lack. Proceedings of the National Academy of Sciences of the United States of America. 1996;**93**:9493-9498.

[65] Zhukova NV, Imbs AB, Yi LF. Diet-induced changes in lipid and fatty acid composition of *Artemia salina*. Comparative Biochemistry and Physiology Part B: Biochemistry and Molecular Biology. 1998;**120**(3):499-506. DOI: http://dx.doi.org/10.1016/S0305-0491(98)10036-6

[66] Dalsgaard J, John MS, Kattner G, Müller-Navarra D, Hagen W. Fatty acid trophic markers in the pelagic marine environment. Advances in Marine Biology. 2003;**46**:225-340. DOI: http://dx.doi.org/10.1016/S0065-2881(03)46005-7

[67] Kelly JR, Scheibling RE. Fatty acids as dietary tracers in benthic food webs. Marine Ecology Progress Series. 2012;**446**:1-22.

[68] Bergé JP, Barnathan G. Fatty acids from lipids of marine organisms: molecular biodiversity, roles as biomarkers, biologically active compounds, and economical aspects. In: Le Gal Y, Ulber R, editors. Marine Biotechnology I. Berlin Heidelberg: Springer; 2005. pp. 49-125. DOI: 10.1007/b135782

[69] Fokina NN, Ruokolainen TR, Nemova NN, Bakhmet IN. Alteration of lipid composition of blue mussels *Mytilus edulis* L. as a result of their acclimation to laboratory conditions. Proceedings of KarRC of RAS. 2015;**11**:76-84. DOI: 10.17076/eb235

[70] Newell RIE. Species profiles: life histories and environmental requirements of coastal fishes and invertebrates (North and Mid-Atlantic). Blue mussel. Biological Report. 1989;**82(11.102) TR EL-82-4**:1-34.

[71] Gerasimova AV, Maximovich NV. Age–size structure of common bivalve mollusc populations in the White Sea: the causes of instability. Hydrobiologia. 2013;**706**(1):119-137. DOI: 10.1007/s10750-012-1415-3

[72] Freas W, Grollman S. Ionic and osmotic influences on prostaglandin release from the gill tissue of a marine bivalve, *Modiolus demissus*. Journal of Experimental Biology. 1980;**84**:169-185.

[73] Stanley-Samuelson DW. Physiological roles of prostaglandins and other eicosanoids in invertebrates. The Biological Bulletin. 1987;**173**(1):92-109.

[74] Di Marzo V, Cimino G, Crispino A, Minardi C, Sodano G, Spinella A. A novel multifunctional metabolic pathway in a marine mollusc leads to unprecedented prostaglandin derivatives (prostaglandin 1, 15-lactones). Biochemical Journal. 1991;**273**:593-600. DOI: 10.1042/bj2730593

[75] Bell JG, Sargent JR. Arachidonic acid in aquaculture feeds: current status and future opportunities. Aquaculture. 2003;**218**(1):491-499. DOI: http://dx.doi.org/10.1016/S0044-8486(02)00370-8

[76] Labarta U, Fernández-Reiríz MJ, Babarro JMF. Differences in physiological energetics between intertidal and raft cultivated mussels *Mytilus galloprovincialis*. Marine Ecology Progress Series. 1997;**152**:167-173. DOI: 10.3354/meps152167

[77] Babarro JMF, Fernández-Reiriz MJ, Labarta U. Metabolism of the mussel *Mytilus galloprovincialis* from two origins in the Ría de Arousa (north-west Spain). Journal of the Marine Biological Association of the UK. 2000;**80**(05):865-872.

[78] Fernández-Reiriz MJ, Irisarri J, Labarta U. Flexibility of physiological traits underlying inter-individual growth differences in intertidal and subtidal mussels *Mytilus galloprovincialis*. PLoS ONE. 2016;**11**(2):e0148245. DOI: 10.1371/journal.pone.0148245

[79] Fokina NN, Nefedova ZA, Nemova NN, Khalaman VV. Modulating role of lipids and their fatty acids in adaptation of the White Sea mussels *Mytilus edulis* L. to environmental salinity change. Journal of Evolutionary Biochemistry and Physiology. 2007;**43**(4):379-387. DOI: 10.1134/S0022093007030023

[80] Vlahogianni TH, Valavandis A. Heavy-metal effects on lipid peroxidation and antioxidant defence enzymes in mussels *Mytilus galloprovincialis*. Chemistry and Ecology. 2007;**5**:361-371. DOI: http://dx.doi.org/10.1080/02757540701653285

[81] Geret F, Serafim A, Barreira L, Bebianno MJ. Response of antioxidant systems to copper in the gills of the clam *Ruditapes decussatus*. Marine Environmental Research. 2002;**54**:413-417. DOI: http://dx.doi.org/10.1016/S0141-1136(02)00164-2

[82] Company R, Serafim A, Bebianno MJ, Cosson R, Shillito B, Fiala-Medioni A. Effect of cadmium, copper and mercury on antioxidant enzyme activities and lipid peroxidation in the gills of the hydrothermal vent mussel *Bathymodiolus azoricus*. Marine Environmental Research. 2004;**58**:377-381. DOI: http://dx.doi.org/10.1016/j.marenvres.2004.03.083

[83] Thompson GA Jr. Metabolism and control of lipid structure modification. Biochemistry and Cell Biology. 1986;**64**(1):66-69. DOI: 10.1139/o86-010

[84] Dzogbefia VP, Kling D, Gamble W. Polychlorinated biphenyls in vitro and in vitro modifications of phospholipid and glyceride biosynthesis. Journal of Environmental Pathology and Toxicology. 1978;**1**:841-856.

[85] Capuzzo JM, Leavitt DF. Lipid composition of the digestive glands of *Mytilus edulis* and *Carcinus maenas* in response to pollutant gradients. Marine Ecology Progress Series. 1988;**46**:139-145.

[86] Kato N, Kawai K, Yoshida A. Effects of dietary polychlorinated biphenyls and protein level on liver and serum lipid metabolism of rats. Agricultural and Biological Chemistry. 1982;**46**:703-708. DOI: http://dx.doi.org/10.1080/00021369.1982.10865132

[87] Albergamo A, Rigano F, Purcaro G, Mauceri A, Fasulo S, Mondello L. Free fatty acid profiling of marine sentinels by nanoLC-EI-MS for the assessment of environmental pollution effects. Science of the Total Environment. 2016;**571**:955-962. DOI: http://dx.doi.org/10.1016/j.scitotenv.2016.07.082

[88] Filimonova V, Gonçalves F, Marques JC, De Troch M, Gonçalves AM. Fatty acid profiling as bioindicator of chemical stress in marine organisms: a review. Ecological Indicators. 2016;**67**:657-672. DOI: http://dx.doi.org/10.1016/j.ecolind.2016.03.044

[89] Gladyshev MI, Anishchenko OV, Sushchnik NN, Kalacheva GS, Gribovskaya IV, Ageev AV. Influence of anthropogenic pollution on content of essential polyunsaturated fatty acids in links of food chain of river ecosystem. Contemporary Problems of Ecology. 2012;**5**:376-385. DOI: http://dx.doi.org/10.1134/S1995425512040051

[90] Hannam ML, Bamber SD, Galloway TS, John Moody A, Jones MB. Effects of the model PAH phenanthrene on immune function and oxidative stress in the haemolymph of the temperate scallop *Pecten maximus*. Chemosphere. 2010;**78**:779-784. DOI: http://dx.doi.org/10.1016/j.chemosphere.2009.12.049ICRAM

[91] Filimonova V, Gonçalves F, Marques JC, De Troch M, Gonçalves AM. Biochemical and toxicological effects of organic (herbicide Primextra® Gold TZ) and inorganic (copper) compounds on zooplankton and phytoplankton species. Aquatic Toxicology. 2016;**177**:33-43. DOI: http://dx.doi.org/10.1016/j.aquatox.2016.05.008

[92] Rocchetta I, Pasquevich MY, Heras H, Ríos de Molina MdC, Luquet CM. Effects of sewage discharges on lipid and fatty acid composition of the Patagonian bivalve *Diplodon chilensis*. Marine Pollution Bulletin. 2014;**79**:211-219. DOI: http://dx.doi.org/10.1016/j.marpolbul.2013.12.011

[93] Signa G, Di Leonardo R, Vaccaro A, Tramati CD, Mazzola A, Vizzini S. Lipid and fatty acid biomarkers as proxies for environmental contamination in caged mussels *Mytilus galloprovincialis*. Ecological Indicators. 2015;**57**:384-394. DOI: http://dx.doi.org/10.1016/j.ecolind.2015.05.002

[94] Nechev J, Stefanov K, Popov S. Effect of cobalt ions on lipid and sterol metabolism in the marine invertebrates *Mytilus galloprovincialis* and *Actinia equina*. Comparative Biochemistry and Physiology. Part A. 2006;**144**:112-118. DOI: http://dx.doi.org/10.1016/j.cbpa.2006.02.022

[95] Bakhmet IN, Fokina NN, Nefyodova ZA, Ruokolainen TR, Nemova NN. Blue mussels *Mytilus edulis* L. in the White Sea as bioindicators under diluted oil impact. Proceedings of KarRC of RAS. 2012;**2**:38-46.

[96] Mal Reddy N, Venkateswara Rao P. Copper toxicity to the fresh water snail, *Lymnaea luteola*. Bulletin of Environmental Contamination and Toxicology. 1987;**39**(1):50-55. DOI: 10.1007/BF01691788

Chapter 8

# An Insightful Model to Study Innate Immunity and Stress Response in Deep-Sea Vent Animals: Profiling the Mussel *Bathymodiolus azoricus*

Raul Bettencourt, Inês Barros, Eva Martins,
Inês Martins, Teresa Cerqueira, Ana Colaço,
Valentina Costa, Domitília Rosa, Hugo Froufe,
Conceição Egas, Sergio Stefanni, Paul Dando and
Ricardo S. Santos

Additional information is available at the end of the chapter

http://dx.doi.org/10.5772/68034

### Abstract

Deep-sea environments are, in some cases, largely unexplored ecosystems, where life thrives driven by the geochemical features of each location. Among these environments, chemosynthesis-based ecosystems, in the Mid Atlantic Ridge, have an exclusive combination of high depth, high sulfur, and high methane concentrations. This is believed to modulate the biological composition of vent communities and influence the overall vent animal transcriptional activity of genes involved in adaptation processes to extreme environments. This opens, thus, the possibility of finding gene expression signatures specific to a given hydrothermal vent field. Regardless of the extreme physicochemical conditions that characterize deep-sea hydrothermal vents, the animals dwelling around the vent sites exhibit high productivity and thus must cope with toxic nature of vent surrounding, seemingly deleterious to the animals, while developing surprisingly successful strategies to withstand adverse environmental conditions, including environmental microbes and mechanical stress whether ensuing from animal predation or venting activity. The deep-sea vent mussel *Bathymodiolus azoricus* has adapted well to deep-sea extreme environments and represents the dominating faunal community from hydrothermal vent sites in the Mid-Atlantic Ridge, owing its successful adaptation and high biomasses to specialized exploitation of methane and sulfide sources from venting activity. Its extraordinary capabilities of adapting and thriving in chemosynthesis-based environments, largely devoid of photosynthetic primary production and characterized by rapid geochemical regime changes are due to symbiotic associations with chemosynthetic bacteria within its large gills. In an attempt to understand physiological reactions in animals normally set to endure extreme deep-sea environments, our laboratory has undertaken, for the last few years, a series of investigations, aimed at characterizing molecular indicators of adaptation processes of which components of the host defense system has received most attention. This study reviews recent advances on the characterization of molecules and genes participating in immune reactions, using *in vivo* and *ex vivo* models, to elucidate cellular and humoral defense mechanisms in vent mussels and the strategies they have adopted to survive under extreme environments.

**Keywords:** innate immunity, deep-sea hydrothermal vents, chemosynthetic ecosystems, long-term acclimatization, host-symbionts interactions, endosymbionts, differential gene expression, transcriptomics, mollusc bivalve, hydrothermal vent mussel, *Bathymodiolus azoricus*

## 1. Introduction

Deep-sea hydrothermal vents were discovered 40 years ago in the Galapagos Rift [1] revealing for the first time, to the amazement of the scientific community, unusual life forms that have developed unique biochemical adaptations to high temperatures and toxic chemical nature of vent surrounding, otherwise harmful to life as we know it on the surface of the planet [2–5]. The animals dwelling around the vent sites exhibit high productivity and thus must cope with the seemingly deleterious physical and chemical conditions, while developing surprisingly successful strategies to withstand adverse environmental conditions, including environmental microbes and mechanical stress whether due to animal predation or from deep-sea volcanic eruptions [6–8].

At such depths and in the absence of light, life is thriving in chemosynthesis-based ecosystems where most abundant marine invertebrates have developed mutualistic relationships with chemosynthetic bacteria. These symbiotic interactions are believed to play a crucial role in the survival of hydrothermal vent animals, driving their transcriptional activities, and their successful adaptation strategies to subsist under extreme environmental conditions. They essentially rely on the establishment of endosymbiosis relationships between vent animals and sulfur-oxidizing (SOX) or methane-oxidizing (MOX) bacteria [9–13].

Deep-sea vent mussels of the *Bathymodiolus* genus are dominant members at hydrothermal vents and cold seep habitats. These mussels have the peculiarity of sheltering both endosymbiotic sulfide-oxidizing and methane-oxidizing bacteria in their gills [9–13], supporting thus their endurance within this type of environment. *Bathymodiolus azoricus* is also the dominant species in deep-sea hydrothermal vents in the Azores region and is well adapted to extreme conditions particularly to toxic concentrations of heavy metals, acidic pH, and absence of light [14–17].

In an attempt to understand physiological reactions of animals normally set to endure extreme conditions, in deep-sea environments, our laboratory has undertaken, for the last 6 years, a series of investigations aimed at characterizing molecular indicators of adaptation processes of which components of the immune and stress-related systems have received most of our attention [18, 19]. Central to our studies is the long-term maintenance of vent mussels to atmospheric pressure proven to be a useful model to study unique molecular relationships under which the regulation of gene transcription may be affected by aquaria conditions and by the gradual disappearance of endosymbiont bacteria from gill epithelia [20]. Nonetheless, vent mussels subsist for months at atmospheric pressure in aquaria supplemented with plain sea water in or with artificial diet. This has allowed us to focus on developing experiments to investigate new physiological responses of animals sustaining experimental challenges involving immunological and stress-related reactions and to provide new approaches to assess the effect of natural microorganisms and metal toxicity at vent environments [21, 22]. As a research model, the choice of the vent mussel *B. azoricus* is of great significance given its unique symbiosis with SOX and MOX bacteria. It has provided us with the means to understanding the molecular mechanisms underlying immune reactions in animals normally set to endure extreme deep-sea environments and the role of their symbiotic bacteria in controlling immune gene transcriptional activity.

In line with this, we have investigated main constituents of the vent mussel immune system and demonstrated how immune and stress genes could be modulated upon different experimental challenges in the absence of the characteristic high hydrostatic pressure found at deep-sea vent sites without methane and/or sulfide supplementation [21–23]. The proximity to the nearby hydrothermal vent fields, in the Azores region, has given us a geographical advantage for earning first insight into immediate physiological responses comprising both cellular and humoral responses of live mussels, freshly collected from the hydrothermal vents, which upon arrival, are acclimatized to our aquarium system, LabHorta [23–25]. The maintenance of live mussels in our laboratory is thus a key factor in gaining knowledge into the physiology of vent animals including the study of evolutionary conserved immune, inflammatory, and stress-related factors commonly found in other marine bivalves [19, 26].

## 2. Innate immunity in *Bathymodiolus azoricus*

The interaction between microorganisms and host defense mechanisms is a decisive factor for the survival of marine bivalves. They rely on cell-mediated and humoral reactions to overcome the pathogens that naturally occur in the marine environment [27]. Growing interest in deep-sea vent biology has turned the vent mussel *B. azoricus* into a model organism centered on research activities based on the premise that vent mussels clearly have need for an immune system to overcome microbial challenges in their natural surroundings. For this reason, our research strategies have been focused on the molecular characterization of molecules participating in immune reactions, using *in vivo* and *ex vitro* models, to elucidate cellular and humoral defense mechanisms in vent mussels and its survival strategies under extreme environments. As for other bivalves, the innate immune system of *B. azoricus* is based on cellular constituents and soluble hemolymph (blood) factors, which play a prominent role in protecting the animals against invading microorganisms. The circulating hemocytes or blood cells are mostly found in the hemolymph and extrapallial fluid. They are responsible for cell-mediated defense reactions such as phagocytosis and the activation of a variety of cytotoxic reactions including the release of lysozomal enzymes and antimicrobial peptides [28–30]. Moreover, the generation of highly reactive oxygen intermediates (ROIs) and nitric oxide also plays an important defense role against pathogens [30–33]. Besides their decisive role in protecting the host from microbial assaults, bivalve hemocytes have also been implicated in other important physiological functions, including nutrient transport, digestion, wound healing and shell regeneration and/or mineralization, and excretion [34]. In addition, the hemolymph serum contains humoral defense factors such as lectins and cytokine-like molecules that are directly and indirectly involved in the killing of pathogens and in mediating cell-cell interactions, respectively. Lectins are important mediators of cellular reactions and exhibit opsonin properties, which facilitate the phagocytosis [35–39]. The hemolymph also contains antibacterial factors and lysozomal components that ensure, along with hemocyte phagocytic and cytotoxic processes, the clearance of pathogenic bacteria [38, 39]. Using a combination of light microscopy and staining procedures, three major hemocyte types are discernible in the extrapallial fluid and hemolymph of *B. azoricus*. The most abundant type was identified as granulocyte readily recognizable by their cytoplasmic granules [19]. They appear fairly homogeneous in size and showing a characteristic crescent, or half-moon shape morphology upon adherence to glass slides and before migratory movements. Granulocytes spread well onto the glass surface averaging 30–40 μm in length. In contrast, hyalinocytes presented smoother cytoplasm, i.e., a nongranular appearance due to a lower amount of cytoplasmic granules noticeable under phase contrast and differential interference contrast visualizations [19]. A third less common hemocyte type was also observed. They correspond to hemoblast-like cells and presented a spherical shape appearance with higher nucleus to cytoplasm ratio when compared to granulocytes and hyalinocytes [19]. *In vitro* phagocytic assays carried out with *B. azoricus* hemocytes revealed that 70% of the hemocytes containing more than two zymosan particles were granulocytes and to a lesser extent the percentage of phagocytic cells corresponding to hyalinocytes was 23%. In contrast, the percentage of hemoblasts containing ingested zymosan particles was 5–7%, the lowest revealed in our studies [19].

Along with hemocytes studies, we began to tackle signaling pathways putatively involved in the mediation of cellular responses in the presence of *Vibrio* spp. It was demonstrated that compounds of microbial origin could trigger detectable phosphorylation events in *B. azoricus* hemocyte extracts and likely involving the activation of different classes of mitogen-activated protein kinases (MAPKs). When challenged with a marine bacterium, *Vibrio parahaemolyticus* or a nonmarine bacterium, *Bacillus subtilis*, to stimulate hemocytes, cellular proteins were differently phosphorylated as demonstrated in Western blotting experiments using the MAPK/ERK, p38, and JNK rabbit polyclonal antibodies. Moreover, the differences seen in phosphorylation patterns could be attributed to inherent properties of the bacterial strain used, differences in the mechanisms of binding to hemocytes, or differential activation of cell membrane receptors and signaling pathways, resulting in different patterns of protein phosphorylation. Western blotting analyses suggest that *B. azoricus* hemocytes display receptors with binding affinities toward microbial molecules or to live bacteria [19].

## 3. *Ex vivo* experiments with gill tissues: new insights into tissue immune specificity toward microorganisms

As mytilid species, the deep-sea vent mussel *B. azoricus* exhibits large lamellated gills (ctenidia) arranged as numerous filament structures stacked together through ciliary junctions. Each filament is organized in two coalescent epithelial cell sheets overlaying a central lumen where hemocytes can be found. The thinness of gill filaments allows for the visualization of live hemocytes through the epithelium, and to monitor hemocyte motility directly under light microscopy [19]. Bivalve mollusk gills assume thus a strategic importance at the interface between the external milieu and the internal body cavities of the animal where contact with microorganisms is inevitable during feeding processes inasmuch as host defense responses may incur from interactions with infective pathogens during normal filtration [36–39]. For this reason, a number of typical cellular and humoral immune reactions are likely to take place in gill tissues and observable as hemocyte proliferation and phagocytosis, the activation of immune signaling pathways, and the activation of genes involved in immune, antioxidant and antibacterial responses against invading bacterial pathogens or the presence of metal toxicants [36–39].

To further test the gill's ability to mount immune reactions, a series of *ex vivo* experiments have been performed using gill tissues freshly dissected from vent mussels and subjected to short-term incubations in tissue culture well-plates and under different experimental settings. Different stimuli were carried out to demonstrate the expression of genes in gill tissues exposed to a mixture of endosymbionts previously obtained from gill extracts, and *V. parahaemolyticus*, in comparison with sterile sea water incubations (**Figure 1**). Differential gene expression results indicated that exposure to methanotrophic and thiotrophic endosymbiont preparations led to general upregulation of genes involved in immune recognition reactions, without the addition of hemolymph, while gill incubations with endosymbiont extract and to which hemolymph serum (hemocyte free) was added led to an opposite effect, resulting in a lower expression of immune recognition genes. These results contrasted with incubations performed with *V. parahaemolyticus* or with control sterile sea water. In this case, higher levels of gene expression were achieved when hemolymph was added to gill tissues incubated with *V. parahaemolyticus* or control sterile sea water (**Figure 1**). *Ex vivo* experiments as described bring evidence supporting a yet uncharacterized effect of hemolymph and its humoral constituents over endosymbionts, likely controlling immune gene expression of its host *B. azoricus*. This prompted the question of whether the host gill tissue would be able or not to recognize endosymbiont as self-particles and to which extent the host immune system does not disturb the acquisition of endosymbionts by horizontal transfer during the host larval stages. Moreover, the permanence and survival of endosymbionts within gill tissue would require a fitted control over the host immune system, acting on the transcriptional regulation of immune genes and at the level of pattern recognition receptors (PRRs) expressed by cells of the host innate immune system to detect microbial-associated molecular patterns (MAMPs) present on the surface of microorganisms [40, 41]. This endosymbiont effect over the host immune system would likely require the presence of hemolymph and its humoral constituents, as demonstrated by the gill *ex vivo* experiments (**Figures 1** and **2**). Since the gill tissue does grow over the mussel life's time we also probed different gill sections to determine levels of immune gene expression along the anterior-posterior axis, notably the "budding zone" on the posterior end, considered as the youngest section of the gill and through which endosymbionts are believed to make their entry. It was found that vent mussel gene expressions were markedly lower than other gill tissue sections, suggesting a thigh mechanism of transcriptional regulation of host genes in the presence of endosymbiont bacteria in the gill budding zone.

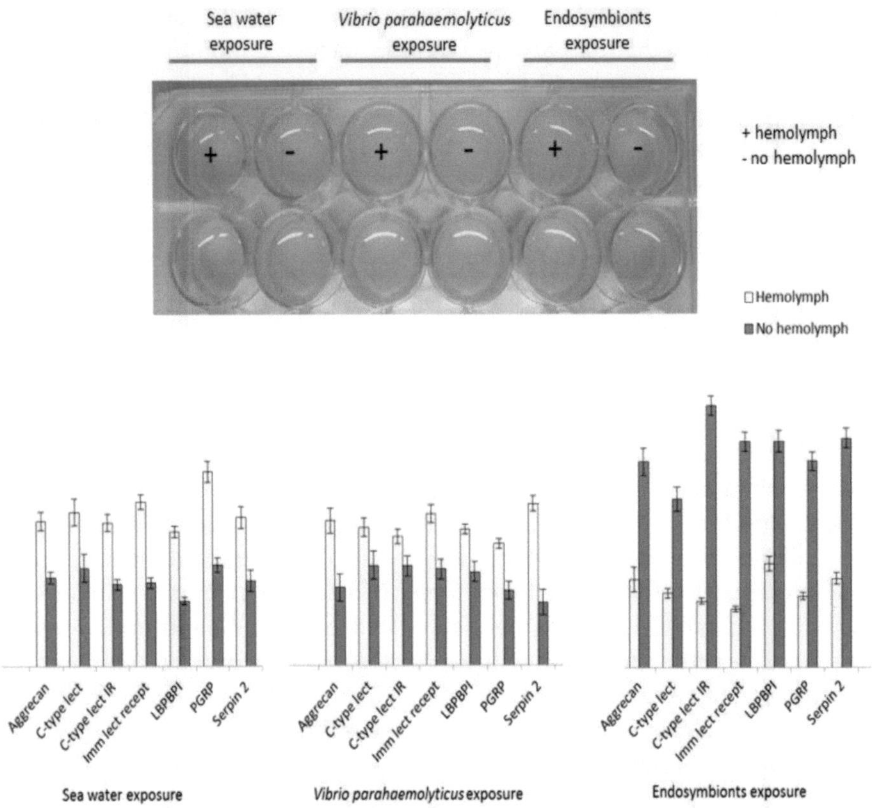


**Figure 1.** *Ex vivo* experiments performed with dissected gills. Gill fragments were incubated with *V. parahaemolyticus* and with an enriched preparation of endosymbionts freshly obtained from gill homogenates and gradient centrifugation. The effect of hemolymph humoral factors was tested by incubating gill fragments with and without hemolymph in the presence of *V. parahaemolyticus* and endosymbiont mixture. Results were compared to control incubations with plain sterile sea water. Gene expression was performed by qPCR targeting the immune recognition genes Aggrecan, C-type lectin, C-type lectin immune receptor, immune lectin receptor, lipopolysaccharide-binding protein-bactericidal/permeability-increasing protein, peptidoglycan recognition protein, and serine protease inhibitor-2 [50]. Incubations performed with endosymbiont preparations distinctively induced immune recognition genes in the absence of hemocyte-free hemolymph.

As filter feeders living most of their lives attached to a substrate, bivalves are exposed to constant biologically available pollutants over an extended period of time [42–44]. They have been studied as biological models to assess the impact of pollution in the environment and used as a biomonitoring "tool" due to their capacity of bioaccumulating high concentrations of trace metals, mostly in soft tissues such as gills and digestive gland [45–47]. The large surface of the gills and their involvement in gas exchange and feeding processes bring bivalves to constant and intimate contact with their environment where pathogens may also find their route of entry and encounter the bivalve first-line immune defense reactions.

While marine bivalves living in sandy, rocky intertidal, and shallow subtidal environments may rely on well-established humoral and cellular immune reactions to counteract pathogenic microorganisms, a new level of molecular intricacy may be seen between endosymbiont-bearing

http://dx.doi.org/10.5772/68034

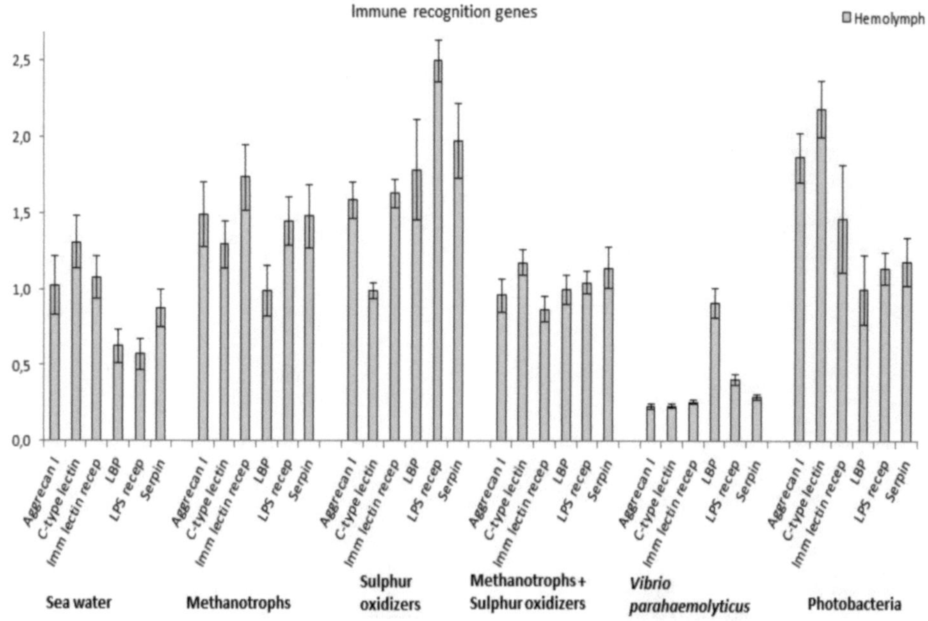


**Figure 2.** *Ex vivo* experiments performed with dissected gills. As in **Figure 1**, gill fragments were incubated with distinct bacterial stimulants: an enriched methanotrophic bacterial preparation; an enriched thiotrophic bacterial preparation, a mixture of methanotrophic and thiotrophic bacterial preparations, *Vibrio parahaemolyticus* and *Photobacteria* bacterium. Results were compared to control incubations with plain sterile sea water. Gene expression was performed by qPCR targeting same immune recognition genes as in **Figure 1**. Separate methanotrophs and thiotrophs preparations induced higher levels of immune gene expressions when compared to a mixture of the two endosymbiont bacteria preparations. Incubations using *Vibrio parahaemolyticus* resulted in drastic downregulation of immune recognition genes.

bivalves living in anaerobic and sulfide-rich environments and the pathogenic microorganisms they encounter. These natural molecular interactions would account for the role of endosymbionts in modulating the host immunity by controlling the transcriptional activity of immune genes. Bivalve associations with chemoautotrophic endosymbionts are now well known and widely distributed across a range of different chemosynthetic environments, including deep-sea hydrothermal vents (*Bathymodiolus* spp., *Calyptogena* sp.); gas seeps, mud volcanoes, and petroleum seeps (*Bathymodiolus* spp., *Calyptogena* sp.); whale and wood falls (*Idas* spp., *Adipicola* spp., *Vesicomya* sp., *Axinodon* sp.); and shallow water anoxic sediments mediated by sulfate reduction (*Solemya* spp., Codaki spp., *Anodontia* spp., *Lucina* spp.) [7].

Our recent results from *ex vivo* gill tissue experiment proven to be a valuable system for the study of tissue-specific immune responses where the thin epithelial cell layers of gill filaments would make it possible to signal pathogen-sensing directly through gill epithelia and affecting adjacent methanotrophic or thiotrophic endosymbionts which in turn would functionally prime host immune cells, the hemocytes, into altering their transcriptional activity (**Figure 3**). The endosymbiont immunomodulatory effect on the host immune system, as discussed in detail further below, is still under current investigations by our research team as the complexity of host-endosymbiont interactions in the deep-sea vent mussel *B. azoricus* remains to be

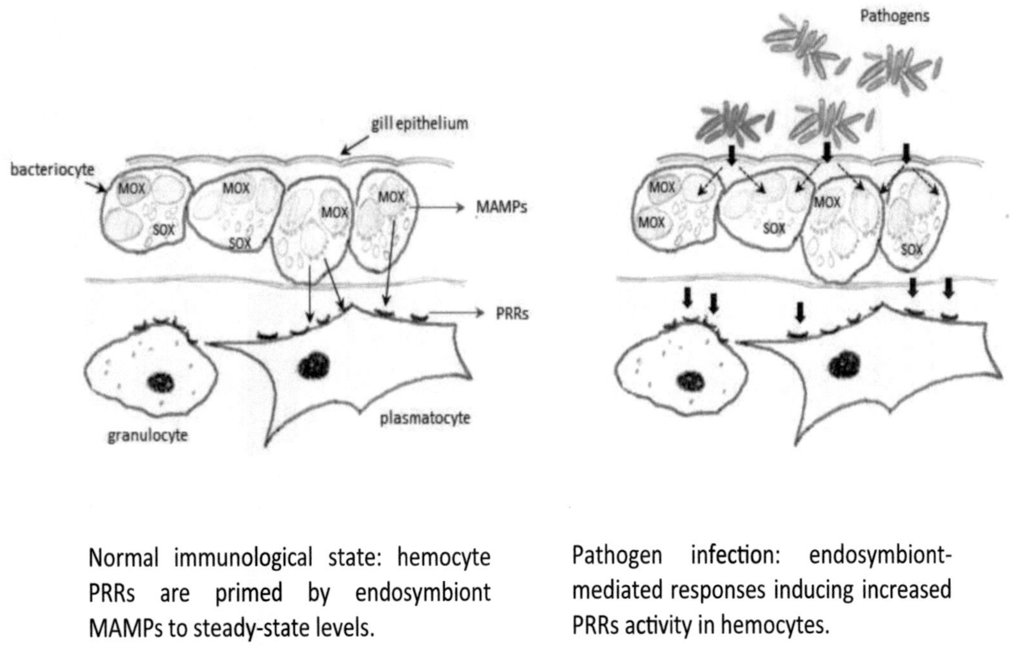


**Figure 3.** Hypothetical model representing the host-endosymbiont-mediated immune responses against pathogens. In a normal immunological state, hemocytes PRRs are being sensitized by host-endosymbiont interactions allowing the vent mussel immune system to remain active and tolerant to the presence of MOX and SOX bacteria. Upon interacting with extracellular pathogens, host-symbiont interactions are altered and incur in higher endosymbiont genes transcriptional activity [74] and subsequently affecting host hemocytes by triggering its immune repertoire via PRRs activation.

fully understood. This model is consistent with the hypothesis that innate immune receptors are required to promote long-term colonization by microbiota. This emerging perspective challenges current paradigms in immunology and suggests that PRRs may have evolved, in part, to mediate the bidirectional cross-talk between microbial symbionts and their hosts [48, 49].

## 4. Posttranscriptomics studies in *Bathymodiolus azoricus*

In 2010 a high-throughput sequencing and analysis of the gill tissue transcriptome from the deep-sea hydrothermal vent mussel *B. azoricus* was reported by our group [50]. It represented the first tissue transcriptional analysis of a deep-sea hydrothermal vent animal, using next generation sequencing technology, enabling the creation of a searchable catalog of genes that provided a direct method of identifying and retrieving vast numbers of novel coding sequences which could then be applied in gene expression profiling experiments, using quantitative polymerase chain reaction (qPCR), from a nonconventional model organism [50]. It provided the most comprehensive sequence resource for identifying novel genes currently available for a deep-sea vent organism, in particular, genes putatively involved in immune and inflammatory reactions in vent mussels. This first transcriptional analysis of gill tissues from the deep-sea hydrothermal vent *B. azoricus* was organized as a searchable catalog of genes providing a direct method of identifying and retrieving vast numbers of novel coding sequences, which can be applied in gene expression profiling experiments. The assembled and annotated sequences were organized in a dedicated database, accessible through the website http://transcriptomics.biocant.pt/deepSeaVent [50].

With an unprecedented high number of gene sequences available from our transcriptomic data, we were able to tackle signaling pathways and compare gene expression profiles in a series of experiments aiming at better understanding innate immunity in animals physiologically programmed to endure deep-sea vent conditions. Responses to bacterial infections with different strains of *Vibrio* wound experiments, long-term acclimatization in aquarium conditions and pressurization experiments with the hyperbaric chamber IPOCAMP [51] became the main focus of our research, setting thus the grounds for more in-depth analyzes revealing distinct gene expression profiles behind unique molecular relationships under which the regulation of gene transcription may be affected by biotic factors including microorganisms, the presence of endosymbiont bacteria and shell damage incurring in opportunistic infections or by abiotic factor as the hydrostatic pressure. The majority of the genes comprising four functional categories as described by Bettencourt et al. [50] and relating to immune recognition, signaling transduction, transcription, and effector molecules mechanisms were analyzed by qPCR.

The long-term aquarium maintenance of vent mussels to atmospheric pressure has long been central to our studies and proven to be a useful model to study unique molecular relationships under which the regulation of gene transcription may be affected by the gradual disappearance of endosymbiont bacteria from gill epithelia [20, 52]. Nonetheless, vent mussels from Menez Gwen hydrothermal vent site subsist for months at atmospheric pressure in aquarium conditions, in plain sea water or supplemented with methane and sulfide. This has allowed us to focus on developing experiments to investigate new physiological responses of vent mussels sustaining experimental challenges involving bacterial pathogens of the *Vibrio* genus, even in the absence of the characteristic high hydrostatic pressure found at deep-sea vent sites and without methane and sulfide supplementation [21, 22].

Earlier results from experimental exposures to *Vibrio splendidus, Vibrio alginolyticus, Vibro anguillarum,* and *Flavobacterium* sp. pointed at the immune discriminatory capacity of *B. azoricus* to distinguish different *Vibrio* strains, and at significant differences of immune gene expression levels between 12 and 24 h exposure times. These studies concluded that the immune gene transcriptional activity was modulated at two levels, i.e., over the course of time and according to the bacterial strain tested, suggesting thus, a selective response toward *Vibrio* spp. when vent mussels were experimentally challenged during 24 h [53, 54]. Additional experiments were carried out with *Vibrio diabolicus* aiming at the analysis of gene expression differences between distinct vent mussel populations from the hydrothermal vent sites Menez Gwen (MG, 800 m depth) and Lucky Strike (LS, 1700 m depth) both located on the Mid-Atlantic region, near the Azores islands. These comparative studies revealed unique immune transcriptional specificities at the gill, digestive gland, and mantle tissues level providing further evidence supporting different usage of transcription factors at the promoter region of immune genes possibly linked to the hydrothermal vent environment Furthermore, Menez Gwen (MG) and Lucky Strike (LS) *B. azoricus* showed significant gene expression differences during *V. diabolicus* challenges over time demonstrating that immune genes are differentially expressed within the same mussel populations regardless of their hydrothermal vent origin suggesting thus site-related tissue-specific gene expression patterns [55]. Moreover, these results also suggested different tissue tolerance to decompression and adaptation to atmospheric pressure not seen so far. Mantle tissues from LS mussels seemed unaffected by deep-sea retrieval showing significantly higher levels of immune gene expressions as compared to MG mantle tissues. Thus, the decompression effect on the animal's internal organs may be evaluated by ways of its ability to respond, at the immune transcriptional level, to *V. diabolicus* challenges. For that reason, mantle tissues from LS animals appear to be decompression-resistant and immune competent toward bacterial challenges. On the other hand, the digestive gland revealed the most increased gene expression levels in MG animals showing how the tissue microenvironment is relevant to *in situ* immune responses. Gill immune transcriptional activity in both MG and LS mussels was not as significantly different as for the other tissues tested which may be attributed to the presence of endosymbiont bacteria in gill epithelia acting as a driving factor likely to affect host-gene expression and the overall physiological statuses of MG and LS vent mussels while interacting with *V. diabolicus*. Even though gill tissues have been the main focus of most of our previous investigations in the deep-sea vent mussel *B. azoricus*, the digestive gland and mantle tissues hold the potential for highlighting specific immune responses in tissues other than gills and how they can modulate the outcome of the animal's overall immune responses [55].

In addition to *ex vivo* experiments and *Vibrio* exposures to live vent mussels, we were able to carry out long-term acclimatization experiments with vent mussels kept in aquaria and at atmospheric pressure. These experiments were devised to assess the effect of such prolonged aquarium conditions on immune and stress-related reactions as mussels were gradually releasing their endosymbiont bacteria from gill bacteriocytes. These studies provided a basis for understanding the interactions between host-immune and endosymbiont gene expressions during postcapture long-term acclimatization in plain sea water and represented an ideal model for investigating *B. azoricus* immune genes transcriptional activity and symbiont bacteria prevalence, in view of changes in the availability of chemical-based energy sources during acclimatization at atmospheric pressure. It also pointed out the relevance of gene expression studies while addressing the swift changes affecting metabolic adaptations and food intake fluctuations, whether induced by or as a result of the gradual loss of endosymbionts and subsequent presence of symbiont bacteria in the aquarium environment, altering thus the physiological homeostasis of *B. azoricus* [56]. These studies demonstrated that the transcriptional activity profiles for immune and bacterial endosymbiont genes followed a time-dependent mRNA transcriptional pattern evidenced at 24 h, 1 week, and 3 weeks acclimatization. Furthermore, after 1 week acclimatization, vent mussels were under the influence of what appears to be a concomitant host-immune and endosymbiont gene expression, possibly indicating a physiological transition point which induces higher levels of transcriptional activity [56]. Under such circumstances, survival of vent mussels will require immune gene repertoire switching involving the differential expression (DE) of recognition, signaling, transcription, and effector genes tied to environmental parameters and to the symbiotic relationships in *B. azoricus*. Metabolic adaptations and food intake changes, whether induced as a result of the gradual loss of endosymbionts and subsequent release in the aquarium environment, are likely to affect gene transcription activities and prevalence of symbionts in gill tissues [56–58].

The geographic proximity to the nearby hydrothermal vent fields, in the Azores region, gave our laboratory a positional advantage for earning first insight into immediate physiological responses comprising both cellular and humoral responses of freshly collected mussels from different hydrothermal vents, which upon arrival, are acclimatized to our aquarium system, LabHorta [23].The maintenance of live mussels from the shallower vent field, Menez Gwen, became thus a key factor in gaining knowledge into the physiology of vent animals including the study of evolutionary conserved immune, inflammatory and stress-related factors commonly found in other marine bivalves [18–22].

Taking advantage of the LabHorta facility, comparisons studies were made possible, with live vent mussels subjected to *V. parahaemolyticus* infection, wound injury, hyperbaric pressurization, and 3 months acclimatization (**Figure 4**). These experiments allowed for the characterization of the differential activation of signaling pathways and the relative quantification of immune genes expressed during each type of stimulation. Differential gene expression results indicated that the four experimental conditions tested were distinctively inducing the immune genes of vent mussels to different levels of transcriptional activity of which the immune and signal transduction genes showed the highest expressions (**Figure 5**).

Of the four challenging conditions *V. parahaemolyticus* infections resulted in the highest number of genes with higher level of expression during this comparison study based on qPCR and selected genes targeting immune recognition, signal transduction, transcription, and synthesis of effector molecules processes (**Figure 5**). Also, cross-talk between signaling pathways may occur in *B. azoricus* individuals subjected to *Vibrio* infections, wound responses, and hyperbaric stimulations, i.e., same immune or pro-inflammatory signaling molecules may serve different signaling pathways whether they are conspicuously more expressed or not during such experiments. Cleary, the activation of signaling pathways involved in *Vibrio* infections was distinct from that of wound and hyperbaric reactions and thus conferring the animal model presented here with the physiological versatility to cope with deep-sea hydrothermal vent environments. These experiments were important to elucidate the molecular mechanisms under which, physiological responses to bacterial infections, would responses, hyperbaric stimulations and long-term maintenance in aquaria conditions, may be involved in *B. azoricus* adaptation processes whether in deep-sea vent environments or at atmospheric pressure. However, in-depth analysis of different signaling genes and pathways involved in such experimental challenges remained fragmentary and elusive.

One the most common goals of RNA Sequencing (RNA-Seq) profiling is to identify genes or molecular pathways that are differentially expressed (DE) between two or more biological conditions [59–63]. Changes in expression can then be associated with differences in physiological reactions, providing clues for further investigation into potential mechanisms of action [64, 65]. In order to gain additional insight into the different signaling genes involved in *Vibrio* infection, wound response, long-term acclimatization, and hyperbaric repressurization, we sequenced the full transcriptome of gill tissues from each of these experimental challenges to which deep-sea vent mussels were subjected and compared their differential

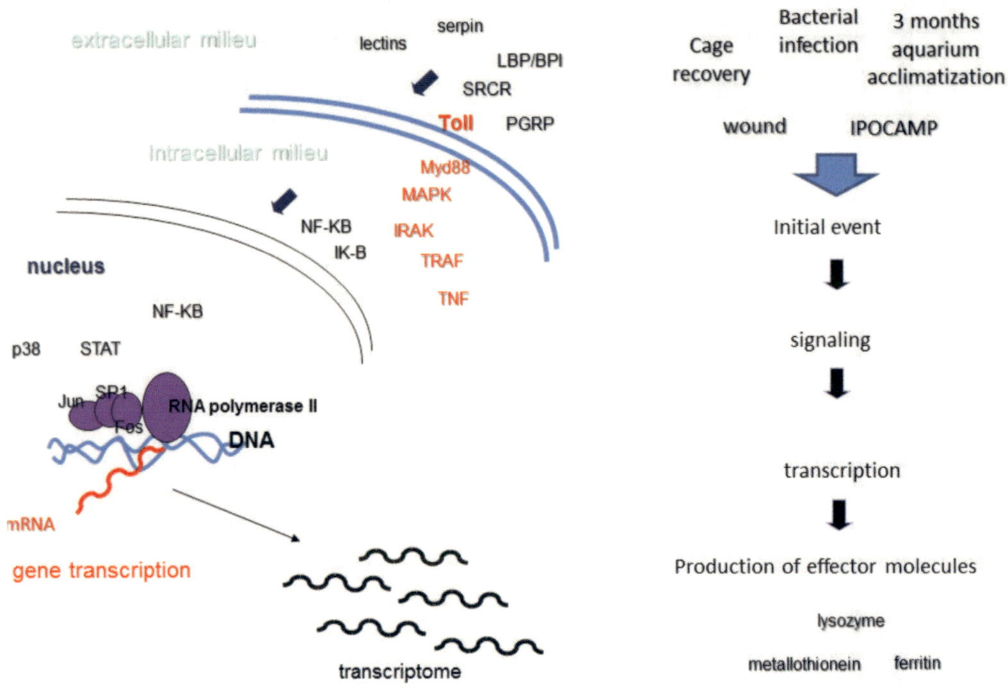


**Figure 4.** Schematic representation of immune signaling activation. After initial events characterized by immune recognition and stress-related reactions, signal transduction pathways are induced into transmitting a series of protein phosphorylation events, through the intracellular milieu, which ultimately result in the translocation of transcription factors into the nucleus that initiates the transcription of genes encoding immune effector molecules, here represented as lysozyme, metallothionein, and ferritin.

gene expression levels with that of gene expression in animals immediately retrieved from the vent sites with the help of acoustically trigged cages that were recovered at the sea surface. Transcript sequences for the five cDNA libraries were obtained from the Illumina RNA-sequencing platform and *de novo* assembly of RNA-Seq transcripts performed with Trinity [66, 67] followed by differential expression (DE) analyses using the edgeR package [68–70]. DE results were presented as Heatmaps clusters (transcriptional cluster report for edgeR DE analysis). The advantage of Heatmaps is that it can display the expression pattern of the genes across all the RNA samples. Visualization of the results is aided by clustering together genes that have correlated expression patterns [68].

Here we present examples of expression plots for some of the most DE genes across the five different experimental conditions referred to as "cage," animals freshly collected with acoustically triggered cages, from the bottom of the deep-sea vent floor; "3 months," same animals as in "cage" acclimatized for 3 months in aquaria environment at 1 atm; "Vibrio," same animals as in 3 months exposed to *V. parahaemolyticus*; "Wound," same animals as in 3 months with shell injury caused by mechanical abrasion to expose the mantle; "IPOCAMP," same animals as in 3 months subjected to 80 bar hydrostatic pressure for 72 h. The top-scoring BLAST hit for each of the gene exemplified is shown on top of the respective expression plot (**Figure 6**).

http://dx.doi.org/10.5772/68034

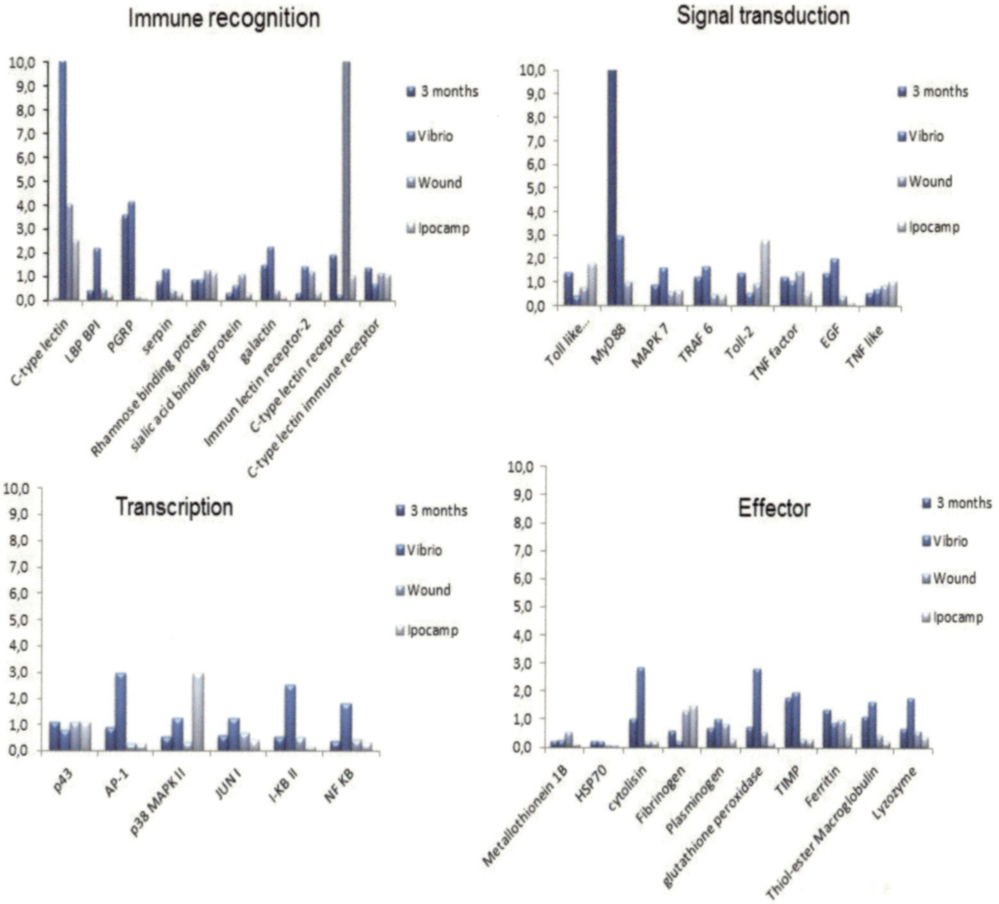


**Figure 5.** Comparative gene expression profiles from vent mussels subjected to 3 months acclimatization in aquaria at atmospheric pressure; *Vibrio parahaemolyticus* exposures; wound injury, and repressurization in the IPOCAMP chamber. Results are presented as relative expression folds calculated by qPCR and targeting immune genes from recognition, signaling transduction, transcription and effector functional gene categories as defined in Bettencourt et al. [50].

Comparison of DE across the five experiment revealed interesting correlations as for "cage" and "3 months" mussels indicating that vent mussels endured well aquarium conditions for as long as 3 months, as demonstrated by similar levels of gene expression. *Vibrio* infections and IPOCAMP pressurization also showed clustering patterns of gene expression which would seem to indicate that once mussels are acclimatized to atmospheric pressure, repressurization stimulus is impacting vent mussels in similar ways as in *V. parahaemolyticus* challenges, suggesting thus the occurrence of stress-related reactions in both types of stimulations. The expression pattern seen for wound injury was particularly distinct as compared to the other four experimental conditions. Wound injury seemed to affect drastically the vent mussel transcriptional activity which some of its genes were severely downregulated probably due to the damaging effect caused by the mechanical abrasion and direct exposure of the

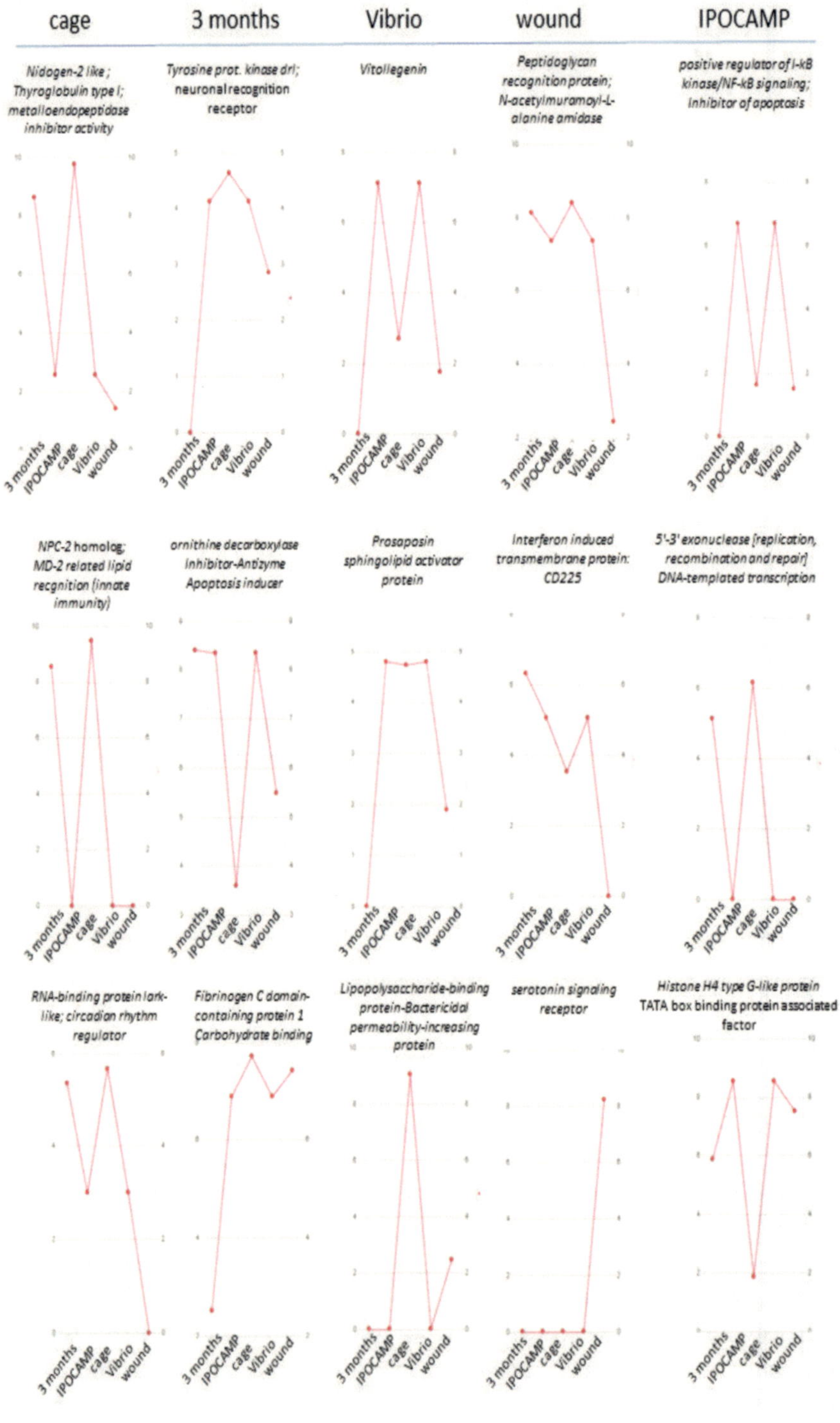


**Figure 6.** Expression plots across the five experimental conditions, "3 months"; "IPOCAMP"; "cage"; "*Vibrio*," and "Wound," representing differential gene expression analyses using EdgeR. The top-scoring BLAST hit for each of the genes exemplified is shown on top of the respective expression plot.

mantle to the aquarium environment. Taken together these experiments proven to be insightful in demonstrating the contrasting behavioral expression of given important physiological transcripts such as the peptidoglycan recognition and LBP-BPI proteins, both involved in innate immune responses, when vent mussels are met with distinct environment factors.

## 5. Host-endosymbiont interactions: implications for host immunity and defense mechanisms against bacterial pathogens

The transcriptome sequencing of gill tissues from the mussel *B. azoricus* revealed a set of genes of bacterial origin, providing a functional insight into the microbial vent community [71]. The transcripts supported a metabolically active microbiome and a variety of bacterial mechanisms and pathways, among which the fixation of carbon, the use of nitrate as a terminal acceptor of electrons and oxidation of sulfur and methane. The bacterial genes ensued from this sequencing work were deemed relevant to evaluate the influence of abiotic and biotic environmental conditions on *B. azoricus* transcriptional activity and also potentially useful to assess symbiont density differences in vent animals originated from distinct hydrothermal vent sites, respectively, to their environmental settings [72]. Keeping in line with the assumption that geographically distinct vent mussels will adopt different physiological statuses in relation to their environmental conditions, we also surmised that the relative abundance of methanotrophic and sulfide oxidizing endosymbiotic bacteria would differ between the Menez Gwen and Lucky Strike mussels as previously reported by other researchers [12, 58, 73]. We hypothesized that geographically distinct *B. azoricus* individuals may be experimentally traced back to their original hydrothermal vent sites based on their bacterial transcriptional activity and bacterial gill densities at the time animals were retrieved from the shallower Menez Gwen and deeper Lucky Strike vent sites. A taxonomical structure of the vent mussel gill's microbiome was also assessed to determine the bacterial community composition of gill tissue from MG and LS mussels to infer the symbiont densities differences between animals from both vent sites. Results from the ribosomal RNA amplicon sequencing of the V6 hypervariable regions, by massive parallel 454 pyrosequencing, indicated that the percentage of sequences obtained was from endosymbiont bacteria at nearly the same proportion between Menez Gwen and Lucky Strike samples. Moreover, comparative analyses based on BLAST searches in the RDP database, using the 16S rRNA OTU sequences, revealed that the thiotrophic endosymbiont represented 90% of all the sequences and methanotrophic endosymbiont almost 5% of the sequences from vent mussel samples originated from the distinct Menez Gwen and Lucky Strike hydrothermal vent fields [72].

While the majority of our experiments using live vent mussels were performed shortly after their retrieval from the Menez Gwen hydrothermal vent, long-term studies with vent mussels acclimatized to atmospheric pressure conditions have hardly been addressed until recently. As above-mentioned, long-term acclimatization experiments in aquarium systems have allowed us to study the expression of bacterial symbionts genes, particularly methanotrophic and thiotrophic bacteria, over time of acclimatization while their mussel host is faced with drastic physiological challenges, metabolic adaptations, and food intake changes in an effort to adapt to an aquarium environment at atmospheric pressure and without supplementation of methane and sulfur [56]. The physiological adaptation to aquarium environment is likely to be aggravated by the expelling of endosymbionts into the aquarium environment, progressively emptying the gill tissue of its autotrophic bacteria, essential for the host vent mussel nutritional sustenance. Long-term aquarium acclimatization represents thus a model study to investigate the presence and maintenance of symbiotic associations between chemosynthetic bacteria and vent animals, which depend on controlled cell-cell communication between host and endosymbionts and the role of the host immune system [56, 74].

Presumably, the loss of endosymbiont induces a dramatic change in host gene expression profiles especially if endosymbiont genes exert some transcriptional control over host gene expression. For this reason, acclimatization studies have been instrumental to further our understanding of *B. azoricus* immune system. These studies have provided insights into physiological principles underlying mechanisms of adaptation to aquarium conditions at sea level pressure while taking advantage of the remarkable capacity of vent mussels to survive well decompression once brought to surface [21, 22, 56]. Furthermore, these studies have allowed analyses using immune challenged mussels comparatively to acclimatized control mussels, maintained under aquarium conditions. In view of our previous experiments performed with live gill tissues and postcapture immune gene expression studies in *B. azoricus* acclimatized to atmospheric pressure, the presence of endosymbiont bacteria is now being under investigation as a driving factor under which host-immune genes may transcriptionally be modulated and reciprocally endosymbiont genes may transcriptionally be modulated by the host [53–56]. Moreover, the impact of aquarium acclimatization on *B. azoricus* immune responses and its capacity to react to *V. diabolicus* challenges was recently evaluated during recurrent incubations with *V. diabolicus* during short periods of time, followed by clean sea water incubations allowing animals to depurate and subsequently be reexposed to the same load of *V. diabolicus* over a period of 3 weeks acclimatization experiment [74]. As previously described, we found a time-dependent immune gene response in *B. azoricus* tied to the endosymbiont presence inside the vent mussel gills. The vent mussel's immune defense capabilities were affected by the gradual loss of symbiont bacteria suggesting a symbiont-mediated defense mechanism under which the transcriptional regulation of host immune genes is directly affected by symbiont density and/or activity. The host-immune system-endosymbiont interactions were actively higher during the first week of acclimatization as a result of *Vibrio* exposures, demonstrating the ability of *B. azoricus* to increase the transcription of immune genes while endosymbiont gene expression also correlated with an increased symbiotic metabolism and prevalence. A synergistic response was proposed to counteract the presence and potential infection by *V. diabolicus* bacterium while modulating *B. azoricus* immune defenses-endosymbiont interactions to an extant, which host-immune and endosymbiont genes are mutually reliant during the first weeks of acclimatization [74]. The evidence presented suggests successful *V. diabolicus* recognition prompting immune genes to increase their levels of transcriptional activity particularly for genes involved in the Toll-like receptor signaling [75, 76] and apoptosis-related pathways [77] during first day of acclimatization in aquarium environments. In agreement with this, *B. azoricus* is presented as a suitable model to study molecular interactions involving host-mediated immune recognition events and adaptation mechanisms, to mitigate apoptosis harmful effects induced by *Vibrio* exposure against which, endosymbionts were prompted to increase their transcriptional activity, evocative of a possible protection role to the host [74]. This work brings to light other questions relating to how the host-immune system regulates the symbiont population within their gills and conversely symbionts avoid being recognized and eliminated by the host. These topics are being further investigated in our group and focused on finding and characterizing the molecular mechanisms underlying the establishment (recognition and acquisition) and functioning of symbiosis between deep-sea vent mussel *B. azoricus* and the methanotrophic and thiotrophic bacteria (gene expression, energy metabolism, regulation of symbiont population).

Interestingly, the study of intricate associations with chemosynthetic symbiont-bacteria living in the gills of deep-sea vent animals led us to the conception of a new pathogenesis model system based on an unconventional host-symbiont model system. This new marine invertebrate model system, as for the ecotoxicological model *Mytilus* spp. [78] relies, instead, on its unique host-immune-symbiont bacteria interactions believed to play a crucial role in counteracting infectious pathogens. The establishment of invertebrate host pathogen systems may serve as suitable and useful models to study pathogenicity. The molecular mechanisms through which pathogens are able to colonize and overtake host's immune system, particularly during the initial phase of infection when molecular recognition of MAMPs is occurring, as the pathogen defines its route of entry, are expected to reveal new molecular strategies that could help developing new therapies in aquaculture diseases. Using the deep-sea vent mussel *B. azoricus* as an alternate invertebrate model system to study pathogenesis brings a new perspective into the search for new drug targets that could directly interfere with pathogen recognition processes and/or with *in situ* inflammatory process where immune cells (i.e., hemocytes) and cytokine-like molecules are being mobilized. Indeed, in such host-endosymbiont model systems, the role of endosymbiont-derived molecules could have an important influence in mediating pathogenesis and in counteracting the deleterious effect of pathogens on the host immune system. From an experimental approach, several genera of bacterial fish pathogens may be used in *B. azoricus,* as infectious agents, e.g., *Vibrio, Flavobacterium, Pseudomonas, Aeromonas, Streptococcus*. Host and endosymbiont gene expression profiles may be studied during infection experiments carried with a given bacteria and genes that are markedly upregulated or downregulated further analyzed and their cDNA sequences determined by traditional sequencing methods.

Particular attention should be given to genes whose encoded proteins are participating in signal transduction pathways directly influencing the outcome of immune effector molecules, as antibacterial peptides; immune recognition lectins and antioxidant products such as superoxide dismutase, ferritin, metalloproteinases, metallothioneins, and heat shock proteins [26]. Synergistic effects resulting from interactions between host immune and endosymbiont activity, in counteracting infectious pathogens, may now be studied at the molecular level, for future therapies design, targeting key steps during pathogen infection processes, for instance, host recognition events; production of the anti-inflammatory factor TNF-alpha and cytokine-like growth factors; enhancement of antibacterial molecules synthesis.

## 6. Concluding remarks

In an attempt to understand physiological reactions of animals normally set to endure extreme conditions, in deep-sea environments, our laboratory has undertaken, for the last 6 years, a series of investigations aimed at characterizing molecular indicators of adaptation processes of which components of the immune and antioxidative stress response systems have received most of our attention. As a research goal, long-term maintenance of vent mussels to atmospheric pressure was instrumental to further our understanding on molecular relationships under which the vent mussel-endosymbiont interactions are affected by aquaria conditions and by the gradual disappearance of endosymbiont bacteria from gill epithelia. Hence, the maintenance of live mussels in our aquarium laboratory system has been a key factor in gaining knowledge into the physiology of vent animals including the study of evolutionary conserved immune, inflammatory, and stress-related factors commonly found in other marine bivalves. *In vivo* and *ex vivo* experiments conducted with live mussels and their excised gill tissues as primary tissue cultures, allowed the specific host-endosymbiont interactions to be revealed, and further characterized in the deep-sea vent model *B. azoricus*, establishing distinct genetic signatures for the expression of endosymbiont genes and host-immune genes in relation to different environmental conditions. Increasing evidence now support the role of gills as a *bone fide* immune-responsive tissue in *B. azoricus*, consistent with a suitable study model for exploring molecular interactions involving host-endosymbiont-mediated immune recognition events and adaptation mechanisms to deep-sea hydrothermal vent environments. Such adaptation mechanisms are likely to be influenced by the microbial community composition surrounding the mussel beds at hydrothermal vents and therefore it is important to continue metatranscriptomic and metagenomic studies [79] from the gill-associated microbial diversity and surrounding hydrothermal vent sediments [80, 81] in view of the broader ecological organization and evolutionary importance of animal-bacterial microbiomes in chemosynthetic-based ecosystems in the deep sea [82, 83].

In recent years, researchers have turned to the human microbiome for its functional role in human health [84] and both composition and alterations in the microbiome have been found associated with diabetes, inflammatory bowel disease, obesity, asthma, rheumatoid arthritis, and susceptibility to infections [85]. Other microbiomes from nonmammalian and nonvertebrate species have also been characterized, for instance in insects where it was found to be highly dependent on the environment, species, and populations and affecting the fitness of species. These fitness effects may have important implications for the conservation and management of species and populations [82, 83]. Given the temporal instability of deep-sea hydrothermal vents and their constant fluctuations of physical and chemical environmental conditions, vent animal-microbiome associations have become critical for our understanding of invasion of nonnative species, responses to pathogens, and responses to chemicals and global climate change in the present and future [82] particularly when deep-sea mining activities are projected to have a major impact on deep-sea vent ecosystems [86].

## Acknowledgements

Acknowledgements are due to IMAR—Marine Institute, Research and Development Unit #531 in the Azores (Thematic Area E) and Portuguese Fundação para a Ciência e Tecnologia

(FCT) through the funding program FCOMP-01-0124-FEDER-007376 and the pluri-annual and programmatic funding schemes—OE-FEDER-POCI2001-FSE and PEst project—OE/EEI/LA0009/2011–2014. We also acknowledge the Regional Government of the Azores and Its Science and Technology Directorate through the pluri-annual funding scheme and its funding support through the Operational Program PROCONVERGENCIA 2007-20013 under the auspices of the European Union Regional strategic fund FEDER and through the research grant IBBA M2.1.2/I/029/2008-BioDeepSea awarded to RB. The Luso-American Foundation FLAD research grants project 600-08/2016 and project L-V-173/2006; the Fulbright Commission through the Fulbright Research Scholar grant awarded to RB (2005-2006) are also acknowledged. This work also received the support from the Fundação para a Ciência e Tecnologia, through the strategic research fund UID/MAR/04292/2013 granted to MARE-Marine and Environmental Sciences Centre; through the research grant IMUNOVENT PTDC/MAR/65991/2006 awarded to RB; through the research grants DEEPFUN-PTDC/MAR/111749/2009 and FCOMP-OE/EEI/LA0009/2011) awarded to AC; through the post-doctoral scholarship SFRH/BDP/14896/2004 awarded to RB; through the post-doctoral scholarship SFRH/BPD/73481/2010 awarded to IM; through the doctoral scholarships SFRH/BD/689511/2010, and EM SFRH/BD/73152/2010 awarded to EM and IB, respectively. The Azorean Science and Technology Directorate doctoral scholarship M3.1.2/F/052/2011 granted to TC is acknowledge. RB was also supported by Fundação para a Ciência e Tecnologia through CIÊNCIA-2007 Research contract (2008-2013). AC was supported by the Program Investigador FCT-IF/00029/2014/CP1230/CT0002 from the Fundação para a Ciência e Tecnologia.

## Author details

Raul Bettencourt[1,2,3]*, Inês Barros[2,3], Eva Martins[1,2], Inês Martins[1,2,3], Teresa Cerqueira[2,4], Ana Colaço[1,2,3], Valentina Costa[2,3,4], Domitília Rosa[4], Hugo Froufe[5], Conceição Egas[5], Sergio Stefanni[6], Paul Dando[7] and Ricardo S. Santos[1,2,3,4]

*Address all correspondence to: raul.s.bettencourt@uac.pt

1 IMAR, Marine Institute, Azores, Horta, Portugal

2 MARE, Marine and Environmental Sciences Centre, Azores, Horta, Portugal

3 OKEANOS, Research Unit, Faculty of Science and Technology, University of the Azores, Horta, Azores, Portugal

4 Department of Oceanography and Fisheries, Faculty of Science and Technology, University of the Azores, Horta, Portugal

5 Next Generation Sequencing Unit, UCBiotech, CNC, Parque Tecnológico de Cantanhede, Cantanhede, Portugal

6 Stazione Zoologica A Dohrn, Napoli, Italy

7 Marine Biological Association of the UK, Plymouth, United Kingdom

## References

[1] Corliss JB, Dymond J, Gordon LI, Edmond JM, von Herzen RP, et al. Submarine thermal springs on the Galapagos Rift. Science. 1979;**203**:1073-1083. DOI: 10.1126/science.203.4385.1073

[2] Childress JJ, Fisher CR. The biology of hydrothermal vent animals: Physiology, biochemistry, and autotrophic symbioses. Oceanography and Marine Biology: An Annual Review. 1992;**30**:337-441. Aberdeen University Press/Allen & Unwin: London. ISSN 0078-3218

[3] Childress JJ. Are there physiological and biochemical adaptations of metabolism in deep-sea animals? Trends in Ecology and Evolution. 1995;**10**(1):30-36. PMCID: PMC40849

[4] Van Dover CL. The Ecology of Deep-Sea Hydrothermal Vents. Princeton, NJ: Princeton University Press; 2000. ISBN: 9780691049298

[5] Salerno JL, Macko SA, Hallam SJ, Bright M, Won YJ, McKiness Z, Van Dover CL. Characterization of symbiont populations in life-history stages of mussels from chemosynthetic environments. Biological Bulletin. 2005;**208**:145-155.DOI:10.2307/3593123

[6] Cavanaugh CM, McKiness ZP, Newton IL, Stewart FJ. Marine chemosynthetic symbioses. In: The Prokaryotes. New York: Springer; 2006. pp. 475-507

[7] Dubilier N, Bergin C, Lott C. Symbiotic diversity in marine animals: The art of harnessing chemosynthesis. Nature Reviews Microbiology. 2008;**6**:725-740.DOI:10.1038/nrmicro1992

[8] Duperron S, Lorion J, Samadi S, Gros O, Gaill F. Symbioses between deep-sea mussels (Mytilidae: Bathymodiolinae) and chemosynthetic bacteria: Diversity, function and evolution. Comptes Rendues Biologies. 2009;**332**:298-310. DOI:10.1016/j.crvi.2008.08.003

[9] Von Cosel R, Comtet T, Krylova EM. *Bathymodiolus* (Bivalvia: Mytilidae) from hydrothermal vents on the Azores triple junction and the Logatchev hydrothermal field, Mid Atlantic Ridge. Veliger. 1999;**42**:218-248. biostor/129413

[10] Distel DL, Lee HKW, Cavanaugh CM. Intracellular coexistence of methanotrophic and thioautotrophic bacteria in a hydrothermal vent mussel. Proceedings of the National Academy of Sciences of the United States of America. 1995;**92**:9598-9602. PMCID: PMC40849

[11] Fiala-Medioni A, McKiness ZP, Dando P, Boulegue J, Mariotti A, Alayse-Danet AM, Robinson JJ, Cavanaugh CM. Ultrastructural, biochemical, and immunological characterization of two populations of the mytilid mussel *Bathymodiolus azoricus* from the Mid-Atlantic Ridge: Evidence for a dual symbiosis. Marine Biology. 2002;**141**:1035-1043

[12] Duperron S, Bergin C, Zielinski F, Blazejak A, Pernthaler A, McKiness ZP, DeChaine E, Cavanaugh CM, Dubilier N. A dual symbiosis shared by two mussel species, *Bathymodiolus azoricus* and *Bathymodiolus puteoserpentis* (Bivalvia: Mytilidae), from hydrothermal vents

along the northern Mid-Atlantic Ridge. Environmental Microbiology. 2006;**8**:1441-1447. DOI: 10.1111/j.1462-2920.2006.01038

[13] Duperron S, Gaudron SM, Rodrigues CF, Cunha MR, Decker C, Olu K. An overview of chemosynthetic symbioses in bivalves from the North Atlantic and Mediterranean Sea. Biogeosciences. 2013;**10**:3241-3267. DOI: 10.5194/bg-10-3241-2013

[14] Charlou JL, Donval JP, et al. Compared geochemical signatures and the evolution of Menez Gwen (37°50′N) and Lucky Strike (37°17′N) hydrothermal fluids, south of the Azores Triple Junction on the Mid-Atlantic Ridge. Chemical Geology. 2000;**171**:49-75

[15] Fisher CR, Brooks JM, Childress JJ, Felbeck H, Hessler RR, Johnson KS, Macko SA, Moe A, Nelson D, Somero GN. Microhabitat requirements of the hydrothermal vent mussel, *Bathymodiolus thermophilus*. American Zoology. 1989;**29**:A81. DOI: http://dx.doi.org/10.1016/0198-0149(88)90049-0

[16] Desbruyères D, Almeida A, Biscoito M, Comtet T, Khripounoff A, Le Bris N, Sarradin PM, Segonzac M. A review of the distribution of hydrothermal vent communities along the northern Mid-Atlantic Ridge: Dispersal vs. Environmental controls. Hydrobiologia. 2000;**400**:201-216. DOI: 10.1023/A:1004175211848

[17] Desbruyères D, Biscoito M, Caprais JC, Colaço A, Comtet T, Crassous Ph, Fouquet Y, Khripounoff A, Le Bris N, Olu K, Riso R, Sarradin PM, Segonzac M, Vangriesheim A. Variations in deep-sea hydrothermal vent communities on the Mid-Atlantic Ridge near the Azores plateau. Deep Sea Research Part I. 2001;**48**:1325-1346. DOI: http://dx.doi.org/10.1016/S0967-0637(00)00083-2

[18] Bettencourt R, Roch P, Stefanni S, Rosa D, Colaço A, Serrão Santos R. Deep sea immunity: Unveiling immune constituents from the hydrothermal vent mussel *Bathymodiolus azoricus*. Marine Environmental Research. **64**:108-127. DOI: 10.1016/j.marenvres.2006.12.010

[19] Bettencourt R, Dando P, Collins P, Costa V, Allam B, Serrão Santos R. Innate immunity in the deep-sea hydrothermal vent mussel *Bathymodiolus azoricus*. Comparative Biochemistry and Physiology Part A: Molecular & Integrative Physiology. 2009;**152**:278-289. DOI: http://dx.doi.org/10.1016/j.cbpa.2008.10.022

[20] Bettencourt R, Dando P, Rosa D, Riou V, Colaço A, Sarrazin J, Sarradin PM, Serrão Santos R. Changes of gill and hemocyte-related bio-indicators during long termmaintenance of the vent mussel *Bathymodiolus azoricus* held in aquaria at atmospheric pressure. Comparative Biochemistry and Physiology Part A. 2008;**150**:1-7. DOI: 10.1016/j.cbpa.2008.02.020

[21] Bettencourt R, Costa V, Laranjo M, Rosa D, Pires L, Colaco A, Lopes H, Serrao Santos R. Out of the deep-sea into a land-based aquarium environment: Investigating physiological adaptations in the hydrothermal vent mussel *Bathymodiolus azoricus*. ICES Journal of Marine Science. 2011;**68**(2):357-364. DOI: 10.1093/icesjms/fsq119

[22] Bettencourt R, Costa V, Laranjo M, Rosa D, Pires L, Colaco A, Sarradin Pierre M, Lopes H, Sarrazin Marie J, Santos Serrao R. Out of the deep-sea into a land-based aquar-

ium environment: Investigating innate immunity in the hydrothermal vent mussel *Bathymodiolus azoricus*. Cahiers De Biologie Marine. 2010;**51**(4):341-350. DOI: http://archimer.ifremer.fr/doc/00019/13042/

[23] Colaco A,Bettencourt R, Costa V, Lino S, Lopes H, Martins I, Pires L, Prieto C, Serrão Santos R. LabHorta: A controlled aquarium system for monitoring physiological characteristics of the hydrothermal vent mussel *Bathymodiolus azoricus*. ICES Proceedings ICES Journal of Marine Science. 2011;**68**(2):349-356. DOI: 10.1093/icesjms/fsq120

[24] Kadar E, Bettencourt R, Costa V, Santos RS, Lobo-Da-Cunha A, Dando P. Experimentally induced endosymbiont loss and re-acquirement in the hydrothermal vent bivalve *Bathymodiolus azoricus*. Journal of Experimental Marine Biology and Ecology. 2005;**318**:99-110. DOI: 10.1016/j.jembe.2004.12.025

[25] Kadar E, Powell JJ. Post-capture experimental investigations using hydrothermal vent macro-invertebrates to study adaptations to extreme environments. Reviews in Environmental Science and Bio/Technology. 2006;**5**:193-201. DOI: 10.1007/ 978-1-40 20-6 285-8_18

[26] Martins I, Romão CV, Goulart J, Cerqueira T, Santos RS, Bettencourt R. Activity of antioxidant enzymes in response to atmospheric pressure induced physiological stress in deep-sea hydrothermal vent mussel *Bathymodiolus azoricus*. Marine Environmental Research. 2016;**114**:65-73. DOI: 10.1016/j.marenvres.2016.01.003

[27] Canesi L, Pruzzo C. Specificity of innate immunity in bivalves: A lesson from Bacteria. In: Ballarin L, Cammarata M, editors. Lessons in Immunity: From Single-Cell Organisms to Mammals. Academic Press; Elsevier Inc; 2016. ISBN: 978-0-12-803252

[28] Allam B, Raftos D. Immune responses to infectious diseases in bivalves. Journal of Invertebrate Pathology. 2015;**131**:121-136. DOI: 10.1016/j.jip.2015.05.005

[29] Song L, Wang L, Zhang H, Wang M. The immune system and its modulation mechanism in scallop. Fish and Shellfish Immunology. 2015;**46**(1):65-78. DOI: 10.1016/j.fsi.2015.03.013

[30] Gerdol M, Venier P. An updated molecular basis for mussel immunity. Fish and Shellfish Immunology. 2015;**46**(1):17-38. DOI: 10.1016/j.fsi.2015.02.013

[31] Richard G, Guérard F, Corporeau C, Lambert C, Paillard C, Pernet F. Metabolic responses of clam *Ruditapes philippinarum* exposed to its pathogen *Vibrio* tapetis in relation to diet. Developmental Comparative Immunology. 2016;**60**:96-107. DOI: 10.1016/j.dci.2016.02.026

[32] Wang Q, Cao R, Ning X, You L, Mu C, Wang C, Wei L, Cong M, Wu H, Zhao J. Effects of ocean acidification on immune responses of the Pacific oyster *Crassostrea gigas*. Fish and Shellfish Immunology. 2016;**49**:24-33. DOI: 10.1016/j.fsi.2015.12.025

[33] Ben Cheikh Y, Travers MA, Morga B, Godfrin Y, Rioult D, Le Foll F. First evidence for a *Vibrio* strain pathogenic to *Mytilus edulis* altering hemocyte immune capacities. Developmental & Comparative Immunology. 2016;**57**:107-119. DOI: 10.1016/j.dci.2015.12.014

[34] Donaghy L, Lambert C, Choi KS, Soudant P. Hemocytes of the carpet shell clam (*Ruditapes decussatus*) and the Manila clam (*Ruditapes philippinarum*): Current knowledge and future prospects. Aquaculture. 2009;**297:**10-24. <hal-00460793>

[35] Wootton EC, Dyrynda EA, Ratcliffe NA. Bivalve immunity: Comparisons between the marine Mussel (*Mytilus edulis*), the edible cockle (*Cerastoderma edule*) and the razor Shell (*Ensis siliqua*). Fish and Shellfish Immunology. 2003;**15**(3):195-210

[36] Allam B, Ashton-Alcox A, Ford SE, Flow cytometric measurement of haemocyte viability and phagocytic activity in the clam, *Ruditapes philippinarum.* Journal of Shellfish Research. 2002;**21**:13-19

[37] Allam B, Paillard C, Ford SE. Pathogenicity of *Vibrio tapetis,* the etiological agent of brown ring disease (BRD) in clams. Diseases of Aquatic Organisms. 2002;**48**:221-231

[38] Canesi L, Gallo G, Gavioli M, Pruzzo C. Bacteria-hemocyte interactions and phagocytosis in marine bivalves. Microscopy and Research Technique Journal. 2002;**57**(6):469-476

[39] Ciacci C, Citterio B, Betti M, Canonico B, Roch P, Canesi L. Functional differential immune responses of *Mytilus gallorpovincialis* to bacterial challenge. Comparative Biochemistry and Physiology Part B: Biochemistry and Molecular Biology. 2009;**153**(4):365-371

[40] Takeuchi O, Akira S. Pattern recognition receptors and inflammation. Cell. 2010;**140**(6): 805-820. DOI: 10.1016/j.cell.2010.01.022

[41] Vance RE, Isberg RR, Portnoy DA. Patterns of pathogenesis: Discrimination of pathogenic and nonpathogenic microbes by the innate immune system. Cell Host & Microbe. 2009;**6**. DOI: 10.1016/j.chom.2009.06.007

[42] Zuykov M, Pelletier E, Harper DAT. Bivalve mollusks in metal pollution studies: From bioaccumulation to biomonitoring. Chemosphere. 2013;**93**(2):201-208

[43] El-Sikaily A, Khaled A, El Nemr A. Heavy metals monitoring using bivalves from Mediterranean sea and Red sea. Environmental Monitoring and Assessment. 2004;**98**(1): 41-58

[44] Spada L, Annicciarico C, Cardellicchio N, Giandomenico S, Di Leo A. Heavy metals monitoring in the mussel *Mytilus galloprovincialis* from the Apulian coast (Southern Italy). Mediterranean Marine Science. 2013;**14**(1):99-108

[45] Martins I, Cosson RP, Riou V, Sarradin P-M, Sarrazin J, Santos Serrão R, Colaço A. Relationship between metal levels in the vent mussel *Bathymodiolus azoricus* and local microhabitat chemical characteristics of Eiffel Tower (Lucky Strike). Deep Sea Research Part I: Oceanographic Research Papers. 2011;**58**(3):306-315. DOI: 10.1016/j.dsr.2011.01.002

[46] Cosson RP, Thiébaut E, Company R, Castrec-Rouelle M, Colaço A, Martins I, Sarradin PM, Bebianno MJ. Spatial variation of metal bioaccumulation in the hydrothermal vent mussel *Bathymodiolus azoricus*. Marine Environmental Research. 2008;**65**(5):405-415. DOI: 10.1016/j.marenvres. 2008.01.005

[47] Lino AS, Galvão PMA, Longo RTL, Azevedo-Silva CE, Dorneles PR, Torres JPM, Malm O. Metal bioaccumulation in consumed marine bivalves in Southeast Brazilian coast. Journal of Trace Elements in Medicine and Biology. 2016;**34**:50-55. http://dx.doi.org/10.1016/j.jtemb.2015.12.004

[48] Chu H, Mazmanian SK. Innate immune recognition of the microbiota promotes host microbial symbiosis. Nature Immunology. 2013;**14**(7):668-675. DOI:10.1038/ni.2635

[49] Littman DR, Pamer EG. Role of the commensal microbiota in normal and pathogenic host immune responses. Cell Host Microbe. 2011;**10**(4):311-323. DOI: 10.1016/j.chom.2011.10.004

[50] Bettencourt R, Pinheiro M, Egas C, Gomes P, Afonso M, Shank T, Santos RS. High-throughput sequencing and analysis of the gill tissue transcriptome from the deep-sea hydrothermal vent mussel *Bathymodiolus azoricus*. BMC Genomics. 2010;**11**:559. DOI: 10.1186/1471-2164-11-559

[51] Shillito B, Jollivet D, Sarradin P, Rodier P, Lallier F, et al. Temperature resistance of *Hesiolyra bergi*, a polychaetous annelid living on deep-sea vent smoker walls. Marine Ecology Progress Series. 2001;**216**:141-149. DOI: http://archimer.ifremer.fr/doc/00000/811/

[52] Riou V, Halary S, Duperron S, Bouillon S, Elskens M, Bettencourt R, Santos RS, Dehairs F, Colaço A. Influence of $CH_4$ and $H_2S$ availability on symbiont distribution, carbon assimilation and transfer in the dual symbiotic vent mussel *Bathymodiolus azoricus*. Biogeosciences. 2008;**5**:1681-1691. DOI:10.5194/bg-5-1681-2008

[53] Martins E, Queiroz A, Serrão Santos R, Bettencourt R. Finding immune gene expression differences induced by marine bacterial pathogens in the Deep-sea hydrothermal vent mussel *Bathymodiolus azoricus*. Biogeosciences. 2013;**10**(11):7279-7291. DOI: 10.5194/bg-10-7279-2013

[54] Martins E, Figueras A, Novoa B, Serrão Santos R, Moreira R, Bettencourt R. Comparative study of immune responses in the Deep-sea hydrothermal vent mussel *Bathymodiolus azoricus* and the Shallow-water mussel *Mytilus galloprovincialis* challenged with *Vibrio* bacteria. Fish and Shellfish Immunology. 2014;**40**(2):485-499. DOI: 10.1016/j.fsi.2014.07.018

[55] Martins E, Serrão Santos R, Bettencourt R. *Vibrio diabolicus* challenge in *Bathymodiolus azoricus* populations from Menez Gwen and Lucky Strike hydrothermal vent sites. Fish Shellfish Immunology. 2015;**47**:962-977. DOI: http://dx.doi.org/10.1016/j.fsi.2015.10.038

[56] Barros I, Divya B, Martins I, Vandeperre F, Santos RS, et al. Post-capture immune gene expression studies in the deep-sea hydrothermal vent mussel *Bathymodiolus azoricus* acclimatized to atmospheric pressure. Fish Shellfish Immunology. 2014;**42**:159-170. DOI: 10.1016/j.fsi.2014.10.018

[57] Boutet I, Jollivet D, Shillito B, Moraga D, Tanguy A. Molecular identification of differentially regulated genes in the hydrothermal-vent species *Bathymodiolus thermophilus* and *Paralvinella pandorae* in response to temperature. BMC Genomics. 2009;**10**:222

[58] Boutet I, Ripp R, Lecompte O, Dossat C, Corre E, Tanguy A, Lallier FH. Conjugating effects of symbionts and environmental factors on gene expression in deep-sea hydrothermal vent mussels. BMC Genomics. 2011;**12**:530. DOI: 10.1186/1471-2164-12-530

[59] Wang Z, Gerstein M, Snyder M. RNA-Seq: A revolutionary tool for transcriptomics. Nature Reviews Genetics. 2009;**10**(1):57-63. DOI: 10.1038/nrg2484

[60] Wall PK, et al. Comparison of next generation sequencing technologies for transcriptome characterization. BMC Genomics. 2009;**10**:347. DOI: 10.1186/1471-2164-10-347

[61] Hahn DA, Ragland GJ, Shoemaker DW, Denlinger DL. Gene discovery using massively parallel pyrosequencing to develop ESTs for the flesh fly *Sarcophaga crassipalpis*. BMC Genomics. 2009;**10**:234. DOI: 10.1186/1471-2164-10-234

[62] Emrich SJ, Barbazuk WB, Li L, Schnable PS. Gene discovery and annotation using LCM-454 transcriptome sequencing. Genome Research. 2007;(1):69-73

[63] Jarvie T, Harkins T. Transcriptome sequencing with the genome sequencer FLX system. Nature Methods. 2007;**4**

[64] Grabherr MG, et al. Full-length transcriptome assembly from RNA-seq data without a reference genome. Nature Biotechnology. 2011;**29**(7):644-652. DOI: 10.1038/nbt.1883

[65] Chen Y, Lun ATL, Smyth GK. From reads to genes to pathways: Differential expression analysis of RNA-Seq experiments using Rsubread and the edgeR quasi-likelihood pipeline. F1000Research. 2016;**5**:1438

[66] Brown ]Stuart M. Sequencing-by-Synthesis: Explaining the Illumina Sequencing Technology [Internet] 2012. Available from: http://bitesizebio.com/13546/sequencing-by-synthesis-explaining-the-illumina-sequencing-technology/ [Accessed: August 30, 2012]

[67] Hass BJ, et al. De novo transcript sequence reconstruction from RNA-seq using the Trinity platform for reference generation and analysis. Nature Protocols. 2013;**8**:1494-1512. DOI: 10.1038/nprot.2013.084

[68] Robinson MD, McCarthy DJ, Smyth GK. edgeR: A bioconductor package for differential expression analysis of digital gene expression data. Bioinformatics. 2010;**26**(1):139-140

[69] http://trinityrnaseq.sourceforge.net/analysis/diff_expression_analysis.html

[70] http://manuals.bioinformatics.ucr.edu/home/R_BioCondManual

[71] Egas C, Pinheiro M, Gomes P, Barroso C, Bettencourt R. The transcriptome of *Bathymodiolus azoricus* gill reveals expression of genes from endosymbionts and free-living deep-sea bacteria. Marine Drugs. 2012;**10**:1765-1783. DOI: 10.3390/md10081765

[72] Bettencourt R, Rodrigues M, Barros I, Cerqueira T, Freitas C, Costa V, Pinheiro M, Egas C, Serrão Santos R. Site-related differences in gene expression and bacterial densities in the mussel *Bathymodiolus azoricus* from the Menez Gwen and Lucky Strike deep-sea hydrothermal vent sites. Fish and Shellfish Immunology. 2014;**39**(2):343-353. DOI: 10.1016/j.fsi.2014.05.024

[73] Duperron S, Nadalig T, Caprais J-C, Sibuet M, Fiala-Médioni A, Rudolf A, Dubilier N. Dual symbiosis in a *Bathymodiolus* sp. mussel from a methane seep on the Gabon continental margin (Southeast Atlantic): 16S rRNA phylogeny and distribution of the symbionts in gills. Applied and Environmental Microbiology. 2005;**71**(4):1694-1700

[74] Barros I, Mendes S, Rosa D, Serrão Santos R, Bettencourt R. *Vibrio diabolicus* immunomodulatory effects on *Bathymodiolus azoricus* during long-term acclimatization at atmospheric pressure. Journal of Aquaculture and Development. 2016,**7**:12 DOI: 10.4172/2155-9546.1000464

[75] Kawai T, Akira S. Signaling to NF-kB by Toll-like receptors. Trends in Molecular Medicine. 2007;**13**(11). DOI: 10.1016/j.molmed.2007.09.002

[76] Toubiana M, Rosani U, Giambelluca S, Cammarata M, Gerdol M, Pallavicini A, Venier P, Roch P. Toll signal transduction pathway in bivalves: Complete cds of intermediate elements and related gene transcription levels in hemocytes of immune stimulated *Mytilus galloprovincialis*. Developmental Comparative Immunology. 2014;**45**(2):300-312. DOI: 10.1016/j.dci.2014.03.021

[77] Canesi L, Ciacci C, Bergami E, Monopoli MP, Dawson KA, Papa S, Canonico B, Corsi I. Evidence for immunomodulation and apoptotic processes induced by cationic polystyrene nanoparticles in the hemocytes of the marine bivalve *Mytilus*. Marine Environmental Research. 2015;**111**:34-40. DOI: 10.1016/j.marenvres.2015.06.008

[78] Schmidt W, Power E, Quinn B. Seasonal variations of biomarker responses in the marine blue mussel (*Mytilus* spp). Marine Pollution Bulletin. 2013;**74**(1):50-55. DOI: 10.1016/j.marpolbul.2013.07.033

[79] Simon C, Daniel R. Metagenomic analyses: Past and future trend. Applied and Environmental Microbiology.2011;**77**(4):1153-1161. DOI: 10.1128/AEM.02345-10

[80] Cerqueira T, Pinho D, Egas C,Froufe H, Altermark B, Candeias C, Santos RS, Bettencourt R. Microbial diversity in deep-sea sediments from the Menez Gwen hydrothermal vent system of the Mid-Atlantic Ridge. Marine Genomics. 2015. DOI: 10.1016/j.margen.2015.09.001

[81] Cerqueira T, Pinho D, Froufe H, Serrão Santos R, Bettencourt R, Egas C. Sediment microbial diversity of three deep-sea hydrothermal vents southwest of the Azores. Microbial Ecology. 2017. DOI 10.1007/s00248-017-0943-9

[82] Christian N, Whitaker BK, Clay, K. Microbiomes: Unifying animal and plant systems through the lens of community ecology theory. Frontiers in Microbiology. 2015;**6**:869. DOI: 10.3389/fmicb.2015.00869

[83] Bahrndorff S, Alemu T, Alemneh T, Lund Nielsen J. The microbiome of animals: Implications for conservation biology. International Journal of Genomics. 2016; 5304028. DOI: 10.1155/2016/5304028

[84] Turnbaugh PJ, Ley RE, Hamady M, Fraser-Liggett CM, Knight R, Gordon JI. The human microbiome project. Nature. 2007;**449**(7164):804-810. DOI: 10.1038/nature06244

[85] Cho I, Blaser MJ. The human microbiome: At the interface of heath and disease. Nature Reviews Genetics. 2012;**13**(4):260-270. DOI: 10.1038/nrg3182

[86] Van Dover CL. Impacts of anthropogenic disturbances at deep-sea hydrothermal vent ecoystems: A review. Marine Environmental Research. 2014;**102**:59-72. DOI: 10.1016/j.marenvres.2014.03.008

Chapter 9

# Gut Microbiome Analysis of Snails: A Biotechnological Approach

Mudasir A. Dar, Kiran D. Pawar and
Radhakrishna S. Pandit

Additional information is available at the end of the chapter

http://dx.doi.org/10.5772/68133

**Abstract**

Mollusks are a diverse group of animals not only at the species level but also with respect to their habitat and behavior. Gastropods comprise 80% of the mollusks with approximately 62,000 living species including snails. Over the period of time, snails have evolved into marine, freshwater and terrestrial forms with a transitional shift in their feeding habits. From prehistoric times, mollusks have established an intimate relationship with humans. These animals are used as food, medicine, offering to gods and are also responsible for economic losses in the form of agricultural pests. As most of these animals feed on plant biomass, their guts have evolved to digest such lignocellulosic biomass with extraordinary efficiency. The plant fiber digestion in their guts depends predominantly on the metabolic activities of the gastro-intestinal microflora. Besides digestive functions, the seasonal dynamic and spatial distribution of bacterial gut community largely influences cold hardiness and many other metabolic properties in snails. Here, we assessed an overview of the various bacterial populations dwelling in digestive tracts of snails. This chapter provides insights into the gut microbiome of various snails that can be exploited for various industrial applications such as biomass degradation, production of biofuel, paper, wine and laundry detergents.

**Keywords:** Mollusca, snails, gut microbiome, symbiont, bacteria, industrial uses

## 1. Introduction

The phylum Mollusca is one of the most diverse groups of animals on earth that comprises 50,000 living species. Mollusks are soft-bodied animals that inhabit almost every kind of habitat. These are dominantly free-living metazoans that possess a calcareous exoskeleton to provide structural support for a muscular foot and enclose mantle cavity which is generally used for feeding, respiration and sometimes locomotion [1]. It constitutes the second largest, and most variable, invertebrate phylum. The living species of the mollusks are divided into seven classes, that is aplacophora, polyplacophora, monoplacophora, gastropoda, cephalopoda, pelecypoda and scaphopoda [2]. Gastropods are the largest group of mollusks, comprising about 80% of the living mollusks with ca. 62,000 living species. The first gastropods originated during the late Cambrian period and over 500 million years ago. Since then, gastropods have radiated into marine, freshwater and terrestrial environments, changing their food preferences from herbivorous to carnivorous, endo-parasitism or symbiont-mediated chemoautotrophy [3].

The class gastropoda is the most speciose among animals to inhabit a variety of habitats such as oceans, rivers, etc. and are the ones that have inhabited the land among mollusks [4]. The aquatic forms have adapted to benthic forms while others remained pelagic. The life span ranges from months to decades [5, 6] and in some cases life is marked by varying periods of dormancy [7]. All gastropods are commonly called head-foot or cephalopodium which is a typical character of all gastropods because the head and foot arise from the same region making it very difficult to differentiate where the head ends or the foot begins [8]. The head of gastropods typically has two or four sensory tentacles with eyes and a ventral large foot, which gives them their name (in Greek, *gaster* is stomach and *poda* is feet). The anterior division of the foot, that is, propodium, is used for crawling. The shell in the larval stage is called protoconch. Most gastropods have a shell that typically opens on the right-hand side. Several species have operculum that is used to close the shell opening.

Most species of gastropoda include slugs and snails where the snails possess coiled shells on their body. The term snail is often used to describe marine and freshwater snails, along with terrestrial ones. More generally, the term is applied to land snails than to those from the sea or freshwater [9]. Snails generally thrive in habitats rich in calcium, limestone, marl and places with concrete and cement. They are hermaphrodite but reciprocal copulation is required to produce viable eggs. Eggs are laid 8 days after copulation producing about 400 to 1000 eggs per year [10]. Cool and moist soil is necessary for the egg hatching producing juvenile snails that eat their egg shells and remain burrowed for 2 weeks. The juveniles feed on tender shoots of plants while the adult can also digest detritus. Under unfavorable conditions, snails can bury themselves under soil and remain inactive from months to years [11].

The terrestrial snails like *Achatina fulica, Achatina achatina* and *Archachatina marginata* are large-sized terrestrial mollusks that can grow up to 20 cm in length and 10 cm in diameter. In these snails, the brownish shell having dark stripes generally covers half of the body [12]. Among these, the shell of *A. fulica* is smaller that can grow up to 3–4 inches, while *A. achatina* has a larger shell size of 10–11 inches [9].

Snails are both ecologically as well as economically very important animals. In the modern era of technology, the utility of snails is largely neglected, particularly in developed countries. Since snails dwell in a variety of niches, they could harbor a militia of micro-biota which could be exploited for various biotechnological purposes. This work provides insights into the microbiome of various snails. Furthermore, for the first time, we assessed the probable applications of snails in general and their gut micro-biota in particular for various biotechnology-based industries.

## 2. Origin and distribution of snails

The families Lymnaeae and Planorbidae originated from the common ancestor approximately 250 million years ago during the Permian period. Some fossils belonging to family Bulininae and Planorbinae of the upper cretaceous have been obtained from Africa and India [13, 14]. The first fossil record for the family Achatinidae was obtained from the Pleistocene in Africa [15, 16] but the family clearly evolved much earlier. In the 1950s, Mead described the earliest achatinids that originated in Cameroon and Gabon, northward of the river Zambezi in Africa, which later spread to both arid and the sub-arid areas of the southern continent and other moist parts east of the great watershed [17, 18]. This indicated that temperate species were directly evolved from tropical ancestors. Nonetheless, little is known about the evolutionary history of the achatinids.

The habitats of terrestrial snails range from dense tropical forests in Africa to the fringing riparian forests of Savannah [19, 20]. The members of the family Achatinidae comprise more than 200 species in 13 genera that are native to Africa. Several species have attained pest nature within their native African range when the habitat was modified by human activities and cropping. Furthermore, due to the increased mobility of humans and globalization of trade and travel, several alien species have been accidentally or purposefully transported to areas outside of the African continent. In these new areas, Achatinidae have caused significant economic and ecological impacts [21]. Due to its invasive capacity, *Achatina fulica* has spread from East Africa to many regions around the globe including rainforests, tropics, subtropics, etc. Apart from anthropogenic activities [22], the higher adaptability of this snail to variety of habitats is often contributed by its gut micro-biota that it selectively chooses from the favorable environments for successful dispersion [23]. However, terrestrial species have a great capacity of adaptation, survival and may contain an intriguing micro-biota serving in the efficient degradation [24] of ingested lignocellulosic plant biomass into many useful products. Due to its fast distribution and voracious feeding, this species is now considered as the most destructive terrestrial gastropod [12]. The *A. fulica* has been blamed as an intermediate vector of many worms and microorganisms, causing a variety of ailments [25]. The species was introduced to the USA in 1939, to India in 1947 near Kolkata and to Brazil in 1980s.

The widespread distribution of *A. fulica* is caused by a number of factors [26]. Sometimes, it has been deliberately introduced by humans as pet and in some cases as a source of food or for ornamental and medicinal purposes (**Figure 1**). It is also transported unintentionally with agricultural, horticultural and other commercial products or in containers in which they are shipped. They were also transported accidentally with military equipment in many countries [27]. The land snail, *A. fulica,* spreads extensively along rivers and streams, either on floating mats of vegetation or by surviving long enough in the water to float downstream.

The pulmonate snails are native to Africa but are currently found in Asia, the Pacific, Madagascar, Indonesia, Australia, the Caribbean Basin, the United States and South America (Colombia, Venezuela, Ecuador, Brazil and Argentina) [28, 29].

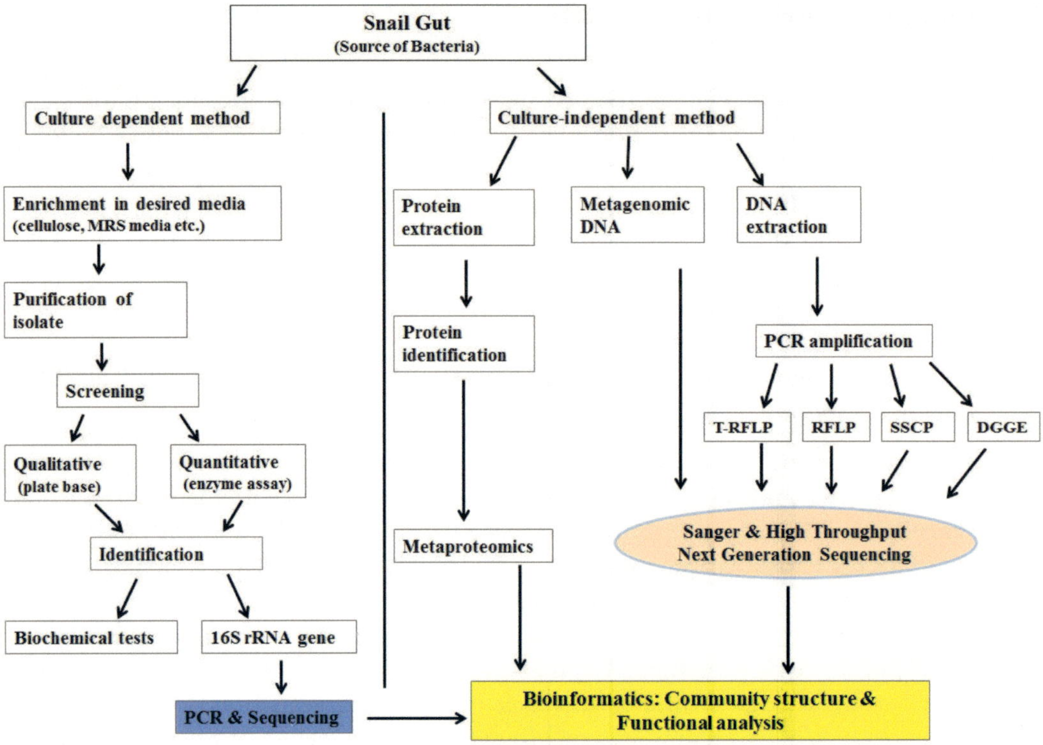


**Figure 1.** Different methods involved in the isolation and identification of microbes of the snail gut.

## 3. Uses of snails

From prehistoric times, it is quite evident that mollusks have a precarious relationship with humans. Many snails are known to damage the wooden ships and poison scuba divers. Researchers have also found that snails actually harbor a secret that could help humans to stay healthy and pain free. Even some authors quoted that guts of mollusks contain a unique set of microorganisms that might save human lives. During evolution, snails have also coevolved with ancient bacteria that reside in their guts. In return, the bacteria also express some drug-like molecules that help the snail's proper functioning and ward off diseases [30]. For example the *Leuconostoc mesenteroides* strain isolated from the gut of *Cornu aspersum* produced some bacteriocins-like substance that inhibited the growth of the pathogen *Propionibacterium acnes* [31].

### 3.1. Snails as food

The gastropods, particularly snails, have been used both as food as well as treatment for a variety of human diseases. The fossil remains of prehistoric shellfish found in caves indicate that snails have served as a delicacy for humans for thousands of years. The snails are easy to culture and majorly composed of muscles. Snails are a rich source of proteins containing high amounts of essential amino acids [32]. From the twentieth century, the food qualities of snails were so appreciated that it was a highly sought-after food. They are preferred as a food source in certain parts of Africa, Asia and South America. In recent decades, the snail's consumption has increased throughout European countries, which consequently lead to their gradual disappearance from freely dwelling areas. This decrease in population contributed to predation of the species and introduction of pathogens that harmed the productivity of snails [33, 34]. The inedible parts of snails are also used in animal-feed preparations as shown in **Figure 1**.

### 3.2. Medicinal uses

Hippocrates reportedly said that crushed snails can be used to relieve inflamed skin and pain. Two decades ago, slime of the Chilean snail was reported to quickly heal the skin lesions with no scars. This innovation later led to the production of "*Elicina*" which is snail slime-based fairness products. Recently in 2010, Missha, a USA-based cosmetic company launched a branded fairness cream, "Aqua Cell Renew Snail Cream", containing 70% slime. The company also claimed that this cream reduces pigments, acne, scars and combats wrinkles [35]. Though snail slime contains unusual crystals of calcite, it may find some use in orthopedics also. This is because scientists at the Herriot-Watt University stated that calcite may be used for the development of bone cement by using inorganic crystals in organic matrix [36].

### 3.3. Religious importance of snails

In southern Miami, snail invasion is very severe because they are linked to religious rituals. In Candomblé religion, coloration of the shell is considered very important for offerings to their gods, Orishas, and symbolizes the personality of an Orisha (e.g., red indicates fire and fury, white indicates tranquility and age while yellow is for prosperity and wealth). However, the color preference can vary between nations of different areas of the religion. For example, sacrificial animals or their parts that are offered to Obatala (white Orisha) should be completely white such as the white blood of *A. fulica* [37].

### 3.4. Ecological importance

Some snails that climb the trees rasping on the surface of leaves can influence biosphere community succession and nutrient cycling. Snails also provide some antimicrobial barriers to the plants by secreting the wax layer which contains antimicrobial compounds [38]. Mucus secreted by the gastropods has been shown to have selective antimicrobial properties as well [39, 40]. Moreover, some snails are also used for monitoring the environmental pollution. Such is the species of *Arianta arbustorum*, which can tolerate higher concentrations of the heavy metals, like cadmium, lead and copper, indicating elevated levels of metal pollution in their niches [41].

## 4. Impact of snails on agriculture

Many researchers have reviewed the impacts of invasive mollusks on agriculture [42, 43] biodiversity and human health. However, the annual costs associated with damage to the environment and agriculture due to alien species in the USA have been recently estimated to be

US$120 billion. The combined costs associated with damage for the United States, the United Kingdom, Australia, South Africa, India and Brazil have been estimated as US$314 billion per year [43]. In the tropics, the loss caused by the snails is threefold. Primarily, there is loss of the agricultural products followed by the cost of labor and materials associated with the management of such pests. Lastly, there is opportunity losses related to the changes in agricultural practice such as cultivation of pest-resistant species only.

Among mollusks, the giant African land snail, *A. fulica* tops the list of agricultural pests. *A. fulica* (*Lissachatina fulica*) is a herbivore, feeding primarily on vascular plants [21] and plant tissues containing high protein and calcium content [44, 45]. All *Achatinidae* species need calcium for the formation of shell and reproduction. Thus, environments rich in calcium carbonate, such as limestone landscapes having a pH of 7.0–8.0, and urban areas with abundant concrete are preferred [28].

The adult snail of *A. fulica* daily consumes large quantity of plant material approximately 10% of its weight [46]. The seedling stage of plants is most preferred and vulnerable. The extent of damage is based on the chemical composition of the plant and varies spatially as well as temporarily [47]. Many researchers have stated that infestations by snails to the nursery stage are so severe that demands change in cultivation practice. For example, in Malaysia, Guam and Indonesia, during the season of peak infestations of *A. fulica*, it is almost impossible to grow vegetables [27, 48, 49].

*A. fulica* is considered the most damaging land snail in the world as it can dwell on over 500 different crop species. It is a non-host specific pest of crops like peanuts, beans, peas, cucumbers and melons. If fruits and vegetables are not available, snails can feed on variety of ornamental plants, tree barks and even paint on houses [21]. The snail also allies with other soil invertebrates to decompose the leaf litter [50] and is the most destructive pest; it is ranked second among the 100 worst alien invasive species [51]. It affects tropical and subtropical areas, causing large damages to farms, commercial plantations and domestic gardens. It can also be found on trees, decaying materials and next to garbage deposits [17]. In urban areas, the deposition of solid waste by humans is primarily responsible for the proliferation of pests [12]. This species has attained pest status also due to its voracious feeding, competing for physical space with the native fauna resulting in disequilibrium of biodiversity [12]. Apart from being an agricultural nuisance, snails can thrive in cities, crawl up the walls of buildings and skid cars on highways [27].

## 5. Control strategies for snails

Snails are important both ecologically as well as economically due to a variety of factors. The prolific breeder *A. fulica*, soon after the introduction to a new habitat, reproduces at alarming rates making the control strategy very difficult. The control strategy of the pest is based on physical, chemical as well as biological methods. The physical control includes collection and destruction of snails and their eggs from the infested site or campaigns organized by local agencies voluntarily supported by health service officials, local people, students and teachers. After collection, snails are crushed and buried deep into the soil, covered with kaolin. Eradication of the species involves a huge amount of chemicals, hand collection and extensive public awareness programs like posters, documentaries, etc. Metaldehyde is the principal component of molluscicides and is indiscriminately used for the control of the snail *A. fulica,* consequently causing loss of productivity of local crops. For example, in Sao Paulo, farmers unknowingly used the molluscicide "metaldehyde" in banana fields to target snails, which killed many species including bats, skunks, lizards and small rodents which were beneficial as natural control agents of agricultural pests [12]. The physical methods are very time-consuming and tedious while the chemicals have resistance problems, killing the non-target flora and fauna. Therefore, biological control is the option that seems very fruitful and ecofriendly. But predatory snails (e.g., rosy wolf snail: *Euglandina rosea*) and flatworms have also failed to control some species such as *A. fulica* [52, 53]. As snails are ecologically and economically important due to the pest nature, the bacterial flora present in the gastrointestinal (GI) tract of snails may have an important role in digestion. These functionally specialized GI tract regions may be unique microenvironments and could harbor unusual bacterial communities.

## 6. Process of digestion in snails

In an ecosystem, the ability to procure enough food is pivotal for the survival of an organism. Feeding is necessary for the maintenance of metabolism, growth and reproductive success of animals. The process of digestion is characterized by a specific set of enzymes that often break the refractory food substances [54]. The alimentary tract of land snails is remarkably simple, possibly because of terrestrial life styles. The alimentary canal is usually divisible into buccal mass, esophagus, crop, stomach, intestine and rectum along with appendages like salivary and digestive glands (hepatopancreas) [55]. In *A. fulica,* like other gastropods, the food scraped by radula and ingested by the buccal mass is mixed with the secretions of the salivary gland and accumulates in the crop (ingluvius), a distensible muscular compartment. The crop and stomach are filled via two cannaliculi with the juice produced by the digestive glands. The medial part of the gut is surrounded by the digestive gland, which secretes more enzymes into the mid-gut lumen and also absorbs nutrients. The epithelium of the digestive tube is ciliated along almost its entire length, allowing the food to mix with digestive juices and helping to transport the alimentary mass. The ciliated epithelia also allow the microbial flora to anchor on the digestive tube [56]. The gut of the giant African land snail, *A. fulica,* is large enough to act as a fermentation vessel where a number of metabolic reactions are mediated by the host symbionts. The unabsorbed part of the alimentary mass (bolus) is compacted and passed directly into the rectum. The snail's digestion is primarily extracellular [55].

## 7. Role of the gut bacteria in snails

The gastrointestinal tracts of animals are modified as per their food requirement and physiological adaptations. All the herbivores that feed on lignocellulosic feed stock share two common features, that is enlarged digestive tract and gut micro-biota. Digestive tract is usually long enough having different regions such as esophagus, crop, rumen, caecum and rectal paunch while gut microbes provide the host with a unique set of necessary enzymes for the digestion of plant materials [57, 58]. The guts of herbivores that largely feed on lignocellulosic rich plant materials act as natural bioreactors for the degradation of plant biomass making them efficient sources of industrially important bacteria [59]. In many herbivores and omnivores, the digestion of the plant biomass is of immense importance for the energy capture [60]. Therefore, bacterial flora present in the GI tract of these animals may have an important role in digestion. These functionally specialized GI tract regions may be unique microenvironments and could harbor unusual bacterial communities.

### 7.1. Abundance of bacterial symbionts in snails

During the past century, scientists have focused on microbes that secrete the cellulose hydrolyzing enzymes. For instance, Seillière [61] pioneered the isolation of bacterial cellulases from the gut of the terrestrial gastropod *H. pomatia*. Similarly, Florkin and Lozet [62] studied the cellulases, whereas Jeuniaux [63] observed that chitinases from *H. pomatia*, of microbial origin, played a major role in the digestion of plant components in all phytophagous snails.

Charrier et al. [64] observed that density of bacteria in *C. aspersum* and *H. pomatia* was up to $5.10^9$ CFU $g^{-1}$ fresh tissue in the distal intestine, while in proximal region it was from 10 to 1000 fold lower than in the distal part. The *H. pomatia* was the least colonized by bacteria. The *C. aspersum* that fed on carboxymethyl cellulose (CMC) harbored approximately $10^7$ $g^{-1}$ bacteria and while those fed on native cellulose contained $10^6$ $g^{-1}$ [65]. In another study carried out in aerobic and anaerobic conditions by the same authors, it revealed that gram-positive bacteria were in the range of $1.57 \times 10^9 \pm 0.10 \times 10^9$ CFU$g^{-1}$ in the intestine. Although the score of gram-negative aerobic bacteria accounted for $5.77 \times 10^8 \pm 1.35 \times 10^8$ CFU$g^{-1}$ in the intestine, but it comprised only 27% of the total bacterial load in *H. aspersa* [66]. However, Simkiss observed only $0.71 \times 10^6$ CFU $g^{-1}$ body weight in *H. aspersa* [67]. In a similar report, researchers [68] noted less than $10^6$ $g^{-1}$ bacteria growing on sterile paper. In the intestine of *Tegula funebralis*, the number of culturable bacteria was $25 \times 10^5$ only [69].

Several strains growing on chitin have been isolated from different species of snails such as *C. gillenii*, *B. agrestis*, *B. noackiae* and *E. malodoratus*. The presence of chitinolytic bacteria in *H. pomatia* has been reported by Jeuniaux [63] where he observed the bacterial density in the range of $10^6$ CFU$g^{-1}$ of the tissue. By culture-dependent method, Pawar with his coauthors [70] enumerated from $10^3$ to $10^6$ CFU from the whole GI tract of *A. fulica*. Koleva et al. [31], while studying the gut bacteria of *C. aspersum*, stated that bacterial diversity varies with the different stages of life cycle and accounted for maximum $1.6 \times 10^9$ CFU $ml^{-1}$ gut extract during the active stage. Since more than 95% of the bacteria in any environment including guts of animals are un-culturable, their composition and community structure cannot be studied completely by culture-dependent approaches. As most of these studies were done using culture-dependent approaches, they might have not revealed much of bacterial composition and community structure in the GI tract of snails. More research is needed to study the bacterial diversity of snails by using advanced *in-silico* and meta-genomic approaches, harnessing the vast diversity of microbes in the snail guts. Very few studies have been carried out to analyze the bacterial populations in snails by using metagenomic methods. The complete details of the processes and protocols involved in the isolation and identification of the gut microbes are beyond the scope of this chapter, however, briefing the outline of most of these methodologies would be helpful. The brief outline of all these methodologies is given in **Figure 2**.

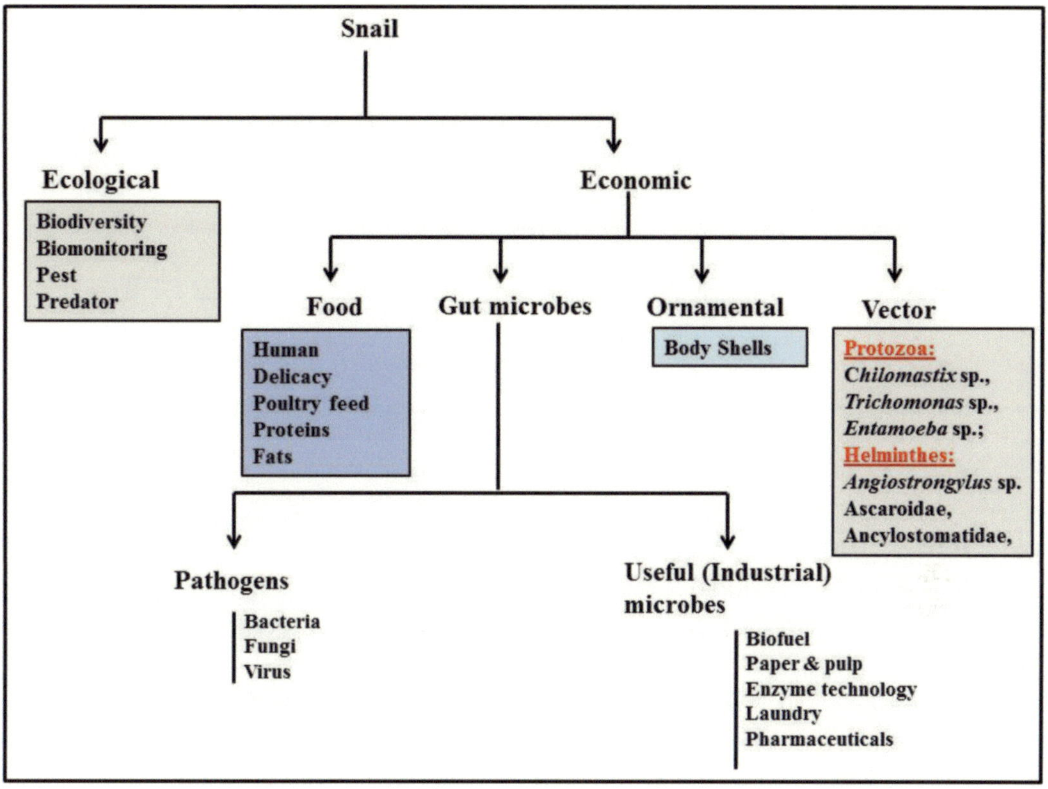


**Figure 2.** Ecological, economic and industrial utility of snails.

## 8. Host-symbiont interactions

Recent evidence for the presence of various kinds of bacteria in the snails suggested that a symbiotic relationship is developed between the host and the microbes during the course of evolution. Hitherto, a large number of eukaryotic symbionts have been isolated from snails in the families particularly, Achatinidae, Ampullariidae, Helicidae, Planorbidae, etc. as given below in **Table 1** [71]. Further, identification of the isolated gut bacteria has been done in vetigastropods of the genus *Haliotis* and in several other pulmonates. Among pulmonates, representatives of the genera *Biomphalaria, Bulinus, Helisoma* [72], *Helix, Cornu* [64, 66] and *Achatina* [70, 73, 74] have also been studied.

The advanced techniques like meta-genomics have proved that the gut bacteria perform many beneficial activities for the host. These resident bacteria help the host in processes such as digestion of complex molecules into simpler forms, generating energy, synthesis of cofactors, amino acids for basic metabolism as well as preventing the growth of pathogens. Some of the bacteria isolated from the snail caused the fermentation of sugars like glucose, lactose, mannitol, rhamnose, arabinose, maltose, etc. showing the positive interaction of the snails with their gut flora [75]. Some authors [40] reported the presence of several bacterial OTUs belonging

| Sr. No. | Snail | Habitat of snail | Family | Microbes studied | Methodology | References |
|---|---|---|---|---|---|---|
| 1 | *Achatina achatina* | Terrestrial | Achatinidae | Bacteria | Biochemical | [9] |
| 2 | *Achatina fulica* | Terrestrial | Achatinidae | Bacteria, fungi, virus, protozoa | 16Sr RNA/ metagenomics/ Microscopic | [23, 70, 73, 74, 137] |
| 3 | *Achatina mustelina* | Terrestrial | Achatinidae | Bacteria and fungi | Metagenomics | [40] |
| 4 | *Archachatina marginata* | Terrestrial | Achatinidae | Bacteria and fungi | Biochemical | [119, 138] |
| 5 | *Batillus cornutus* | Marine | Turbinidae | Bacteria | 16S rRNA | [139] |
| 6 | *Helix aspersa/Cornu aspersum* | Terrestrial | Helicidae | Bacteria and yeast | 16S rRNA/ Biochemical | [31, 67, 68, 140, 141] |
| 7 | *H. pomatia* | Terrestrial | Helicidae | Bacteria | Metagenomics | [131] |
| 8 | *Indoplanorbis exustus* | Freshwater | Planorbidae | Bacteria | Biochemical | [75] |
| 9 | *Lymnaea stagnalis* | Freshwater | Lymnaeidae | Bacteria | Biochemical | [75] |
| 10 | *Pomacea canaliculata* | Freshwater | Ampullariidae | Bacteria | 16S rRNA | [71] |
| 11 | *Pila globosa* | Freshwater | Ampullariidae | Bacteria | 16S rRNA | [142] |
| 12 | *Pila ovata* | Freshwater | Ampullariidae | Bacteria | Biochemical | [115] |
| 13 | *Tegula funebralis* | Marine | Tegulidae | Bacteria | Biochemical | [69] |
| 14 | *Trochus niloticus* | Marine | Tegulidae | Bacteria | Biochemical | [143] |

**Table 1.** Species of snails that have been used for isolation of microorganisms.

to oceanospirillales, enterobacteriaes, alteromonadales, along with $\alpha$-Proteobacteria and Rhizobiales in the fecal samples of *Achatinella mustelina*. Some snails thrive in toxic habitats like deep sea vents due to energy provided by the bacteria. The scaly foot snail, *Chrysomallon squamiferum,* discovered from the Kairei vent of Indian Ocean, flourishes by using a similar strategy, exploiting energy harnessed by the gut symbionts. That is why this snail can grow to up to 45 mm in size, when most of its close relatives did not grow beyond 15 mm in the absence of endosymbionts [76].

The physiology and diet of the host are the main components that determine the community structure of an organism. The gut microbiome of many animals including snails has been characterized recently [23, 70]. Animals are known to choose their gut microbes selectively/ functionally, and the microbial cells outnumber their hosts by many folds [77, 78]. Snails, like other invertebrates, eat soil to get the useful microbes that may augment in digestion. In turn, micro-biota provides important implications to the host's immune system [79] preventing invasion by exogenous pathogenic microbes [80, 81]. This in other words indicates that changes in microbial flora of the snail could have a negative impact such as without which they may stop feeding and ultimately die [82].

### 8.1. Cellulose-degrading bacteria

The plant biomass is comprised of three major components that is cellulose (50%), hemicellulose (30%) and lignin (20–30%). All herbivores do not possess the ability to digest plant polysaccharides and instead depend on their gut symbionts to derive the nutritionally important compounds from the ingested material [83–85]. Therefore, many researchers have extrapolated the gut microbiomes of many animals by using meta-genomics approach. Such studies have revealed that the gut of herbivores is a home to a consortium of microbes that have evolved to efficiently degrade and ferment the plant cellulose ingested by the host [86, 87]. These organisms possess a complex enzyme system known as cellulosome, and the complete enzymatic system includes three different enzyme types, that is exo-β-1, 4-glucanases (EC 3.2.1.91), endo-β-1, 4-glucanases (EC3.2.1.4) and β-1, 4-glucosidase (EC 3.2.1.21) along with several cofactors [88]. Cellulases act by hydrolyzing the β-1, 4 bonds in cellulose, releasing some small chains of oligosaccharides which are concurrently broken into monosaccharides by β-glucosidases [89]. The hydrolysis of lignin occurs due to the concomitant action of a specific set of enzymes such as laccase, lignin peroxidase, etc. In lignin degradation, the ligninolytic enzymes primarily alter the structural conformation of lignin by breaking several stable bonds resulting in production of free radicals [90]. From application point of view, bacteria are generally preferred over the fungi due to their higher growth rate allowing fast production of recombinant proteins [91]. Additionally, some glycoside hydrolases (GHs) of bacterial nature form multi-enzyme complexes called cellulosome provide increased synergy, stability and catalytic efficiency [92], while others are multifunctional, harboring both endoglucanase and xylanase activities [93]. A list of different groups of bacteria can be isolated from snails and thereby exploited for industrial applications. Therefore, enzymes of bacterial origin could offer specific biotechnological interests due to their less dependency on mediators. However, the lignocellulose-hydrolyzing enzymes secreted by bacteria are inducible, extracellular and cell associated [90]. Recently, Chang and his team [94] isolated a *Bacillus* strain that has a repertoire to remove lignin from rice straw; this biomass can be subsequently treated with lactic acid bacteria (LAB) to improve the sugars yield. These sugars can be further utilized for the production of bioethanol, biogas and bio-hydrogen by fermentations [70].

Some of the microbes such as bacteria *Fibrobacter succinogenes*, *R. flavefaciens* and *R. albus* [95] and some fungi are primarily responsible for degradation of plant cell walls. *R. albus*8 is anaerobic, fibrolytic and gram-positive bacterium present in herbivores and can degrade both cellulose and hemicellulose [60, 96]. But *R. flavefaciens* and *R. champanellensis* are very efficient cellulose degraders due to their cellulosome secretion which is lacking in case of *R. albus*8 [97].

The symbiotic bacteria from the gut of gastropods are considered to participate in the digestion of carbohydrates, such as cellulose and hemicellulose comprising the major part of the plants (**Table 2**). Recently, we reported the presence of lignocellulolytic bacteria in the GI tract of *A. fulica* [73]. However, Koch et al. [71] reported that *P. canaliculata* can survive till 56 days on a cellulose-rich diet and concluded the existence of bacterial endoglucanases that helps the snail to utilize cellulose polymer. Earlier studies [65, 68] showed that *H. aspersa* contains very few cellulose-degrading bacteria though some authors [64] claimed the complete absence of these bacteria in the gut. Many authors have demonstrated the degradation of native cellulose, mannan and laminarin by the snails [98, 99], thereby a large set of bacteria producing hydrolytic enzymes may be involved. The cellulases of animal origin were first studied by

| Sr. No. | Snail species | Bacteria | NCBI accession no. (16S rRNA) | Gram stain | References |
|---|---|---|---|---|---|
| 1 | *Achatina fulica* | *Klebsiella pneumoniae* | AB680060 | −ve | [23, 73, 74] |
| 2 | | *Sphingobacterium mizutaii* | NR042134 | −ve | |
| 3 | | *Sphingobacterium multivorum* | FJ459994 | −ve | |
| 4 | | *Microbacterium* sp. | AB646581 | +ve | |
| 5 | | Uncultured *Flavobacterium* sp. | DQ168834 | −ve | |
| 6 | | *Aeromonas punctata* | NR029252 | −ve | |
| 7 | | *Microbacterium* sp. | AB646581 | +ve | |
| 8 | | *Klebsiella variicola* | NR025635 | −ve | |
| 9 | | *Aeromonas caviae* | AB626132 | −ve | |
| 10 | | *Aeromonas caviae* | JF920485 | −ve | |
| 11 | | *Streptomyces kunmingensis* | NR043823 | +ve | |
| 12 | | *Cellulosimicrobium* sp. | AB188217 | +ve | |
| 13 | | *Cellulosimicrobium funkei* | JQ659848 | +ve | |
| 14 | | *Klebsiella* sp. | AB114637 | −ve | |
| 15 | | *Enterobacter* sp. | JQ396391 | −ve | |
| 16 | | *Stenotrophomonas* sp. | DQ242478 | +ve | |
| 17 | | *Cellulosimicrobium cellulans* | AB166888 | +ve | |
| 18 | | *Cellulosimicrobium* sp. | HM367604 | +ve | |
| 19 | | *Agromyces allii* | NR_04393 | +ve | |
| 20 | | *Nocardiopsis* sp. | HQ433551 | +ve | |
| 21 | | *Microbacterium binotii* | JQ659823 | +ve | |
| 22 | | *Bacillus subtilis* | | +ve | |
| 23 | | *Ochrobactrum* sp. | KJ669202 | −ve | |
| 24 | | *Achromobacter xylosoxidans* | KJ669206 | −ve | |
| 25 | | *Klebsiella* sp. | KJ669189 | −ve | |
| 26 | | Enterobacter sp. | KJ669197 | −ve | |
| 27 | | *Enterobacter cloacae* | KJ669195 | −ve | |
| 28 | | *Bacillus.* sp. | KR866144 | +ve | |
| 29 | *Archachatina marginata* | *Bacillus subtilis* | NA | +ve | [119] |
| 30 | | *E. casseliflavus* | NA | +ve | |
| 31 | | *Streptococcus faecalis* | NA | +ve | |
| 32 | | *Staphylococcus aureus* | NA | +ve | |
| 33 | *Pomacea canaliculata* | *Nostoc* sp. | NA | −ve | |

http://dx.doi.org/10.5772/68133

| Sr. No. | Snail species | Bacteria | NCBI accession no. (16S rRNA) | Gram stain | References |
|---|---|---|---|---|---|
| 34 | *Helix aspersa* | *Pseudomonas* sp. | NA | −ve | [68, 133] |
| 35 | | *Xanthomonas* sp. | NA | −ve | |
| 36 | | *Acinobacter* sp. | NA | −ve | |
| 37 | | *Vibrio* sp. | NA | −ve | |
| 38 | | *Enterobacteriae* sp. | NA | −ve | |
| 39 | | *Bacillus* sp. | NA | +ve | |
| 40 | | *Staphylococcus* sp. | NA | +ve | |
| 41 | | *Micrococcus* sp. | NA | +ve | |
| 42 | *Bulinus africanus, Biomphalaria pfeifferi, Helisoma duryi* | Chloroacidobacteria | NA | −ve | [72] |
| 43 | | Chryseobacterium | NA | −ve | |
| 44 | | Comamonadaceae | NA | −ve | |
| 45 | | *Bacillus* spp. | NA | +ve | |
| 46 | | *Aeromonas* spp. | NA | −ve | |
| 47 | | Verrucomicrobiae spp. | NA | −ve | |
| 48 | *Batillus conutus* | *Bacillus* sp. JMP A | HM776393 | +ve | [139] |
| 49 | | Bacillus sp. JMP B | HM776394 | +ve | |
| 50 | | *Staphylococcus* sp. JMP-C | HM776395 | +ve | |
| 51 | *Pila globosa* | *Klebsiella oxytoca* | KF017601 | −ve | [142] |

NA: not available.

**Table 2.** Cellulose degrading bacteria isolated from the digestive tract of different snails.

Biedermann and Moritz [100], in *Helix* spp., at the end of nineteenth century. Further, snails possess a micro-biota specialized in a variety of functions, thus contributing to an extraordinary (up to 80%) efficiency to digest plant biomass [24]. The abundance of carbohydrate-secreting bacteria and the rate of enzyme activity in various parts of the herbivorous guts are inversely proportional to each other, therefore, bacteria have become complementary for digestion of food. However, Payne et al. [101] also reported that wherever the enzyme production is less or nil, the enzymes released by the gut microflora would be of much help for digestion. The bacterial glycoside hydrolase (GH) genes and carbohydrate-binding modules (CBMs) are abundant in the digestive tract of animals [84, 102–106] which suggest the potential role of microbial symbionts in the hydrolysis of plant material to help extract nutrients [107]. The metagenomic and *in silico* studies have proved that gut symbionts perform useful functions to the host such as production of amino acids, energy generation and act as a barrier against diseases [108]. Recent works by researchers [23, 72] using advanced microbiological techniques elucidated that snails contain a vast array of microbial diversity within their guts.

### 8.2. Lactic acid bacteria

The lactic acid bacteria (LAB) comprise a significant proportion of the gut-bacterial communities of many animals including pigs, fowls, rodents, chicken, horses, gastropods and insects. These bacteria are vital for the host as they behave as protagonists in maintaining the ecological equilibrium between the different species of microorganisms inhabiting these environments. This microbial community takes part in the fermentation of the food, providing energy to the host [64]. Koleva et al. [31] isolated 55 strains of LAB from the gut of *C. aspersum* (**Table 3**). Based on 16S rRNA sequencing, *Lactobacillus* (18), *Enterococcus* (17), *Lactococcus*

| Sr. No. | Snail | Bacteria | NCBI accession no. (16S rRNA) | Gram stain | References |
|---|---|---|---|---|---|
| 1 | *Helix pomatia* | *Buttiauxella agrestis* | DQ223869 | −ve | [64] |
| 2 | | *Citrobacter gillenii* | DQ223882 | −ve | |
| 3 | | *Buttiauxella agrestis* | DQ223871 | −ve | |
| 4 | | *Lactococcus lactis* | DQ223875 | +ve | |
| 5 | | *Kluyvera intermedia* | DQ223868 | −ve | |
| 6 | | *Lactococcus sp.* | DQ223877 | +ve | |
| 7 | | *Obesumbacterium proteus* | DQ223874 | −ve | |
| 8 | | *Enterobacter amnigenus* | DQ223879 | −ve | |
| 9 | | *Enterococcus raffinosus* | DQ223885 | +ve | |
| 10 | | *Enterococcus malodoratus* | DQ223886 | +ve | |
| 11 | *Cornu aspersum,* | *Buttiauxella noackiae* | DQ223870 | −ve | [66] |
| 12 | | *Clostridium* sp. | DQ223883 | +ve | |
| 13 | | *Raoultella terrigena* | DQ223873 | −ve | |
| 14 | | *Enterobacter amnigenus* | DQ223878 | −ve | |
| 15 | | *Citrobacter gillenii* | DQ223881 | −ve | |
| 16 | | *Enterococcus casseliflavus* | DQ223887 | +ve | |
| 17 | | *Citrobacter sp.* | DQ223880 | −ve | |
| 18 | *Helix aspersa* | *Lactobacillus brevis* | NA | +ve | [31] |
| 19 | | *Lactobacillus plantarum* | NA | +ve | |
| 20 | | *Lactococcu lactis* | NA | +ve | |
| 21 | | *Weissella confusa* | NA | +ve | |
| 22 | | *Lactobacillus* curvatus | NA | +ve | |
| 23 | | *Enterococcus* mundtii | NA | +ve | |
| 24 | | *E. faecium* | NA | +ve | |

NA: not available.

**Table 3.** List of lactic acid bacteria used by snails in fermentation of digested food.

(12) and *Leuconostoc* (7) accounted for 33, 32, 21 and 13% of the bacterial diversity, respectively, including the strains belonging to genus *Weissella*. Among these genera, *Enterococcus* and *Lactococcus* exhibited the lactic acid activity, thereby indicating their role in the digestive physiology of the snail. However, the LAB are also reported to have a stimulatory response in a marine gastropod *Nassarius obsoletus* [109]. The epiphyte enterococci being the dominant lactic acid bacterium in the snail's intestine is quite interesting. *Lactococcus lactis* is a nonpathogenic bacterium that has been extensively used in the dairy industry for the manufacture of buttermilk, yogurt and cheese. These microbes are also routinely used in the fermentation process of wines, beer, bread and pickles.

*Enterococcus*, a LAB, inhabiting the gut of many herbivores, is considered as beneficial for the hosts because it forms a biofilm-like structure on the gut epithelium which could prevent the host gut from colonization of pathogenic microbes [110]. The members of the genus *Enterococcus* also produce some bacteriocins. The synergistic effect of this biofilm formation and production of antimicrobial compound probably impedes the entrance and establishment of perilous pathogens in the snail gut [111, 112].

### 8.3. Proteolytic bacteria

Proteases are enzymes that perform proteolysis, that is, hydrolysis of peptide bonds between two amino acids of a polypeptide chain. Protease enzymes are ubiquitous [113] in nature. Some proteases determine the lifetime of functional molecules like hormones, antibodies, or other enzymes that are very important for physiological processes. In the present era of advanced technology, more research is being done on eco-friendly products replacing the chemical processes by using enzymatic methods. Proteases have a high demand in industries like bread and meat industry, pharmaceuticals and agro-waste disposal management [114]. They are widely used in the film industry for recovery of silver from X-ray films, in the chemical industry for peptide synthesis, in the feed and food industry for production of protein hydrolysates, by waste processing companies, in the field of textile processing for degumming of silk and processing of wool and in the manufacture of detergents, pharmaceuticals and leather [115].

Though produced by many microorganisms, that is fungi, yeast, actinomycetes and molds, the proteases of bacterial origin are considered as most significant [116] because bacteria can be manipulated genetically to generate new enzymes with desired properties for the specific applications [117]. The bacterial proteases constitute about two-thirds of the industrially important enzymes and account for about 60% of the total worldwide sale in markets. Protease-producing bacteria are also useful for the ecosystem as these microbes decompose the dead and decaying animal or plant matter that is primarily composed of proteins. They can create pollution-free environment and are responsible for the recycling of nutrients.

Ariole and Ilega [115] isolated the proteolytic *Pseudomonas aeruginosa* from the gut of freshwater snail, *Pila ovata*. They concluded that this bacterium augmented the snail in degradation of nutrients showing a maximum proteolytic activity of 372 U/ml at pH 9. The saprophagous nature of *H. pomatia* suggests that its gut can be a site for protein digestion [118]. Proteolytic activity contributed by the bacteria was also reported by Koleva et al. [31] in the gut of *C. aspersum* during the actively feeding stage.

In the African snail, *A. marginata,* the five-cellulase-and-protease-positive bacteria, belonging to genus *B. subtilis, S. aureus, S. casseliflavus* and *S. faecalis,* have been studied [119]. Few researchers have reported the protein digestion augmented by the gut symbionts in case of gastropods [120–122], with a 32-kDa protease present in gut lumen and midgut gland of *P. canaliculata.*

Snails are cheap, easy to rear and collect and contain copious microbes in their guts that can be exploited for various industrial purposes. The industrially important enzymes, like cellulases and proteases, can be isolated, extracted and purified from the gut microbes of snails thereby reducing the cost of imported materials. These enzymes are not only used in biofuel production but also harvested for other industries like pharmaceutical, waste disposal and detergent industries [119].

### 8.4. Chitinolytic bacteria

The omnivorous snails feed on insects that are a rich source of chitin, and in some cases, traces are often detected in gastropod feces. The body of phytophagous gastropods consists of 10% nitrogen, while food plants dined by snails contain only 4% of nitrogen. Chitin and its derivatives like chitosans could serve as a readily available nitrogen source for the gut bacteria and ultimately their host can take advantage of chitin-derived products [123].

Functional studies described extensively the importance of bacterial gut flora for the snail's digestion and nutrient supply [124]. Since the endogenous enzymatic activity in the intestine of the snail is very low, the snails may use their allochthonous and autochthonous bacteria for organic matter degradation [23, 99]. The digestive tract also harbors bacteria with special functions like metal chelation [67] and fermentation activity [64, 66], particularly on chitin and soluble cellulose, thereby providing nitrogen, lactate and acetate that are used as precursors as well as energy sources [70]. The DGGE fingerprinting technique along with NMDS analysis have revealed that intestine of the land snail *H. pomatia* harbors a unique set of bacterial flora. These authors also stated that sequences related to Pseudomonadaceae and Enterobacteriaceae spp. dominated the intestinal and digestive gland of snail populations. However, Kiebre-Toe et al. [125] and Charrier et al. [64] also reported the dominance of *Pseudomonas* sp., *Pantoea* sp. and *Buttiauxella* sp. in the intestine of *Helix* sp.

Lesel et al. [65] isolated the chitinolytic bacteria from the *H. pomatia* where chitinolytic bacteria were 10 times more abundant in the stomach and intestine than in the crop. In *Redix peregra,* the chitobiase activity was reduced when fed on antibiotic-treated diet, which also resulted in the loss of bacteria. This dual reduction indicates the synthesis of chitobiase by the bacteria inhabiting the gut [54]. Same conclusion was recounted by the Jeuniaux [126] and Donachie et al. [127] for the pulmonate *H. pomatia* and krill (*Megunyctiphunes norvegicu*) by showing a reduction in the enzymatic activity of the gut after the treatment of antibiotics.

### 8.5. Sulfate-reducing bacteria

Snails are copper-dependent animals as they use copper for the formation of the respiratory pigment haemocyanin. They also contain pore cells that can recycle the copper within the body. The sulfate-reducing bacteria increase the availability of copper to their snail hosts possibly by the effect of their metal-chelating activities [67]. The sulfate-reducing bacteria *Desulvibrio* sp. found in the crop of *H. aspersa* chelates the metals like Cu, Zn, Fe and Ni and make them ready for absorption. Similarly, some authors [128] concluded that digestive gland of the pulmonate *H. aspersa* acts as the store of Pb, Zn and Cd, which would represent a detoxification system. On the other hand, Simkiss [67] demonstrated the presence of sulfate-reducing bacteria in the crop of the snail *C. aspersum*.

Recently, Koch et al. [72] isolated the *Pseudomonas, Enterobacter* and *Lactococcus* bacterial species that were capable to degrade uric acid. However, in snails, uricase is found in several tissues, shuts down during estivation and does not participate in uric acid oxidation during arousal from this state [129]. However, tissue uricase along with bacterial uricase plays a role in nitrogen recycle of animals. In *P. canaliculata,* many bacteria not only help in digestion but also take part in recycling of uric acid like in arthropods.

## 9. Effect of gut physiology on the bacteria

The community structure of the microbes inhabiting the gut is predominantly altered by physiological states like hibernation and aestivation of the host [126, 130]. The physiological states like aestivation or hibernation are characterized by marked decrease in bacterial diversity due to expulsion of gut contents where some phylotypes are intentionally eliminated from the body. This gut clearance and other physico-chemical modifications may be responsible for the restructuration of the bacterial community like absence of mollicutes and $\alpha$-proteobacteria in *H. pomatia* [131]. The snails also choose their gut biota as per physiological requirements. At the beginning of hibernation, certain groups are reduced and disappear while those that were meager during active stage may gain in space and become dominant. Further, during aestivation, the snails also lose large quantities of water, which may affect the viability of the gut bacteria and eventually their number and metabolism [31]. This could also be reason for the loss of allochthonous bacterial populations. During hibernation, there is a noticeable reduction of water content of the body along with reduction of food and low temperatures, which induce the snail to select the psychotropic bacteria only. These studies indicate that the gut flora is altered by different life stages and related physiological processes of the snails [132].

Though the bacteria survive during different physiological states like starvation, aestivation and hibernation of the snails, there is always a reduction in their number [64, 68] and these bacteria can be considered as autochthonous members of the snail gut. During these stages, mucous ribbon acts as the main nutritive medium for the bacterial growth [133]. In *C. aspersum,* amylolytic bacteria are adopted by vertical transmission [31] whereas proteolytic and cellulolytic bacteria were seen only during the adult stages of the animal. The higher cellulolytic and proteolytic activity within the snail were predominately exhibited in active stage only indicating the transient nature of these bacteria, that is being ingested with the food from the environment thereby augmenting and improving digestion processes [65]. However, proteolytic bacteria were completely absent during hibernation, aestivation and in juvenile stages. The hibernation was marked with the decline of cellulolytic bacteria.

In *H. pomatia*, $\gamma$-proteobacteria and $\alpha$-proteobacteria were the most abundant classes in all populations of snails. Only one phylotype of firmicutes has been reported during hibernation of snail populations. In non-hibernating snails, firmicutes were found only in the proximal intestine and digestive gland. In active snails, firmicutes were observed in distal intestine, with Mollicute specimen established abundantly in all three gut regions. However, they were restricted to the distal intestine and digestive gland at the beginning of hibernation [131].

The changes in the pH of the gut have serious effects on the microbial community. During anaerobioses, these bacteria in turn change the pH of the gut through fermentative reactions [119] producing end products that affect the acid-base balance of the digestive tract. But Churchill and Storey [134] postulated that in dormant snails, there is no accumulation of end-products (lactate and succinate) in dormant snails.

Besides all these functions that are contributed by the bacteria to their hosts, they also influence cold hardiness in their hosts. In snails such as *H. pomatia* and *C. aspersum*, the gut bacteria participate in ice-nucleating activity thereby reducing the cold hardiness in these snails [131, 135]. *H. pomatia* is known to decrease its supercooling point ca. by 3°, from –2 during its active state to –7°C in hibernation depending on the geographic location [136]. Lastly, enzymes secreted by the gut microbial community are very suitable for various biotechnological applications within the food, pharmaceutical and chemical industries along with detoxification of many hazardous chemicals.

In conclusion, snails present a vast diversity among mollusks with inherent industrial importance. Snails provide benefits not only as food for humans but are also routinely used in agriculture for the control of many insect pests. Though there are pros and cons associated with mollusks, a key need is better knowledge of the basic biology of these useful animals, with rigorous documentation of their habitats for the possible conservation. Little is known about the composition of snail micro-biota because a large number of species have been underestimated. Understanding the microbial ecology of snails may illustrate many useful processes like development of medicines from mucus or utilization of gut symbionts to challenge the emerging issues of environmental pollution and energy crisis. There is a dire need to explore more and more diversity of microbes that is encrypted in extreme environments like digestive tracts of snails. To accomplish this, many advanced techniques like high throughput next generation sequences (NGSs) along with other metagenomic techniques can be employed to unleash the role of these microbes in the host physiology.

## Acknowledgements

Mudasir A. Dar acknowledges UGC, New Delhi, India, for providing the Maulana Azad National fellowship.

http://dx.doi.org/10.5772/68133

## Author details

Mudasir A. Dar[1], Kiran D. Pawar[2] and Radhakrishna S. Pandit[1*]

*Address all correspondence to: rspandit@unipune.ac.in

1 Department of Zoology, Savitribai Phule Pune University, Ganeshkhind, Pune, Maharashtra, India

2 School of Nanoscience and Biotechnology, Shivaji University, Vidyanagar, Kolhapur, Maharashtra, India

## References

[1] Runnegar B, Pojeta Jr. Molluscan phylogeny: The paleontological viewpoint. Science. 1974;**186**:311-317

[2] Stasek CR. The molluscan framework. In: Florkin M, Sheer BT, editors. Chemical Zoology. Vol. 7. Mollusca: Academic Press New York and London; 1972. pp. 1-44

[3] Loker ES. Gastropod immunobiology. In: Kenneth S, editors. Invertebrate Immunity. USA: Springer Science; 2010. pp. 17-33

[4] Holthuis BV. Evolution between marine and freshwater habitats: A case study of the gastropod suborder Neritopsina [thesis]. Washington, USA: University of Washington; 1995

[5] Heller J. Longevity in mollusks. Malacologia. 1990;**31**:259-295

[6] Curtis LA. Spatial heterogeneityin size and parasitism: How it arises in an estuarine snail population. Journal Experimental Marine Biology and Ecology. 2007;**352**:317-330

[7] Storey KB. Life in the slow lane: Molecular mechanisms of estivation. Comparative Biochemistry and Physiology Part (A), Molecular and Integrative Physiology 2002;**133**:733-754

[8] Nordsieck R. The living word of Molluscs.The Locomotion of Gastropods. 2011. http://molluscs.at/gastropoda/terrestrial.html. Accessed on [22-12-2016]

[9] Ugoh SC, Ugbenyo AJ. Studies on the isolation of enteropathogens associated with the intestines of Giant African land snails (*Achatina* and *Archachatina*) species sold in Gwagwalada, FCT, Abuja–Nigeria. Researcher. 2013;**5**:56-60

[10] Skelley PE, Dixon WN, Hodges G. Giant African Land Snail and Giant South American Snails: Field Recognition. Florida: Pest Alert, FDACS; 2010

[11] USDA–APHIS-PPQ. New Pest Response Guidelines. Giant African Snails: Snail Pests in the Family Achatinidae. USDA–APHIS–PPQ–Emergency and Domestic Programs–Emergency Planning,Riverdale,Maryland. 2007. www.aphis.usda.gov/import_export/plants/manuals/emergency/downloads/nprg_gas.pdf [Accessed on [22-12-2016]

[12] Ohlweiler FP, Guimarães MCA, Takahashi FY, Eduardo JM. Current distribution of *Achatina fulica*, in the state of São Paulo including records of *Aelurostrongylus abstrusus* (Nematoda) larvae infestation. Revista do Instituto Medicina Tropical de Sao Paulo. 2010;**52**:211-214. http//DOI: 10.1590/S0036-46652010000400009

[13] Newton RB. On some freshwater fossils from central South Africa. Annals and Magazine of Natural History. 1920;**9**:241-249

[14] Smith AG, Smith DG, Funnell BM. Atlas of Mesozoic and Cenozoic coastlines. Cambridge, UK: Cambridge University Press; 1994. p. 114

[15] Solem A. Biogeographic significance of land snails, Paleozoic to Recent. In: Gray J, Boucot AJ, editors. Historical Biogeography, Plate Tectonics, and the Changing Environment. Corvallis: Oregon State University Press; 1979a. pp. 277-287

[16] Solem, A. A theory of land snail biogeographic patterns through time. In: van der Spoel S, Van Bruggen AC, Lever J, editors. Pathways in Malacology. The Hague: Bohn S, Holkema U, and Junk W; 1979b. pp. 225-249

[17] Mead AR. Anatomy, phylogeny, and zoogeography in the African land snail family Achatinidae. In: Proceedings of the 12th International Malacological Congress, Vigo, 1995; pp. 422-423

[18] Mead AR. Comparative anatomy establishes correlativity in distributional direction and phylogenetic progression in the Achatinidae. In: Bieler R, Mikkelsen PM, editors. Abstracts, World Congress of Malacology, Washington D.C. Field Museum of Natural History, Chicago, for UnitasMalacologica, 1998; p. 214

[19] Ajayi SS, Tewe SO, Milligan JK. Influence of seasonality on aestivation and behaviour of the forest African giant land snail, *Archachatina marginata* (Swaison). Bulletin of Annual Health Proc. 2009;**28**:328

[20] Odaibo, AB. Snail and snail farming. Nigeria edible land snails. Ibadan: Stirling-Horden Publishers. 2007;**1**:1-11

[21] Rauth SK, Barker GM. *Achatina fulica* (Bowdich) and other Achatinidae as pest in tropicultural agriculture. In: Hamilton LR, editors. Mollusks as Crop Pests. New Zealand: KABI International Publishing New Zealand; 2002. p. 474

[22] De Winter AJ. New records of Achatina fulica (Bowdich) from the Côte d'Ivoire. Basteria. 1989;**53**:71-72

[23] Cardoso AM, Cavalcante JJV, Vieira RP, Lima JL, et al. Gut bacterial communities in the giant land snail *Achatina fulica* and their modification by sugarcane based diet. PLoS One. 2012;**7**:e33440

[24] Charrier M, Daguzan J. Food consumption: Production and energy budget in *Helix aspersa* Müller (Gastropoda: Pulmonata). Annales de la Nutrition et de l'Alimentation. 1980;**34**:147-166

[25] Lv S, Zhang Y, Liu H-X, Hu L, Yang K, et al. Invasive snails and an emerging infectious disease: Results from the First National Survey on *Angiostrongylus cantonensisin* China. PLoS Neglected Tropical Disease. 2009;**3**:368-375

[26] Cowie RH, Robinson DG. Pathways of introduction of nonindigenous land and freshwater snails and slugs. In: Ruiz G, Carlton JT, editors. Invasive Species: Vectors and Management Strategies. Washington, DC: Island Press; 2003. pp. 93-122

[27] Mead AR. The Giant African Snail: A Problem in Economic Malacology. Chicago: University of Chicago Press; 1961

[28] USDA.United States Department of Agriculture.National Invasive Species Information Center 2008. *Achatina fulica* Bowdich; Giant African Land Snail). Available from: http://www.invasivespeciesinfo.gov/animals/africansnail.shtml [Accessed 22-01-2017]

[29] Gutiérrez DE, Núñez V, Vogler R, Rumi A. Invasion of the Argentinean Paraense rainforest by the Giant African snail Achatina fulica. American Malacological Bulletin. 2011;**29**:135-137

[30] MacRae M. Engineering Drugs from Marine Mollusks. 2016. https://www.asme.org/engineeringtopics/articles/bioengineering/engineeringdrugsfrommarinemollusks). [Accessed on 22-01-2017]

[31] Koleva ZV, Kizheva YK, Tishkov SH, Dedov IK, Kirova EL, et al. Dynamics of bacterial community in the gut of *Cornu aspersum*. Journal of BioScience and Biotechnology. 2015;**4**:263-269

[32] Abadom M. Digestive enzymes of the West African giant land snail, *Archachatina marginata*. Biochemical Society Transactions. 1991;**19**:334S

[33] De la Piedra R. Biología del caracol (*Helix aspersa* Muller) y propuesta de instalación de uncriaderomixto modificado. Tesina de MédicoVeterinario, Universidad Nacional Mayor de San Marcos, Lima. 2005:**88**

[34] Benito M. Evaluacióntécnicaeconómica de unacrianzaintensiva de Caracoles (*Helix aspersa*). Tesis de IngenieroAgrónomo. Pontificia Universidad Católica, Santiago de Chile. 2004;**62**

[35] Reporter DM. Snail Slime Hailed Latest Beauty Wonder Product, Promising to Clear Acne, Reduce Scarring and Beat Wrinkles'. Daily Mail. 2012. URL:http://www.dailymail.co.uk/femail/article2216457/Snailslimehailedlatest-beautywonder productpromisingclearacnereducescarringbeatwrinkles.html [Accessed on 22-12-2016]

[36] Aitken R. Snail Slime' Could Mend Bones'. BBC News. 2015. URL: http://news.bbc.co.uk/1/hi/health/900869.stm. [Accessed on 22-12-2016]

[37] Léo Neto NA, Brooks SE, Alves RRN. From Eshu to Obatala: Animals used in sacrificial rituals at Candomblé "terreiros" in Brazil. Journal of Ethnobiology and Ethnomedicine. 2009;**5**:1-10. DOI:10.1186/1746-4269-5-23

[38] Thomas S. Medicinal Use of Terrestrial Molluscs (Slugs and Snails) with Particular Reference to Their Role in the Treatment of Wounds and Other Skin Lesions. 2013. http://www.worldwidewounds.com/2013/July/Thomas/slug-stevethomas.html [Accessed on 15-11-2016]

[39] Kubota Y, Watanabe Y, Otsuka H, Tamiya T, Tsuchiya T, Matsumoto JJ. Purification and characterization of an antibacterial factor from snail mucus. Comparative Biochemistry and Physiology 1985;**82**

[40] Rorke RO, Cobian GM, Holland BS, Price MR, Costello V, et al. Dining local: the microbial diet of a snail that grazes microbial communities is geographically structured. Environmental Microbiology. 2015;**17:**1753-1764

[41] Berger B, Dallinger R. Terrestrial snails as quantitative indicators of environmental metal pollution. Environmental Monitoring and Assessment. 1993;**25**:65-84. DOI:10.1007/BF00549793

[42] Godan D. Pest Slugs and Snails. Berlin, Heidelberg, New York: Springer Verlag; 1983

[43] Cowie RH, Dillon RT, Robinson DG, Smith JW. Alien non-marine snails and slugs of priority quarantine importance in the United States: A preliminary risk assessment. American Malacological Bulletin. 2009;**27**:11-132

[44] Iglesias J, Castillejo Y. Field observations of the land snail *Helix aspersa* Müller. Journal of Molluscan Studies. 1999;**65**:411-423. DOI:10.1093/mollus/65.4.411

[45] Chevalier L, Coz-Bouhnik M Le, Charrier M. Influence of inorganic compounds on food selection by the brown garden snail *Cornu asperses* (Müller) (Gastropoda: Pulmonata). Malacologia. 2003;**45**:125-132

[46] Schreurs J. Investigations on the biology, ecology and control of giant African snail in West New Guinea. Report, Manokwari Agricultural Research Station; 1963; 18 p

[47] Raut SK, Ghose KC. Pestiferous land snails of India. Zoological Survey of India No. 11, Calcutta: Bani Press; 1984. 151 p

[48] South FW. The giant snail (*Achatina fulica*, Fer.) in Malaya. Malayan Agricultural Journal. 1926;**14**:231-240

[49] Kondo Y. Report on the *Achatina fulica* Investigation on Palau, Pagan and Guam. Washington, DC, USA: Pacific Science Board of the National Research Council, Invertebrate Consultants Committee for the Pacific; 1950. p. 94

[50] Hatziioannou M, Eleutheriadis N, Lazaridou-Dimitriadou M. Food preferences and dietary overlap by terrestrial snails in Logos Area (Edessa, Macedonia, Northern Greece). Journal of Molluscan Studies. 1994;**60**:331-341

[51] Anonymous. Global Invasive Species Database, 2013. *Achatina fulica* (Mollusc). Available from: http://www.issg.org/database/species/ecology.asp? [Accessed on 13-11-2016]

[52] Cowie RH. Evolution and extinction of Partulidae, endemic Pacific island land snails. Philosphical Transactions of the Royal Society of London Series B. 1992;**335**:167-191

[53] Hadfield MG, Miller SE, Carwile AH. The decimation of endemic Hawaiian [sic] tree snails by alien predators. American Zoologist. 1993;**33**:610-622

[54] Brendelberger H. Bacteria and digestive enzymes in the alimentary tract of *Radix peregra* (Gastropoda:Lymnaeidae). Limnology and. Oceanography. 1997;**42**:1635-1638

[55] Ghose KC. The alimentary system of *Achatina fulica*. Transactions of the American Microscopical Society. 1963;**82**:149-167

[56] Zbinden M, Pailleret M, Ravaux J, Gaudron SM, Hoyoux C, et al. Bacterial communities associated with the wood-feeding gastropod *Pectinodonta* sp. (Patellogastropoda, Mollusca). FEMS Microbiology Ecology. 2010;**74**:450-463

[57] Distel DL, DeLong EF, Waterbury JB. Phylogenetic characterization and in situ localization of the bacterial symbionts of shipworms (Teredinidæ: Bivalvia) by using 16S rRNA. Applied and Environmental Microbiology. 1991;**57**;2376-2382

[58] Breznak JA, Brune A. Role of microorganisms in the digestion of lignocellulose by termites. Annual Review of Entomology. 1994;**39**:453-487

[59] Sun JZ, Scharf ME. Exploring and integrating cellulolytic systems of insects to advance biofuel technology. Insect Science. 2010;**17**:163-165

[60] Devendran S, Abdel-Hamid AM, Evans AF, Iakiviak M, Kwon IH, et al. Multiple cellobiohydrolases and cellobiosephosphorylases cooperate in the ruminal bacterium *Ruminococcus albus* 8 to degrade cellooligosaccharides. Scientific RepoRts. 2016;**6**:35342. DOI: 10.1038/srep35342

[61] Seillière G. Sur uncasd'hydrolysediastasique de la cellulose du cotton, après dissolution dans la liqueur de Schweitzer. Comptes Rendus Des Seances De La Societe De Biologie Et De Ses Filiales. 1906;**61**:205-206

[62] Florkin M, Lozet F. Originebactérienne de la cellulase du contenu intestinal de l'escargot. Archives Internationales De Physiologie et De Biochimie. 1949;**57**:201-207

[63] Jeuniaux C. Production dexochitinase par des streptomyces. Comptes Rendus des Séances de la Société de Biologieet Filiales. 1955a;**149**:1307-1308

[64] Charrier M, Fonty G, Gaillard-Martinie B, Ainouche K, Andant G. Isolation and characterization of cultivable fermentative bacteria from the intestine of two edible snails, *Helix pomatia* and *Cornu aspersum* (Gastropoda: Pulmonata). Biological Research. 2006;**39**:669-681

[65] Lesel M, Charrier M, lesel R. Some characteristics of the bacterial flora housed by the brown garden snail *Helix aspersa* (Gastropoda: Pulmonata) preliminary results. In: Lesel R, editors. Proceedings of the International Symposium on Microbiology in Poecilotherms. Amsterdam: Elsevier Sciences. 1990; pp. 149-152

[66] Charrier M, Yannick CB, Ollivier B. Bacterial flora in the gut of *Helix aspersa* (Gastropoda: Pulmonata): Evidence for a permanent population with a dominant homolactic intestinal bacterium *Enterococcus casseliflavus*. Canadian Journal of Microbiology. 1998;**44**:20-27

[67] Simkiss K. Prokaryote-eukaryote interactions in trace element metabolism *Desulfovibrio* sp. in *Helix aspersa*. Experientia. 1985;**41**:1195-1197

[68] Watkins B, Simkiss K. Interactions between soil bacteria and the molluscan alimentary tract. Journal of Molluscan Studies. 1990;**56**:267-274

[69] Galli DR, Giese AC. Carbohydrate digestion in a herbivorous snail, *Tegula funebralis*. Journal of Experimental Zoology Part A. 1959;**140**:415-440. DOI: 10.1002/jez.1401400305

[70] Pawar KD, Banskar S, Rane SD, Charan SS, Kulkarni GJ, et al. Bacterial diversity in different regions of gastrointestinal tract of Giant African Snail (*Achatina fulica*). Microbiology Open. 2012;**1**:415-426

[71] Koch E, Lozada M, Dionisi H, Alfredo CV. Uric acid-degrading bacteria in the gut of the invading apple snail *Pomacea canaliculata* and their possible symbiotic significance. Symbiosis. 2014;**63**:149-155

[72] Van Horn DJ, Garcia JR, Loker ES, Mitchell KR, Mkoji GM, et al. Complex intestinal bacterial communities in three species of planorbid snails. Journal of Molluscan Studies 2011;**78**:74-80

[73] Dar MA, Pawar KD, Jadhav JP, Pandit RS. Isolation of cellulolytic bacteria from the gastro-intestinal tract of *Achatina fulica* (Gastropoda: Pulmonata) and their evaluation for cellulose biodegradation. International Biodeterioration Biodegradation. 2015;**98**:73-80. http://dx.doi.org/10.1016/j.ibiod.2014.11.016

[74] Pinheiro GL, Correa RF, Cunha RS, Cardoso AM, Chaia C. Isolation of aerobic cultivable cellulolytic bacteria from different regions of the gastrointestinal tract of giant land snail *Achatina fulica*. Frontiers in Microbiology. 2015;**6**:860. DOI: 10.3389/fmicb.2015.00860

[75] Mahejabin NS, Tarannum TS. Biochemical study of bacterial strains isolated from snail gut collected from Rauzabagh, Maulana Azad College, Aurangabad, Maharashtra, India. Journal of Environmental Research and Development. 2015;**9**:577-584

[76] Van Dover CL, Humphris SE, Fornari D, Cavanaugh CM, Collier R, et al. Biogeography and ecological setting of Indian Ocean hydrothermal vents. Science. 2001;**294**:818-23

[77] Turnbaugh PJ, Ley RE, Hamady M, Fraser-Liggett CM, Knight R, et al. The human microbiome project. Nature. 2007;**449**:804-810

[78] Zilber-Rosenberg I, Rosenberg E: Role of microorganisms in the evolution of animals and plants: the hologenome theory of evolution. FEMS Microbiology Reviews. 2008;**32**:723-735

[79] MCFall-Ngai M. Adaptive immunity: Care for the community. Nature. 2007;**445**:153

[80] Dillon RJ, Dillon VM. The gut bacteria of insects: Nonpathogenic interactions. Annual Review of Entomology. 2004;**49**:71-92

[81] Dong YM, Manfredini F, Dimopoulos G. Implication of the mosquito midgut microbiota in the defense against malaria parasites. PLoS Pathogens. 2009;**5**:e1000423

[82] Nisbet RH. The Life of Achatinidae in London, Presidential address. Proceeding of the malacological Society of London. 1974;**41**:171

[83] Warnecke F, Luginbuhl P, Ivanova N, Majid G, Richardson TH, et al. Metagenomic and functional analysis of hindgut microbiota of a wood-feeding higher termite. Nature. 2007;**450**:560-565

[84] Brulc JM, Antonopoulos DA, Miller ME, Wilson MK, Yannarell AC, et al. Gene-centric metagenomics of the fiber-adherent bovine rumen microbiome reveals forage specific glycoside hydrolases. Proceedings of the National Academy of Sciences USA. 2009;**106**:1948-1953

[85] Suen G, Scott JJ, Aylward FO, Adams SM, Tringe SG, et al. An insect herbivore microbiome with high plant biomass-degrading capacity. PLoS Genetics. 2010;**6**:e1001129. DOI:10.1371/journal.pgen.1001129

[86] Iakiviak M, Mackie RI, Cann IK. Functional analyses of multiple lichenin-degrading enzymes from the rumen bacterium *Ruminococcus albus* 8. Applied Environmental Microbiology. 2011;**77**:7541-7550

[87] Lee SS, Ha JK, Cheng KJ. Relative contributions of bacteria, protozoa, and fungi to *in vitro* degradation of orchard grass cell walls and their interactions. Applied Environmental Microbiology. 2000;**66**:3807-3813

[88] Wilson DB, Irwin DC. Genetics and properties of cellulases. Advances in Biochemical Engineering/Biotechnology: Recent Progress in Bioconversion. 1999;**65**:1-21

[89] Bhat MK, Bhat S. Cellulose degrading enzymes and their potential industrial applications. Biotechnology Advances. 1997;**15**:583-620

[90] Priyadarshinee R, Kumar A, Mandal T, Dasguptamandal D. Unleashing the potential of ligninolytic bacterial contributions towards pulp and paper industry: key challenges and new insights. Environmental Science and Pollution Research. 2016;**23**:23349-23368. DOI:10.1007/s11356-016-7633-x

[91] Maki M, Leung KT, Qin W. The prospects of cellulase-producing bacteria for the bioconversion of lignocellulosic biomass. International Journal of Biological Sciences. 2009;**5**:500-516. DOI:10.7150/ijbs.5.500

[92] Waeonukul R, Kyu KL, Sakka K, Ratanakhanokchai K. Isolation and characterization of a multienzyme complex (cellulosome) of the *Paenibacillus curdlanolyticus* B-6 grown on Avicel under aerobic conditions. Journl of Bioscience and Bioengineering. 2006;**107**:610-614. DOI:10.1016/j.jbiosc.2009.01.010

[93] Pérez-Avalos O, Sánchez-Herrera LM, Salgado LM, Ponce-Noyola T. A bifunctional endoglucanase/endoxylanase from *Cellulomonas flavigena* with potential use in industrial processes at different pH. Current Microbiology. 2008;**57**:39-44. DOI: 10.1007/s00284-008-9149-1

[94] Chang YC, Choi DB, Takamizawa K, Kikuchi S. Isolation of *Bacillus* sp. strains capable of decomposing alkali lignin and their application in combination with lactic acid bacteria for enhancing cellulase performance. Bioresource Technology. 2014;**152**:429-436

[95] Wolin MJ, Miller TL. Interactions of microbial-populations in cellulose fermentation. Federation Proceedings. 1983;**42**:109-113

[96] Devillard E, Goodheart DB, Karnati SK, Bayer EA, Lamed R, et al. *Ruminococcus albus*8 mutants defective in cellulose degradation are deficient in two processive endocellulases, Cel48A and Cel9B, both of which possess a novel modular architecture. Journal of Bacteriology. 2004;**186**:136-145

[97] Cann I, Bernardi RC, Mackie RI. Cellulose degradation in the human gut: *Ruminococcus champanellensis* expands the cellulosome paradigm. Environmental Microbiology. 2016;**18**:307-310

[98] Charrier M, Rouland C. The digestive osidases of the snail *Helix aspersa*: Localizations and variations according to the nutritional status. Canadian Journal Zoology. 1992;**70**:2234-2241

[99] Flari V, Charrier M. Contribution to the study of carbohydrases in the digestive tract of the edible snail *Helix lucorum* L. (Gastropoda: Pulmonata: Stylommatophora) in relation to its age and its physiological state. Comparative Biochemistry and Physiology. 1992;**102**:363-372

[100] Biedermann W, Moritz P. Beiträgezurver gleichenden Physiologie der Verdauung II. Uebereincelluloselo¨sendesEnzymimLebersecret der Schnecke (*Helix pomatia*). Pflüg Archiv. 1898;**73**:219-287

[101] Payne DW, Thorpe NA, Donaldson EM. Cellulolytic activity and study of the bacterial population in the digestive tract of *Scrobicularia plana* (Decosta). Proceedings of the Malacological Society of London. 1972;**40**:147-160

[102] Ley RE, Backhed F, Turnbaugh P, Lozupone CA, Knight RD, et al. Obesity alters gut microbial ecology. Proceedings of the National Academy of Sciences USA. 2005;**102**:11070-11075. DOI: 10.1073/pnas.0504978102

[103] Pope PB, Denman SE, Jones Tringe SG, Barry K, et al. Adaptation to herbivory by the Tammar wallaby includes bacterial and glycoside hydrolase profiles different from other herbivores. Proceedings of the National Academy of Sciences USA. 2010;**107**:14793-14798. DOI: 10.1073/pnas.1005297107

[104] Martens EC, Lowe EC, Chiang H, Pudlo NA, Wu M, et al. Recognition and degradation of plant cell wall polysaccharides by two human gut symbionts. PLoS Biology. 2011;**9**:e1001221

[105] Zhu LF, Wu Q, Dai JY, Zhang S, Wei F. Evidence of cellulose metabolism by the giant panda gut microbiome. Proceedings of the National Academy of Sciences USA. 2011;**108**:17714-17719

[106] Engel P, Martinson VG, Moran NA. Functional diversity within the simple gut microbiota of the honey bee. Proceedings of the National Academy of Sciences USA. 2012;**109**:11002-11007

[107] Hess M, Sczyrba A, Egan R, Kim TW, Chokhawala H, et al. Metagenomic discovery of biomass-degrading genes and genomes from cow rumen. Science. 2011;**331**:463-467

[108] Guarner F, Malagelada JR. Gut flora in health and disease. Lancet. 2003;**361**:512-519

[109] Carr WE. Chemoreception in the mud snail, *Nassarius obsoletus*. II. Identification of stimulatory substances. The Biological Bulletin. 1967;**133**:106-127

[110] Shao Y, Arias-Cordero E, Guo H, Bartram S, et al. In vivo Pyro-SIP assessing active gut microbiota of the cotton leafworm, Spodopteralittoralis. PLoS One. 2014;**9**:e85948. DOI:10.1371/journal.pone.0085948

[111] Dantur KI, Enrique R, Welin B, Castagnaro AP. Isolation of cellulolytic bacteria from the intestine of Diatraeasaccharalis larvae and evaluation of their capacity to degrade sugarcane biomass. AMB Express. 2015;**5**:15

[112] Ruiz-Rodriguez M, Valdivia E, Martin-Vivaldi M, Martin-Platero AM, et al. Antimicrobial activity and genetic profile of Enteroccoci isolated from hoopoes uropygial gland. PLoS One. 2012;**7**

[113] Kohei O. New families of carboxyl peptidases: serine-carboxyl peptidases and glutamic peptidases. Journal of Biochemistry. 2012;**151**:13-25. DOI: 10.1093/jb/mvr129

[114] Barrett AJ, Rawlings ND, Woessner FJ, editors. Handbook of Proteolytic Enzymes. 2nd ed. London, UK: Elsevier Academic Press; 2004

[115] Ariole, CN, Ilega E. Alkaline protease production by *Pseudomonas aeruginosa* isolated from the gut of *Pila ovata*. Journal of Global Biosciences. 2013;**2**:126-131

[116] Wellingta CADN, Meire LLM. Production and properties of an extracellular protease from thermophilic *Bacillus* sp. Brazilian Journal of Microbiology. 2004;**35**:91-96

[117] Chu WH. Optimization of extracellular alkaline protease production from species of *Bacillus*. Journal of Industrial Microbiology Biotechnology. 2007;**34**:241-245

[118] Charrier M, Brune A. The gut microenvironment of helicid snails (Gastropoda: Pulmonata): in situ profiles of pH, oxygen and hydrogen determined by microsensors. Canadian Journal of Zoology. 2003;**81**:928-935

[119] Oyeleke SB, Egwim EC, Oyewole OA, John EE. Production of cellulase and protease from microorganisms isolated from gut of *Archachatina marginata* (Giant African Snail). Science and Technology. 2012;**2**:15-20

[120] Cockburn T, Reid R. Digestive tract enzymes in two aeolidnudi branchs (Opisthobranchia: Gastropoda). Comparative Biochemistry and Physiology Part B: Comparative Biochemistry. 1980;**65**:275-281

[121] Evans WAL, Jones EG. A note on the proteinase activity in the alimentary tract of the slug *Arion ater* L. Comparative Biochemistry and Physiology. 1962;**5**:223-225

[122] Godoy MS, Castro-Vazquez A, Vega IA. Endosymbiotic and host proteases in the digestive tract of the invasive snail *Pomacea canaliculata*: diversity, origin and characterization. PLos One. 2013;**8**:e66689

[123] Speiser B. Food and feeding behavior. In: Barker GM, editors. The Biology of Terrestrial Molluscs. New York: CAB International Publishing; 2001. pp. 259-288

[124] Myers FL, Northcote DJ. Partial purification and some properties of a cellulase from *Helix pomatia*. Biochemical Journal. 1959;**71**:749-756

[125] Kiebre-Toe MB, Borges E, Maurin F, Richard Y and Kodjo A. Etude de la flore-bactérienneaérobie à Gram négatif de l'escargotd'élevage (*Helix aspersa*). Revue MédicaleetVétérinaire. 2003;**154**:605-610

[126] Jeuniaux C. La flore bactérienne chitinolytique intestinale de l'escargot (*Helix pomatia* L): Analyse quantitative et qualitative. Bulletin de la Société Royale des Sciences de Liège. 1955b;**24**:254-270

[127] Donachie SI, SaborowskiR, Peters G and E Buchholz. Bacterial digestive enzyme activity in the stomach and hepatopancreas of Meganyctiphanesnorvegica (M. Sars, 1857). Journal of Experimental Marine Biology and Ecology. 1995;**188**:151-165

[128] Coughtrey PJ, Martin MH. The uptake of lead, zinc, cadmium and copper by the Pulmonate Mollusc, *Helix aspersa* Miiller, and its relevance to The monitoring of heavy metal contamination of the environment. Oecologia. 1977;**27**:65-74

[129] Giraud-Billoud M, Abud MA, Cueto JA, Vega IA, Castro-Vazquez A. Uric acid deposits and estivation in the invasive apple-snail, *Pomacea canaliculata*. Comparative Biochemistry and Physiology Part A. 2011;**158**:506-512

[130] Pereira CRD, Breckenridge WR. A histo-physiological study of the alimentary system of *Achatina fulica* (Gastropoda: Pulmonata) with particular reference to glands in the tract. Ceylon Journal of Science. 1981;**14**:152-192

[131] Nicolai A, Rouland-lefèvre C, Ansart A, filser J, Lenz R, et al. Inter-population differences and seasonal dynamic of the bacterial gut community in the endangered land snail *Helix pomatia* (Gastropoda: HelIcIdae). MalacologIa. 2015;**59**:177-190. DOI.org/10.4002/040.059.0101

[132] Nicolai A. The Impact of Diet Treatment on Reproduction and Thermos-Physiological Processes in the Land Snails *Cornu aspersum* and *Helix pomatia*. Thèse en Co-tutelle. France: Universität Bremen/Universitè de Rennes; 2010

[133] Charrier M. Evolution, during digestion, of the bacterial flora in the alimentary system of Helix aspersa (Gastropoda:Pulmonata): A scanning electron microscope study. Journal of Molluscan Studies. 1990;**56**:425-433

[134] Churchill AT, Storey BK. Intermediary energy metabolism during dormancy and anoxia in the land snail *Otala lactea*. Physiological Zoology. 1989;**62**:1015-1030

[135] Ansart A, Vernon P, Daguzan J. Photoperiod is the main cue that triggers supercooling ability in the land snail, *Helix aspersa* (Gastropoda:Helicidae). Cryobiology. 2001;**42**:266-273

[136] Nicolai A, Filser J, Lenz R, Bertrand C, Charrier M. Quantitative assessment of haemolymph metabolites in two physiological states and two populations of the land snail *Helix pomatia*. Physiological and Biochemical Zoology. 2012;**83**:274-284

[137] Morocoima A, Rodríguez V, Rivas R, Coriano H, Rivero S. *Achatina fulica* Bowdich, 1822 (Mollusca: Gastropoda) carrier of Helminthes, Protozoa and Bacteria in northeast Venezuela. Boletín De Malariología Y Salud Ambiental. 2014;**54**(2):174-185

[138] Odo GE, Ekeh FN, Ekechukwu NE, Agwu EJ, Aguzie IOE. Modification of pathogenic microbiota and histology of gastrointestinal tract of *Archachatina marginata* (Swainson, 1821) by *Carica papaya* seed meal. African Journal of Microbiology Research. 2016;**10**:1612-1617. DOI: 10.5897/AJMR2016.8017

[139] Gomare S, Kim HA, Ha JH, Lee MW, Park JM. Isolation of the polysaccharidase-producing bacteria from the gut of sea snail, *Batillus cornutus*. Korean Journal of Chemical Engineering. 2011;**28**:1252-1259. DOI: 10.1007/s11814-010-0506-y

[140] González-Núñez C, Calvo-Torras MA. Comparison of the microbiota of snails (*Helix aspersa*) of different weights and its evolution over time. Journal of Veterinary Science and Technology. 2015;**6**:2. DOI:10.4172/2157-7579.1000216

[141] Caullan LP, Vila GG, Angulo EA, Calvo A, Marcelo JA. Microbiota from *Helix aspersa* Müller in Barcelona Area (Spain). Advances in Microbiology. 2014;**4**:604-608. DOI:10.4236/aim.2014.410066

[142] Imran M, Saida B, Ghadi SC, Verma P. Shouche YS. The gut-associated *Klebsiella* sp. of the apple snail produces multiple polysaccharide degrading enzymes. Current Science. 2016;**110**:2170-2172

[143] Thilaga RD, Jayaprabha D, Sivakumar V, Tamilselvi M. Digestive enzymes, gut micro floral study of top shell *Trochus niloticus*. World Journal of Fish and Marine Sciences. 2010;**2**:495-498

Chapter 10

# The Uniqueness of *Achatina fulica* in its Evolutionary Success

Sandip Mukherjee, Shuvasree Sarkar,
Chayan Munshi and Shelley Bhattacharya

Additional information is available at the end of the chapter

http://dx.doi.org/10.5772/68134

**Abstract**

The increasing load of environmental pollutants poses a serious threat over the globe. In this vulnerable situation, it is essential to have alternative sources of medicines, may be from invertebrates. Among invertebrates, although molluscs are known for their consumption as food and ethno-medicinal use, the importance of these animals is still overlooked. Presently attention has been geared toward molluscs including *Achatina fulica* which are now considered as one of the most evolutionary successful animals. During the last few decades, researchers are trying to decipher their complex immune system to harvest valuable molecules to treat human diseases. In the present review, the existence of important immunological factors in *Achatina* is discussed addressing the coagulation system, innate immune molecules, bioactive proteins and lastly the enigmatic C-reactive proteins.

**Keywords:** *Achatina fulica*, innate immunity, antibacterial activity

## 1. Introduction

Extensive research on invertebrate immune system for the last few decades, including molluscs, revealed that invertebrates contain peptides which are endowed with anti-microbial activity [1]. These peptides can trigger specific anti-bacterial reaction by producing different isoforms specific for each bacterial species. Among immunological molecules of invertebrates, Toll-like receptor 4 (TLR4) gained much attention, though its essentiality happens to be more pronounced in vertebrates [1]. Gastropod diversity is well documented, recording 40,000–150,000 species with size variance of 1 mm to 1 m and indicating a strong immune system in gastropods [1–3].

The giant African snail, *Achatina fulica*, is one of the large and most widely distributed land snails, considered as an agrihorticultural pest [4].Since *Achatina* develops rapidly and produces large numbers of offspring, it is now listed as one of the top 100 invasive species in the world [5]. Moreover, *Achatina* is a unique species and maintains three different life cycle stages in the same individual, surviving in the environment for millions of years. Apart from maintaining the critical life cycle stages, *A. fulica* has survived successfully, consequently, gaining the disrepute as an agricultural pest in India. In addition, these snails are considered as bio-indicators of ecosystem health. Although they do not possess immunoglobulins, they have evolved unique modalities to detect and respond to microbial surface antigens such as lipopolysaccharides (LPS), lipoteichoic acids, lipoproteins, peptidoglycans and (1→3) β-D-glucans [6].

Terrestrial snails are well known for accumulating heavy metals in their tissues and serve as a pertinent species for monitoring trace metals, agrochemicals, urban pollution and electromagnetic exposures [7]. The effect of accumulated heavy metals in different molluscan tissues and possible use of such alterations as biomarkers of exposure to xenobiotics has been investigated in some detail [8, 9]. Although snails are considered as alleged pest they are used by humans for various purposes including vigorous consumption of mollusc meat in several countries around the globe, including tribal and urban populations of India and Bangladesh [10]. Another important aspect is the ethno-medicinal use of several mollusc species highlighted by several authors [11, 12]. Pharmacological application of different body parts of mollusc are used to treat several diseases which suggests its potential to act as a source of drug [12]. In the present chapter, various characters of *Achatina* will be described including their unique immune system that contributes toward the evolutionary success of *A. fulica* in the terrestrial ecosystem.

## 2. Molecules in the Innate Immune System of *A. fulica*

### 2.1. Coagulation system in *A. fulica*

Invertebrates are not able to synthesize immunoglobulins, rather they have developed a potential defense system against microbial surface antigens such as lipopolysaccharides (LPS)/endotoxins and glucans [13]. Among various kinds of innate immune mechanisms in invertebrates, two types of coagulation mechanisms are on record: (i) in crustaceans such as lobster, crayfish [14] and insects [15] clotting occurs through Ca-dependent transglutaminase, (ii) serine protease zymogens dependent coagulation system is reported which is similar to mammalian system [13]. In *Limulus polyphemus*, commonly known as the horseshoe crab, endotoxins are sensed by amoebocytes. In invertebrates, amoebocytes are known to be associated in both hemostasis and innate or nonadaptive immune responses against microbial infections [16]. Amoebocytes behave like macrophages in mammals and can either bind pathogens directly or recognize and engulf pathogens that have been opsonized by serum proteins. This direct recognition plays a major role in host defense [17]. It has been proposed that activation of the innate immune system is initiated when pathogens bind to nonclonally distributed pattern recognition receptors on immune cells [18]. In *Limulus*, the ancient horse shoe crab, the components, termed as Factor C [19, 20], Factor B [21] and pro-clotting enzyme [22], undergo endotoxin-mediated sequential limiting proteolysis/activation followed by irreversible conversion of clottable protein (coagulogen) into insoluble gel (coagulin) [23]. A similar pattern of coagulation system is also reported in *A. fulica* akin to endotoxin-mediated coagulation system in the circulating amoebocytes. An endotoxin-sensitive factor (ESF) available in the *Achatina* amoebocyte lysate (AAL) has been purified and characterized to be a serine protease type. These factors undergo a series of events such as aggregation and rapid degranulation leading to coagulation [24]. The aggregation mechanism causes bacterial sequestration, while degranulation results in secretion of serine protease zymogens [16]. Although the molecular basis of coagulation in *A. fulica* and further characterization of AAL remains to be determined, amoebocytes are considered as one of the primary immune cells in innate immune system in *A. fulica*.

## 3. Acharan sulfate, the new glycosaminoglycan from *A. fulica*

Acharan sulfate, a glycosaminoglycan isolated from *A. fulica*, has a major disaccharide repeating unit of 2-acetyl,2-deoxy-a-D-glucopyranose-2-sulfo-a-L-idopyranosyluronic acid, which is structurally related to both heparin and heparin sulfate. Acharan sulfate is known to be a polydisperse, with an average molecular mass of 29 kDa that contain un-sulfated iduronic acid. This glycosaminoglycan was found to be located in the body of this species and considered to be the major constituent of the mucus and the structure and compartmental distribution of acharan sulfate in the snail body [25]. Different populations of acharan sulfate having charge and/or molecular mass heterogeneities were isolated from *Achatina* whole body, mucus and the organic shell matrix. A minor glycosaminoglycan fraction was also purified which appeared to be susceptible to degradation by nitrous acid confirming the presence of N-sulfated glycosaminoglycan molecules. Furthermore, application of histochemical techniques of metachromatic staining and histoautoradiography (following metabolic radiolabeling with [$^{35}$S] sulfate) was evident that acharan sulfate is of wide distribution in the snail body.

## 4. Anti-bacterial protein from mucus of *A. fulica*

Achacin is an antibacterial glycoprotein obtained from the mucus present on the body surface of *A. fulica*. Achacin is known not only to inhibit growth of both Gram-positive and Gram-negative bacteria [26], but also appeared to attack the bacterial plasma membranes [27]. It is hypothesized that achacin is an active molecule although its role in controlling innate immunity warrants further research. However, the sequence of achacin has reported [28] its ability to catalyze oxidative deamination producing ketoacids, hydrogen peroxide ($H_2O_2$) and ammonia ($NH_3$). The antibacterial activity of achacin was found to be dependent on $H_2O_2$ production which is produced by the oxidative deamination reaction. Interestingly, LAOs in vertebrates also have antibacterial activity [29] which effects are most likely due to $H_2O_2$ formation. However, the concentration of achacin-generated $H_2O_2$ in the culture medium was not sufficient to inhibit bacterial growth [28]. Bacteria in their growth phase appeared to play an important role in the antibacterial activity of achacin. These data illustrate that when snails are infected by pathogens, achacin should bind to the plasma membranes of those that are proliferating. Achacin may attack pathogens during other growth phases too by increasing the local concentration of $H_2O_2$ so as not to harm neighboring host cells. Thus, LAOs, which are widely distributed in living organisms, appeared to be of import in both vertebrate and invertebrate host defenses.

## 5. Role of Snail Hemocytes in Innate Immunity

Circulating blood cells known as hemocytes represent the main cellular component of the molluscan immune system. Hemocytes are composed of a mixture of different subpopulations of cells, for example, flow cytometric analyses of hemocytes from the freshwater snails *Biomphalaria glabrata* [30] and *Planorbarius corneus* [31] confirmed two types of circulating cells with two distinct functions [31]. Large granular hemocytes of *B. glabrata,* characterized by the absence of the monoclonal antibody BGH1 surface marker [32], are highly phagocytic in nature, while the $BGH1^+$ is nonphagocytic. *Lymnaea stagnalis* also possess two subtypes having specific surface epitopes such as the mature LS1 and nondifferentiated $LS1^+$ hemocytes [29]. It is presumed that hemocyte subpopulations that differ both chemically and functionally are regulated in their activities or behaviors through specific receptors and the signals conveyed by their interaction with appropriate ligands. It was further concluded [33] that there are five types of cells in the hemolymph of *B glabrata* and *Biomphalaria straminea* which contributes to the knowledge base for studies on hemocytes and their involvement in controlling *Schistosoma mansoni* infection.

If attention is focused on the functional attributes of hemocytes, several reports in this direction revealed diverse immunological functions such as phagocytosis [34], cytotoxicity [35], aggregation [36] and pathogen encapsulation [37, 38]. In addition to hemocytes, hemolymph, the humoral component of the molluscan immune system, is reported to exhibit the activities of superoxide dismutase [39], catalase [40] and acid [41] and alkaline phosphatases [42]. Total hemocyte count in mollusc has been considered as an important immune parameter [43]. Elevation of the total hemocyte count indicates augmentation of immunity of invertebrates [44]. Phagocytosis is an established strategy of immune defense in invertebrates including mollusc. It is considered as the major immunological activity evidenced in many molluscan species [45]. Major cytotoxic molecules such as superoxide anion and nitric oxide generated by the circulatory hemocytes of molluscs are functionally associated with the destruction of pathogens [46, 47]. Phenoloxidase is reported to be functionally associated with phagocytosis, self-nonself discrimination, cytotoxicity and melanization response [48]. Superoxide dismutase and catalase play a significant antioxidation role in the cellular physiology of molluscs. In addition, glutathione-S-transferase is functionally associated with general detoxification response of xenobiotics and anti oxidation activity [49]. All these enzymes are involved in scavenging and deactivating the toxic oxidative radicals and protect the tissue from oxidative damage [46]. Acid and alkaline phosphatases are functionally involved in pathogen destruction in phagolysosome which bear immunological significance [50]. Several reports also demonstrated a range of receptors which bind carbohydrates, extracellular matrix proteins, hormones, growth factors and cytokines resulting specific immunocyte signals not only in vertebrates but also in molluscs [37]. Thus, it can be surmised that signaling systems are evolutionary conserved functions of immunocytes in the animal kingdom.

Apart from the above-mentioned defense mechanisms, snails also undergo starvation and aestivation under any stress condition. Though several reports are available on starvation and aestivation of snails, information on immune-related parameter of Indian mollusc is scant. In *Helix pomatia*, antioxidant enzymes are stimulated during aestivation [51] and physiological correlation exists between antioxidant defense and metabolic depression [52]. Starvation is reported to compromise the immunological activity of a land snail, *Helix aspersa* [53]. As per these reports, several immunological parameters are shown to be influenced by nutrition; some of these parameters are hemocyte count, phenol oxidase activity and phagocytosis. One of the elegant reports [54] in this perspective showed modulation of the innate immune parameters during experimental aestivation and starvation in *Parashorea globosa*. The parameters studied by this group included generation of cytotoxic molecules like superoxide anion, nitric oxide and phenoloxidase and the activities of superoxide dismutase, catalase, glutathione-S-transferase, acid phosphatase, alkaline phosphatase and total protein in hemocytes and hemolymph of *P. globosa* during activity, aestivation, arousal and starvation. This finding appears to be important in the field of comparative immunity and physiology for *P. globosa* which is considered as a commercially important mollusc in India.

### 5.1. C-reactive protein (CRP), a multifunctional player in Achatina

C-reactive protein (CRP) was first discovered in Oswald Avery's laboratory at the Rockefeller Institute for Medical Research [55]. CRP has evolved conservatively, and homologous proteins with similar functional attributes have been found in many other species. The stable preservation of this protein during evolution implies some biological significance. Thus, CRP is an ancient molecule discovered in humans only about 82 years ago. It belongs to a protein family called pentraxin (from the Greek words "penta" five and "ragos," berries) that constitutes a phylogenetically ancient family of proteins exhibiting a remarkable conservation of structure and binding reactivities. The presence of CRP has been reported from a wide range of different animals such as monkey, dog, goat, rabbit, rat, mice, domestic fowl, fish, shark and lumpsucker among vertebrates and horseshoe crab [56] and *A. fulica* [57] among the invertebrates. The finding that CRP is a major blood constituent of primitive animals, for example, horseshoe crab, *L. polyphemus* and dogfish argues strongly for an important role of this protein.

In *A. fulica*, induction of C-reactive protein (CRP) synthesis was triggered by exogenous administration of the steroid 4-androstenedione (4 AD) [58].Further, it has been suggested that the hepatopancreas is the main site of CRP expression and the CRP gene in the hepatopancreas is acutely responsive to Gram-negative bacterial infection [59]. Previously, a question had been raised on whether CRP is inducible in *Limulus* [60]. A search of the *Limulus* CRP promoter for the IL-6 response element and the Drosophila heat shock element in the human CRP promoter [61] revealed an absence of these cis-elements which led to the conclusion that *Limulus* CRP expression is constitutive [60]. However in mammals, hepatic CRP is soluble in nature which is released into circulation [62] induced by proinflammatory cytokines. Recently, in an interesting study, the evolutionary significance of TNF, IFNγ and iNOS in immune response has been amply demonstrated in two Indian mollusc species [63]. Besides assessing different toxicological parameters, anti-bacterial property of the innate immune molecule, namely C-reactive protein (CRP) isolated from *A. fulica*, was also determined. CRP is a prototypic acute phase reactant, which is a phylogenetically conserved protein expressed in invertebrates such as arthropods [56], molluscs [58] and also in all vertebrates [64]. In *Limulus*, an arthropod, CRP acts as a main front-line innate immune molecule [59] which may be the key to a powerful defense mechanism of these animals against microbial infections that are potentially lethal in other organisms. Moreover, the presence of high level of endogenous CRP (2–4 mg/ml) in the hemolymph of *A. fulica* [58] might be the sole reason behind their effective survival in the environment.

Several authors reported that CRP can protect mice from infections caused by both Gram-positive *Streptococcus pneumoniae* [65] and Gram-negative *Neisseria elactamica* [66] and *Haemophilus influenzae* [67] bacteria via direct binding with repetitive phosphorylcholine moieties on the lipoteichoic acid or the lipopolysaccharide (LPS) of these pathogens, respectively. The level of CRP also increases dramatically during periods of immunological challenge and boosts the bactericidal activities of monocytes and neutrophils by enhancing the release of reactive oxygen intermediates [68]. CRP also induces oxidative stress in vitro in endothelial cells, smooth muscle cells and monocyte-macrophages [69, 70]. Although there are many reports on properties of CRP in a wide range of in vitro and in vivo model systems, clear understanding of the actual biological functions of this phylogenetically ancient and highly conserved molecule remains elusive.

It is also noted that bacterial cells are strongly dependent on metabolic cycles for their survival and pathogenicity [71, 72]. Therefore, effect of *Achatina* CRP (ACRP) on these bacterial metabolic cycles comprising key metabolic enzymes such as phosphofructokinase 1(PFK1) in glycolysis, isocitrate dehydrogenase (ICDH) and isocitrate lyase (IL) in TCA cycle and fructose-1,6-bis phosphatase (FBP1) in gluconeogenesis was also investigated. Various authors have reported the existence of eukaryote-like programmed cell death and the involvement of caspase-3-like proteins in bacteria [73]. Based on this information, it was attempted to delineate the anti-bacterial property of ACRP in terms of inhibition of salient metabolic enzymes which decrease bacterial infection accompanied by ROS generation and apoptosis-like phenotypes during bacterial cell death.

Several authors [74] reported potentiality of human CRP to inhibit superoxide ($O_2^-$) generation and delay apoptosis in neutrophils [64]. Recently, it has been reported that immune-potent CRP modulates antioxidant and anti-inflammatory effects in LPS-stimulated human macrophages [75]. The anti-stress property of ACRP was tested in mice which are known to have a very low level of endogenous CRP (~2 μg/mL) even after an inflammatory stimulus [76]. In order to prove this hypothesis, lead nitrate was administered intraperitoneally at an environmentally relevant dose in mice, and the induced oxidative stress was found to be removed when ACRP was administered prior to treating with Pb. Furthermore, in an in vitro study, both native CRP and its subunits were found to accomplish reversal of lead-induced hepatotoxicity in *A. fulica* [77].

In molluscs, several anti-microbial peptide (AMP) genes are triggered during onset of a broad range of pathogenic infections. Furthermore, several categories of immune molecules are extracted from snails including glycosaminoglycans, peptides, proteins (glycoproteins) and enzymes which possess diverse biological activities [78, 79]. Interestingly, evolutionary success of *A. fulica* can be associated, in part, with their relatively simple and effective innate immune system comprised of defense molecules present in their hemolymph such as hemocyanins, lectins, C-reactive proteins and macroglobulins in addition to a large number of granular hemocytes or amoebocytes [78, 79].

It was earlier established that xenobiotics, like heavy metals, are successful in triggering the synthesis of CRP causing inflammatory condition, and in turn, CRP was found to be a very good scavenger in eliminating these heavy metals. In contrast to human and other higher level mammals, the normal fresh water teleost *Channa punctatus* has a high level of CRP [80]. The level of CRP was also found to be significantly high in the snail *A. fulica* [58] during rainy season which is nearer to the level of CRP in the horse-shoe crab, *Limulus*. It was clearly documented by several authors that level of CRP significantly increases in serum during onset of infection or inflammation and thereby CRP acts as an inducible protein in mammals. However, in invertebrates, CRP is constitutively expressed, as for example, *Achatina* hemolymph contains a higher level of CRP which is about 2 mg/ml and showed strong cross reaction with *Limulus* CRP antiserum [58].

Snails accumulate heavy metals more in their tissue inducing numerous acute and sublethal effects [81]. Due to this sensitivity, they are considered as excellent bioindicators of heavy metal contamination [82]. The effect of accumulated heavy metals on different molluscan tissues and possible use of such alterations as biomarkers of exposure to xenobiotics has been investigated [8, 9]. Molluscs have shown considerable promise as biomonitors of metal pollution [83], and an extensive literature has appeared concerning mechanisms of uptake, detoxification and storage of heavy metals [84]. Few studies on several fresh water and marine species further substantiate the role of molluscs as bioaccumulators [85]. Further, ecological and ecophysiological studies suggest that molluscs react to environmental stress and pollution by modifying their behavior [86]. It is reported that terrestrial snails might regulate some metals assimilated from food and xenobiotic exposure [7]. The kinetics of metal accumulation and detoxification are still a subject of discussion, and there is a lack of information regarding metal toxicity in snails [84, 87].

## 6. Conclusion

Presently immunological molecules in mollusc, especially *A. fulica* have gained much attention because they fail to synthesize immunoglobulins but possess a strong innate immune system comprised of several molecules such as microbial surface antigens lipopolysaccharides (LPS)/endotoxins, glucans, acharan sulfate-glycosaminoglycan, achacin-mucus–derived antibacterial protein, hemocyte-derived factors and C-reactive protein. Among these proteins, CRP not only acts as apotent defence molecule but also engages in several vital physiological functions. Recently, elegant studies have clearly indicated the role of *Achatina* CRP in inhibiting growth of both Gram-positive and Gram-negative human pathogenic bacteria [88]. These investigations showed for the first time that CRP itself can trigger apoptotic like cell death in bacterial cells. Another important contribution from this group is that ACRP was found to cross species barrier and reduce metal toxicity in mammals (mice). However, more in depth study is warranted to exploit these molecules for the benefit of human beings.

## Author details

Sandip Mukherjee, Shuvasree Sarkar, Chayan Munshi and Shelley Bhattacharya*

Address all correspondence to: shelleyb38@gmail.com

Environmental Toxicology Laboratory, Department of Zoology, Centre for Advance Studies, Visva Bharati University, Santiniketan, India

## References

[1] Söderhäll K (2011). Invertebrate Immunity. In: Advances in Experimental Medicine and Biology, Spinger Science+ Business Media, LLC, Landes Bioscience, New York, USA.

[2] Marchalonis JJ (1977). Immunity in Evolution. Harvard University Press, Cambridge, MA.

[3] Ponder WF, Lindberg DR (2008). Molluscan Evolution and Phylogeny.In: Phylogeny and Evolution of the Mollusca. University of California Press, Berkley, 1-469.

[4] Ghose KC (1964). The structures of the heart of *Achatinafulica*and *Macrochlamysindica* and pacemaker mechanism and refilling of the auricle in *A. fulica*, with a discussion on the same on the gastropods. *Philipp J Sci*, 93: 219-228.

[5] Lowe S, Browne M, Boudjelas S, De Poorter M (2000). 100 of the World's Worst Invasive Alien Species A selection from the Global Invasive Species Database. Published by The Invasive Species Specialist Group (ISSG) a Specialist Group of the Species Survival Commission (SSC) of the World Conservation Union (IUCN).

[6] Begum N, Matsumoto M, Tsuji S, Toyoshima K and Seya T (2000). The primary host defence system across humans, flies and plants. *Curr Trends Immunol*, 3: 59-74.

[7] Berger B and Dallinger R (1989). Terrestrial snails as quantitative indicators of environmental metal pollution. *Environ Monit Assess*, 25: 65-84.

[8] Triebskorn R, Henderson IF, Martin A and Köhler HR (1996). Slugs as target or non-target organisms for environmental chemical. In: Symposium Proceedings no. 66: Slug and Snail Pests in Agriculture. 65-72.

[9] Marigomez IA, Kortabitarte M and Dussart GBJ (1998). Tissue level biomarkers in sentinel slugs as cost-effective tools to assess metal pollution in soils. *Arch Environ Contam Toxicol,* 34: 167-176.

[10] Baby RL, Hasan I, Kabir KA, Naser MN (2010). Nutrient analysis of some commercially important molluscs of Bangladesh. *J Sci Res,* 2: 390-396.

[11] Prabhakar AK and Roy, SP (2009). Ethnomedicinal uses of some shellfishes by people of Kosi river basin in North Bihar, India. *Ethno Med* 3: 1-4.

[12] Mahawar MM and Jaroli DP (2007). Traditional knowledge on zootherapeutic uses by the Saharia tribe of Rajasthan, India. *J Ethnobiol Ethnomed,* 3: 25.

[13] Muta T and Iwanaga S (1996). The role of hemolymph coagulation in innate immunity. *Curr Opin Immunol,* 8:41-47.

[14] Kopack P, Hall M, Söderhäll K(1993). Characterization of a clotting protein, isolated from plasma of the freshwater crayfish *Pacifastacusleniusculus*. *Eur J Biochem,* 213: 591-597.

[15] Bohn H and Barwig B (1984). Hemolymph clotting in the cockroach *Leucophaeamaderae* (Blattaria). Influence of ions and inhibitors; isolation of the plasma coagulogen. *J Comp Physiol B,* 154: 457-467.

[16] Murer EH, Levin J, Holme R (1975). Isolation and studies of the granules of the amoebocytes of the *Limulus polyphemus*, the horseshoe crab. J Cell Physiol, 86: 533-542.

[17] Pearson AM (1996). Scavenger receptor in innate immunity. *Curr OpinImmunol,* 8:20-28.

[18] Mdzhitov R, Janeway CA Jr (1997). Innate immunity: impact on the adaptive immune response. *Curr OpinImmunol,* 9:4-9.

[19] Ding LJ, Navas MAA III, Ho B (1993). Two forms of factor C from the amoebocytes of *Carcinoscorpiusrotundicauda*: purification and characterization. *Biochem Biophys Acta,* 1202:149-156.

[20] Nakamura T, Morita T, Iwanaga S (1986). Lipopolysaccharide-sensitive serine protease zymogen (factor C) found in *Limulus polyphemus*. *Eur J Biochem,* 154: 511-521.

[21] Nakamura T, Horiuchi T, Morita T, Iwanaga S (1986). Purification and properties of intracellular clotting factor, Factor B, from horseshoe crab (*Tachypleustridentatus*) hemocytes. *J Biochem,* 99: 847-857.

[22] Nakamura T, Morita T, Iwanaga S (1985). Intracellular proclotting enzyme in Limulus (*Tachypleustridentatus*) hemocytes: its purification and properties. *J Biochem,* 7:1561-1574.

[23] Nakamura S, Takagi T, Iwanaga S, Niwa M, Takahashi KA (1976). Clottable protein (coagulogen) of horseshoe crab hemocytes. *J Biochem,* 80:649-652.

[24] Levin J, Bang FB (1964). The role of endotoxin in the extracellular coagulation of Limulus blood. *Bull John Hopkins Hosp,* 115: 265-274.

[25] Vieira TC, Costa-Filho A, Salgado NC, Allodi S, Valente AP, Nasciutti LE, Silva LC (2004). Acharan sulfate, the new glycosaminoglycan from *Achatina fulica*Bowdich 1822. Structural heterogeneity, metabolic labeling and localization in the body, mucus and the organic shell matrix. *Eur J Biochem*, 271:845-854.

[26] Otsuka-Fuchino H, Watanabe Y, Hirakawa C, Tamiya T, Matsumoto JJ, Tsuchiya T (1992). Bactericidal action of a glycoprotein from the body surface mucus of giant African snail. *Comp Biochem Physiol*, 101C: 607-613.

[27] Otsuka-Fuchino H, Watanabe Y, Hirakawa C, Takeda J, Tamiya T, Matsumoto JJ, Tsuchiya T (1993). Morphological aspects of achacin-treated bacteria *Comp Biochem Physiol*, 104C: 37-42.

[28] Ehara T, Kitajima S, Kanzawa N, Tamiya T, Tsuchiya T (2002). Antimicrobial action of achacin is mediated by L-amino acid oxidase activity. *FEBS Lett*, 531:509-512

[29] Stiles BG, Sexton FW, Weinstein SA (1991). Antibacterial effects of different snake venoms: purification and characterization of antibacterial proteins from *Pseudechisaustralis* (Australian king brown or mulga snake) venom. *Toxicon*, 29:1129-1141.

[30] Johnston LA and Yoshino TP (2001). Larval *Schistosoma mansoni* excretory-secretory glycoproteins (ESPs) bind to hemocytes of *Biomphalariaglabrata*(Gastropoda) via surface carbohydrate binding receptors. *J Parasitol*, 87:786-793.

[31] FranceschiC, Cossarizza A, Monti D, Ottaviano E (1991). Cytotoxicty and immunocyte markers in cells from the freshwater *Planorbariuscorneus*(L.) (*Gastropoda pulmonata*) implication on the evolution of natural killer cells. *Eur J Immunol*, 21:489-93.

[32] Dikkeboom R, Tijnagel JM, Van der Knaap WP (1985). Monoclonal antibody recognized hemocyte subpopulations in juvenile and adult *Lymnaea stagnalis*: Functional characteristics and lectin binding. *Dev Comp Immunol*, 12:17-32.

[33] Cavalcanti MGS, Filho FC, Mendonça AMB, Duarte GR, Barbosa CCGS, De Castro CMMB, Alves LC, Brayner FA (2012). Morphological characterization of hemocytes from *Biomphalariaglabrata* and *Biomphalariastraminea*. *Micron*, 43:285-291.

[34] Cheng TC (1977). Biochemical and ultrastructural evidence for the double role of phagocytosis in molluscs: defence and nutrition. *Comp Pathobiol*, 3: 21-30.

[35] AdemaCM, Van Der Knapp WPW, Sminia T (1991). Molluscan hemocyte mediated cytotoxicity: the role of reacting oxygen intermediates. *Rev Aquat Sci*, 4: 201-223.

[36] Chen JH and Bayne CJ (1995). Bivalve molluschemocyte behaviors: characterization of hemocyte aggregation and adhesion and their inhibition in the California mussel (*Mytiluscalifurnianus*). *Biol Bull*, 188: 255-266.

[37] Humphries JE and Yoshino TP (2003). Cellular receptors and signal transduction in molluscanhemocytes: Connections with the innate immunesystem of vertebrates. *Integr Comp Biol*, 43:305-312.

[38] Koutsogiannaki S and Kaloyianni M (2010). Signaling molecules involved in immune responses in mussels. *Invertebrate Surviv J*, 7: 11-21.

[39] MonariH, Serrazanetti GP, Foschi J, Matozzo V, Marin MG, Cattani O (2007). Effects of salinity on the clam *Chamelea gallina*haemocytes. Part II: superoxide dismutase response. *Mar Biol,* 151:1059-1068.

[40] Liu D, Chen Z, Xu H (2008). Effects of leucine-enkephalin on catalase activity and hydrogen peroxide levels in the haemolymph of the Pacific oyster (*Crassostrea gigas*). *Molecules,* 13: 864-870.

[41] Cheng TC and Butler MS (1979). Experimentally induced elevations in acid phosphatase activity in hemolymph of *Biomphalariaglabrata* (Mollusca). *J Invertebr Pathol,* 34: 119-124.

[42] Helal IB, El-Mehlawy MH, Rizk ET, El-Khodary GM (2003). Effect of euphorbia peplus plant extract and the antihelmenthicprazequantel on the defence system of *Biomphalariaalexandria* snail. *Egyptian J Aquat Biol Fisheries,* 7(4): 501-527.

[43] MelloDF, ProencaLA, Barracco MA (2010). Comparative study of various immune parameters in three bivalve species during a natural bloom of Dinophysis*acuminata* in Santa Catarina Island, Brazil. *Toxins,* 2:1166-1178.

[44] Kacsoh BZ and Schlenke TA (2012). High hemocyte load is associated with increased resistance against parasitoids in *Drosophila suzukii,* a relative of *D. melanogaster. PLoS One,* 7: e34721.

[45] Garcia-Garcia E, Prado-Alvarez M, Novoa B, Figueras A, Rosales C (2008). Immune responses of mussel hemocyte sub-populations are differentially regulated by enzymes of the P13-K, PKC and ERK Kinase families. *Dev Comp Immunol,* 32: 637-653.

[46] Manduzio H, Rocher B, Durand F, Gatap, C, Leboulenger F (2005). The point about oxidative stress in molluscs. *Invertebrate Surviv J,* 2: 91-104.

[47] Ray M, Bhunia NS, Bhunia AS, Ray S (2013). A comparative analyses of morphological variations, phagocytosis and generation of cytotoxic agents in flow cytometrically isolated hemocytes of Indian molluscs. *Fish Shellfish Immunol,* 34: 244-253.

[48] Cerenius L and Söderhäll K, (2004). The prophenoloxidase-activating system in invertebrates. *Immunol Rev,* 198: 116-126.

[49] ManduzioH, Monsinjon T, Galap C, Leboulenger F, Rocher B (2004). Seasonal variations in antioxidant defences in blue mussels *Mytilus edulis* collected from a polluted area: major contributions in gills of an inducible isoforme of Cu/Zn-superoxide dismutase and of glutathione S-transferase. *Aquat Toxicol,* 70: 83-93.

[50] Murti R and Shukla GS (1984). Mercuric chloride intoxication in freshwater prawn: II. Effect on phosphatase activity. *Ecotoxicol Environ Saf,* 8: 581-586.

[51] Nowakowska A, Swiderska-Kolacz G, Rogalska J, Caputa M (2009). Antioxidants and oxidative stress in *Helix pomatia* snails during estivation. *Comp Biochem Physiol,* 150C: 481-486.

[52] Hermes-Lima M, Storey JM, Storey KB (1998). Antioxidant defenses and metabolic depression. The hypothesis of preparation for oxidative stress in land snails. *Comp Biochem Physiol,* 120 *B*: 437-448.

[53] AL-Rawadeh A (2010). The role of starvation on selective immunological parameters in land snail *Helix aspersa*. *Adv Environ Biol*, 4: 265-270.

[54] Bhunia AS, Mukherjee S, Bhunia NS, Ray M, Ray S (2016). Immunological resilience of a freshwater Indian mollusc during aestivation and starvation. *Aquaculture Rep*, 3:1-11.

[55] Tillet WS and Francis T (1930). Serological reactions in pneumonae with a non protein somatic fraction of pneumococcus. *J Exp Med*, 52(4): 561-571.

[56] Hokama Y and Nakamura RM (1987). C-reactive protein: Current issues and future perspectives. *J Clin Lab Anal*, 1:15-27.

[57] Agrawal A, Mitra S, Ghosh N, Bhattacharya S (1990). C-reactive protein (CRP) in haemolymph of a mollusc, *Achatina fulica* Bowdich. *Indian J Exp Biol*, 28: 788-789.

[58] Bose R and Bhattacharya S (2000). C-reactive protein in the haemolymph of *Achatina fulica*: Interrelationship with sex steroids and metallothionein. *Comp Biochem Physiol A*, 125: 485-495.

[59] Ng PML, Jin Z, Tan SSH, Ho B, Ding JL (2004). The C-reactive protein: a predominant LPS-binding acute phase protein responsive to Pseudomonas infection. *J Endotoxin Res*, 10(3): 1-12.

[60] Liu TY, Minetti CA, Fortes-Dias CL, Liu T, Lin L, Lin Y (1994). C-reactive proteins, limunectin, lipopolysaccharide-binding protein, and coagulin. *Ann NY Acad Sci*, 712: 146-154.

[61] Li SP, Liu TY, Goldman ND (1990). cis-Acting elements responsible for interleukin-6 inducible C-reactive protein gene expression. *J Biol Chem*, 265: 4136-4142.

[62] Yap SH, Moshage HJ, Hazenberg BP, RoelofsHM, BijzetJ,LimburgPC,Aarden LA and-Van RijswijkMH (1991).Tumor necrosis factor (TNF) inhibits interleukin (IL)-1 and/or IL-6 stimulated synthesis of C-reactive protein (CRP) and serum amyloid A (SAA) in primary cultures of human hepatocytes. *Biochem Biophys Acta*, 1091(3): 405-408.

[63] Eklund CM (2009). Proinflammatory cytokines in CRP baseline regulation. *Adv Clin Chem*, 48: 111-136.

[64] Khreiss T, Jozsef L, Hossain S, Chan JS, Potempa LA and Filep JG (2002). Loss of pentameric symmetry of C-reactive protein is associated with delayed apoptosis of human neutrophils. *J Biol Chem*, 277: 40775-40781.

[65] Volanakis JE and Kaplan MH (1971). Specificity of C-reactive protein for choline phosphate residues of pneumococcal C-polysaccharide. *Proc Soc Exp Biol Med*, 136: 612-614.

[66] Serino L and Virji M (2000). Phosphorylcholine decoration of lipopolysaccharide differentiates commensal *Neisseriae* from pathogenic strains: identification of licA-type genes in commensal *Neisseriae*. *Mol Microbiol*, 35: 1550-1559.

[67] Weiser JN, Pan N, McGowan KL, Musher D, Martin A, Richards J (1998). Phosphorylcholine on the lipopolysaccharide of *Haemophillus influenza* contributes to persistence in the respiratory tract and sensitivity to serum killing mediated by C-reactive protein. *J Exp Med*, 187: 631-640.

[68] Zeller J and Sullivan B (1992). C-reactive protein selectively enhances the intracellular generation of reactive oxygen products by IgG-stimulated monocytes and neutrophils. *J Leukoc Biol,* 52: 449-456.

[69] Venugopal SK, Devaraj S, Jialal I (2003). C-reactive protein decreases prostacyclin release from human aortic endothelial cells. *Circulation,* 108: 1676-1678.

[70] Ryu J, Lee CW, Shin JA, Park C S, Kim JJ, Park SJ, Han KH (2007). FcgammaRIIa mediates C-reactive protein induced inflammatory responses of human vascular smooth muscle cells by activating NADPH oxidase 4. *Cardiovasc Res,* 75: 555-565.

[71] Dougherty TJ and Miller PF (2006). Microbial genomics and drug discovery: exploring innovative routes of drug discovery in the postgenomic era. *IDrugs,* 9: 420-422.

[72] Fang FC, Libby SJ, Castor ME, Fung AM (2005). Isocitratelyase (AceA) is required for *Salmonella* persistence but not for acute lethal infection in mice. *Infect Immun,* 73(4): 2547-2549.

[73] Rice KC and Bayles KW (2008). Molecular control of bacterial death and lysis. *Microbiol Mol Biol Rev,* 72: 85-109.

[74] Ratnam S and Mookerjea S (1998). The regulation of superoxide generation and nitric oxide synthesis by C-reactive protein. *Immunology,* 94: 560-568.

[75] Franco-Molina MA, Mendoza-Gamboa E, Miranda-Hernandez DF, Sierra-Rivera CA, Zapata-Benavides P, Vera-García ME, Tamez-guerra RS and Rodríguez-Padilla C (2011). Anti-inflammatory and antioxidant effects of immunopotent CRP in lipopolysaccharide (LPS)-stimulated human macrophages. *Afr J Microbiol Res,* 5: 3726-3736.

[76] Du Clos TW (2003). C-reactive protein as a regulator of autoimmunity and inflammation. *Arthritis Rheumat,* 48: 1475-1477.

[77] Mukherjee S (2013). Functional properties of C-reactive protein purified from *Achatinafulica* (Bowdich). PhD dissertation, Visva Bharati University, Santiniketan.

[78] Iwanaga S, Kawabata SI, Muta T (1998). New types of clotting factors and defense molecules found in horseshoe crab hemolymph: their structure and functions. *J Biochem,* 123: 1-15.

[79] Iwanaga S and Kawabata S (1998). Evolution and phylogeny of defense molecules associated with innate immunity in horseshoe crab. *Front Biosci,* 3: 973-984.

[80] Ghosh S and Bhattacharya S (1992). Elevation of C-reactive protein in serum of *Channa punctatus* as an indicator of water pollution. *Ind J ExpBiol,* 30(8): 736-737.

[81] Salanki J, Farkas A, Kamardina T, Rozsa KS (2003). Molluscs in biological monitoring of water quality. *Toxicol Lett,* 140-141: 403-410.

[82] Mostafa OM, Mossa AT, El Einin HM (2013). Heavy metal concentrations in the freshwater snail *Biomphalaria alexandrina* uninfected or infected with cercariae of *Schistosoma mansoni* and/or *Echinostomaliei* in Egypt: the potential use of this snail as a bioindicator of pollution. *J Helminthol,* 28: 1-6.

[83] Rainbow PS and Phillips DJH (1993). Cosmopolitan biomonitors of trace metals. *Marine Poll Bull*, 26: 593-601.

[84] Simkiss K and Watkins B (1990). The influence of gut microorganisms on zinc uptake in *Helix aspersa*. *Environ Pollut*, 66: 263-271.

[85] Abd Allah AT and Moustafa MA (2002). Accumulation of lead and cadmium in the marine prosobranch *Neritasaxtilis*: chemical analysis, light and electron microscopy. *Environ Pollut*, 116: 185-191.

[86] Sibly RM and Calow PA (1989). Life-cycle theory of responses to stress. *Biol J Linn Soc*, 37: 101-116.

[87] Laskowski R. and Hopkin SP (1996). Accumulation of Zn, Cu, Pb and Cd in the garden snail (*Helix aspersa*): implications for predators. E*nviron Pollut*, 91: 289-297.

[88] Mukherjee S, Barman S, Sarkar S, Mandal NC, Bhattacharya S (2014). Anti-bacterial activity of *Achatina* CRP and its mechanism of action. *Indian J Exp Biol*, 52: 692-704.